Sustainable Technologies for Value Addition and Management of Plastic Waste

This book provides innovative and sustainable technologies for converting plastic waste into valuable resources. The book is structured into three sections, each focusing on a different aspect of plastic waste management and conversion technologies. The first section delves into sustainable technologies for producing various types of fuels from plastic waste, including liquid fuel, hydrogen, and aviation fuel. The second section examines technologies for converting plastic waste into value-added chemicals and materials, encompassing new materials, chemicals, and new plastics. The third section focuses on the management and recycling of plastic waste, covering treatment methods for general, electronic, micro-, and nano-plastic waste and the application of artificial intelligence in waste management. It also includes a life-cycle assessment of plastic waste management processes. Further, the book examines the broader concepts of circular economy and sustainability in the context of plastic waste management. The chapters discuss the principles and practices of a circular economy, aiming for a sustainable and waste-free future. They also address the role of policies and the implementation of green supply chains in promoting sustainable plastic waste management, emphasizing the importance of integrating policy and technological advancements. This book is a valuable resource for professionals, researchers, and students of environmental sciences and biotechnology.

Key Features

- Emphasizes the sustainable technologies for converting plastic waste into various fuels, including liquid fuel, hydrogen, and aviation fuel
- Highlights the conversion of plastic waste into value-added chemicals and new materials
- Examines advanced management and recycling methods for different types of plastic waste, including general, electronic, micro-, and nano-plastic waste
- Provides a comprehensive life-cycle assessment of plastic waste management processes and evaluates their environmental impact and sustainability
- Discusses the principles and practices of a circular economy for a sustainable and waste-free future through innovative plastic waste management strategies
- Addresses the role of policies and green supply chains in promoting sustainable plastic waste management

Sustainable Industrial and Environmental Bioprocesses

This book series aims to provide a comprehensive collection of books focusing on bioprocesses in industrial and environmental biotechnology. The multidisciplinary content encompasses chemical and biochemical engineering, industrial microbiology, and energy biosciences, all with a central theme of sustainable development and circular economy principles. The books align with the Sustainable Development Goals (SDGs) and offer state-of-the-art information and in-depth knowledge on the subject matter. The books in this series also emphasize the application of emerging tools, such as machine learning and artificial intelligence, for the advancement of bioprocesses. While primarily targeting academicians and researchers, the series is valuable for policy planners and industry professionals, with carefully tailored contents to cater to their specific needs and interests.

Series Editor: Dr Ashok Pandey

Professor Ashok Pandey is currently Executive Director, Centre for Energy and Environmental Sustainability–India, Lucknow. Formerly, he was Distinguished Scientist at the Centre for Innovation and Translational Research, CSIR–Indian Institute of Toxicology Research, Lucknow, India, and Chief Scientist and Head, Centre for Biofuels, CSIR–National Institute for Interdisciplinary Science and Technology, Trivandrum; Research Scientist at Zentral Laboratorium, Sudzucker, West Germany; and Scientist, Biochemistry Division, National Sugar Institute, Kanpur, India. He is HTBS National Innovation Chair (Biotechnology) and has been a Distinguished Professor in UPES, Dehradun; Visiting Professor, Indian Institute of Technology, Roorkee; and Distinguished Adjunct Professor, SIMST, Chennai. His major research and technological development interests are industrial and environmental biotechnology and energy biosciences, focusing on sustainability and circular economy. Professor Pandey is/has been UNESCO Professor and Adjunct/Visiting Professor/Scientist in universities in France, Germany, Brazil, Argentina, Mexico, Canada, China, South Korea, South Africa, Malaysia, Thailand, Switzerland, USA, and the UK.

Professor Pandey is the recipient of many national and international awards and honours, which include Fellow of the World Society for Sustainable Energy Technologies (WSSET); the International Society for Energy, Environment and Sustainability (ISEES); the National Academy of Sciences, India (NASI); and the Indian Chemical Society (ICS). He is a Distinguished Fellow of the Biotech Research Society, India (BRSI) and Academician of European Academy of Sciences and Arts. He has been conferred the Life-Time Achievement Award by several societies and academies, including the BRSI, ISEES, and several other national and international awards. Professor Pandey is the founder of the Biotech Research Society, India (BRSI) and International Bioprocessing Association (IBA).

Microplastic Pollution: Occurrence, Health Risk and Challenges
Navish Kataria, Vinod Kumar Garg, Changseok Han, Eldon Rene

Diatoms and Sustainable Agriculture
Archana Tiwari and Ashok Pandey

Biochar and the Circular Economy: Innovations in Production and Applications
Hrudayanath Thatoi, Shibani Mohapatra, Ashok Pandey, Lam Su Shiung

Sustainable Technologies for Value Addition and Management of Plastic Waste
Abha Kumari, Ashok Pandey, Rohidas Bhoi, Anh Tuan Hoang

For more information about this series, please visit: *https://www.routledge.com/Sustainable-Industrial-and-Environmental-Bioprocesses/book-series/SIEB?srsltid=AfmBOoo33yv3wNbuaZfBUQYyOFSMtnIj2cqH4_d2R8-he_cEFFlJI6Qt*

Sustainable Technologies for Value Addition and Management of Plastic Waste

Edited by

Abha Kumari
Ashok Pandey
Rohidas Bhoi
Anh Tuan Hoang

CRC Press
Taylor & Francis Group
Boca Raton London New York

CRC Press is an imprint of the
Taylor & Francis Group, an **informa** business

Designed cover image: Shutterstock

First edition published 2026
by CRC Press
2385 NW Executive Center Drive, Suite 320, Boca Raton, FL 33431

and by CRC Press
4 Park Square, Milton Park, Abingdon, Oxon, OX14 4RN

CRC Press is an imprint of Taylor & Francis Group, LLC

ISBN: 978-1-032-93946-9 (hbk)
ISBN: 978-1-032-94643-6 (pbk)
ISBN: 978-1-003-58100-0 (ebk)

DOI: 10.1201/9781003581000

Typeset in Times
by Apex CoVanatage, LLC

Contents

Preface

The first time plastic was synthesized for an application as a substitute for shellac, a natural electrical insulator, was by Baekeland to meet the needs of the rapidly electrifying United States. After that, plastic was synthesized to fulfill the requirements of industries and human beings. This resulted in the creation of 220 million tons of plastic waste in 2024. If these trends continue, by 2050, 26 billion metric tons of plastic waste will be produced, almost half of which will be dumped in landfills and the environment. Polystyrene (found in straws, cutlery, coffee cup lids, plastic bottles, plastic bags, medical waste, balloons, and other items); microbeads, packaging films, wrapping materials, shopping and trash bags, fluid containers, apparel, toys, home and industrial goods, and building materials are plastic, which are used every day, leading to pollution after use. Plastic waste refers to plastic objects that have not been recycled or disposed of properly or have been illegally dumped. The World Economic Forum estimates that, in the absence of action, the annual amount of plastic debris entering the oceans of between 1 and 8 million tons will triple by 2030. This causes the generation of micro- and nanoplastics. Waste made of plastic is divided into several categories, including polyethylene terephthalate (PET or PETE), high-density polyethylene (HDPE), low-density polyethylene (LDPE), polypropylene (PP), polystyrene (PS or Styrofoam), and miscellaneous plastics. Careless disposal of plastic waste, inexperienced recycling or reprocessing, and the non-biodegradability of plastic waste create a number of serious environmental problems that also have an impact on human health. Toxic emissions, including carbon monoxide, chlorine, hydrochloric acid, dioxin, furans, amines, nitriles, styrene, benzene 1,3-butadiene, CCl_4, and acetaldehyde, are produced when plastics burn. Plastic production and its unplanned management and disposal have been shown to pollute terrestrial, aquatic, and atmospheric environments. Plastics do not break down and will remain in the environment for a number of years. Plastic waste and petroleum-derived plastics do not decompose and tend to persist in the surrounding environment for a longer time. Plastics can be ingested and accumulate in the tissues of both terrestrial and aquatic animals, which can impede their growth and development. The need of the hour is to have in-depth knowledge of sustainable valorization and scientific plastic waste management.

The rapid increase in plastic waste from both human and industrial activities poses significant challenges to resource management and environmental sustainability. As global populations expand and industrial activities increase, the accumulation of plastic waste endangers ecosystems, worsens pollution, and depletes natural resources. However, this challenge also presents a transformative opportunity: using sustainable technologies, we can convert plastic waste into valuable, marketable products, creating a circular economy that minimizes waste, lessens environmental impact, and promotes resource efficiency.

This book, *Sustainable Technologies for Value Addition and Management of Plastic Waste*, comprehensively explores innovative and sustainable approaches to transform plastic waste into biofuels, valuable materials, and effective waste management systems. It aims to equip researchers, students, policymakers, and industry professionals with a deep understanding of advanced technologies and strategies for converting plastic waste into profitable products, fostering both environmental and economic sustainability.

The book thoroughly examines sustainable production technologies for various value-added products, encompassing advanced methodologies, reactor and fermenter designs, sustainable strategies, commercial feasibility, limitations, prospects, and applications. It provides detailed insights into the physicochemical composition and variations of plastic waste resources, along with sustainable technologies for managing and recycling plastic waste. A key focus is the integration of artificial intelligence (AI) and machine learning (ML) tools to optimize plastic waste management systems. The book's detailed production process flowsheets promote understanding of related challenges and opportunities, motivating contributions toward a more sustainable green future.

The book is organized into three sections, each addressing a critical aspect of plastic waste valorization:

1. **Fuel production**: The first section delves into sustainable technologies for producing various types of fuels from plastic waste, including liquid fuel, hydrogen, and aviation fuel. This underscores biochemical, thermochemical, and catalytic methods, bioreactor engineering, and business viability for less fossil fuel use, in accordance with worldwide energy transition objectives.
2. **High-value chemicals and materials**: The second section examines technologies for converting plastic waste into value-added chemicals and materials, encompassing new materials, chemicals, and new plastics. The conversion of plastic waste into value-added chemicals and new materials is discussed in detail, such as the mechanisms, their properties, characteristics, and applications.
3. **Advanced waste management and recycling**: The third section focuses on the management and recycling of plastic waste, covering treatment methods for general, electronic, micro- and nanoplastic waste, and the application of artificial intelligence in waste management. It also includes a life-cycle assessment of plastic waste management processes.

Further, the book examines the broader concepts of a circular economy and sustainability in the context of plastic waste management. It also discusses the use of AI and machine learning logic for plastic waste management. The chapters discuss the principles and practices of a circular economy, aiming for a sustainable and waste-free future. The book also addresses the sustainable treatment of plastic waste and plastic e-waste and their recycling to value.

With an emphasis on commercially viable production techniques and their potential for industry initiatives, each chapter aims to give readers a thorough grasp of a particular process. By showcasing the vast potential of plastic waste as a valuable resource, this book inspires readers to harness its capabilities for a positive environmental and economic impact. Readers will find this book to be a pivotal source, as it equips them with the information and strategies to foster research, inform policy, and enhance real-world applications in our progression toward a circular, sustainable economy. This book is a useful resource for those working in the field of engineering, environmental sciences, and biotechnology.

We hope *Sustainable Technologies for Value Addition and Management of Plastic Waste* serves as a valuable resource, furnishing its readers with the knowledge and tools to advance research, inform policy, and drive practical implementation in the transition to a circular and sustainable economy. The book is a critical resource, equipping readers with the information and tactics necessary to encourage research, shape policy, and improve the practical application of concepts as we move toward a circular, sustainable economy. This book is a useful resource for those working in environmental sciences and biotechnology.

Abha Kumari
Ashok Pandey
Rohidas Bhoi
Anh Tuan Hoang

The Editors

Abha Kumari is a Professor at the Centre for Biotechnology and Biochemical Engineering, Amity Institute of Biotechnology, Amity University, Noida, India. She received her PhD and MSc (Engg) in Engineering from the Indian Institute of Science (IISc), Bengaluru, India. Her research interests span biochemical engineering and biotechnology, with a particular emphasis on waste-to-value technologies, municipal solid waste-to-biogas and carotenoid production, bio-enzyme production from waste biomass, waste valorization and management, and metal recovery from electronic waste. She has published more than 32 research articles in peer-reviewed international journals and has authored 14 book chapters. She is also the author of one book and the editor of two edited books.

Ashok Pandey is currently Executive Director, Centre for Energy and Environmental Sustainability–India, Lucknow. Formerly, he was Distinguished Scientist at the Centre for Innovation and Translational Research, CSIR–Indian Institute of Toxicology Research, Lucknow, India; Chief Scientist and Head, Centre for Biofuels, CSIR–National Institute for Interdisciplinary Science and Technology, Trivandrum; Research Scientist at Zentral Laboratorium, Sudzucker, West Germany; and Scientist, Biochemistry Division, National Sugar Institute, Kanpur, India. He is HTBS National Innovation Chair (Biotechnology) and has been Distinguished Professor in UPES, Dehradun; Visiting Professor, Indian Institute of Technology, Roorkee; and Distinguished Adjunct Professor, SIMST, Chennai. His major research and technological development interests are industrial and environmental biotechnology and energy biosciences, focusing on sustainability and circular economy.

Rohidas Bhoi is an Assistant Professor at the Malaviya National Institute of Technology, Jaipur. He obtained his PhD in chemical engineering from IIT Bombay. His research areas include waste-to energy, pyrolysis, biofuels, reaction engineering, and catalysis. Dr. Bhoi has published 20 research papers and 7 book chapters.

Anh Tuan Hoang is currently the Vice President of Dong Nai Technology University, Dong Nai, Vietnam. He received his PhD in 2016. His research interests include energy conversion, renewable energy, bioenergy, and alternative fuels. Dr. Hoang has published over 200 papers, 8 book chapters, and 1 book. He has been ranked among the World's Top 2% Scientists in the field of energy from 2019 to 2024. In 2023, he was recognized as a Highly Cited Researcher by Clarivate, Web of Science.

Section I

Sustainable Technologies for Plastic Waste to Fuels

1 Production of Liquid Fuel from Plastic Waste

*Prathwiraj Meena, Shreya Patel, and Rohidas Bhoi**
Department of Chemical Engineering, Malaviya National Institute of Technology Jaipur, Rajasthan, India

1.1 INTRODUCTION

The advancement of technology and societal modernization has led to the proliferation of plastic products, raising concerns about environmental pollution. Researchers worldwide are alarmed by the threat of soil and terrestrial contamination resulting from plastic waste (Nizzetto et al., 2016). Various types of plastics sourced from diverse origins, including industrial and urban landfills, illegal dumping, fibers from clothing, tire wear, airborne plastic particles, and microplastics incorporated into various products, can contaminate soil, water, and air. As per the Organization for Economic Co-operation and Development report, worldwide plastic production continues to escalate and is projected to surpass 500 million tons by 2050. This surge is driven by ever-growing plastic production, along with its rapid disposal and inefficient recycling methods. With a meagre recycling rate of only 10% for plastic waste and 14% being incinerated globally, a substantial 76% ends up in either landfills or the natural environment (Global Plastics Outlook, 2022). Therefore, it is crucial to implement scientifically sound strategies for managing plastic waste to reduce its environmental impact. Plastic has significant energy potential, characterized by a higher heating value (HHV) ranging from 41 to 47 MJ/kg, surpassing that of coal (25–35 MJ/kg) and wood (17–19 MJ/kg), comparable to crude oil (42–47 MJ/kg), which makes it a valuable energy resource, given its derivation from crude oil feedstock (Chang, 2023). Therefore, disposing of energy resources through landfills or waste is undesirable. Utilizing plastic waste for energy production could reduce reliance on fossil fuels, simultaneously addressing the issue of plastic waste management, thus promoting a circular economy.

Pyrolysis is emerging as an advanced thermochemical process that is conducted at elevated temperatures in an oxygen-free environment, capable of converting plastic waste into a range of fuels, including plastic oil, char, and gas, as well as value-added chemicals like alkanes, alkenes, and aromatics (Prajapati et al., 2021). Among the fuels derived from plastic waste pyrolysis, plastic oil emerges as a promising alternative fuel due to its carbon-neutral nature, reduced greenhouse gas emissions compared to traditional fossil fuels, and compatibility with existing equipment and infrastructure for storage and transfer, unlike char and gas.

1.2 PETROCHEMICAL INDUSTRY, PLASTIC PRODUCTION, AND ITS CONVERSION TO PETROCHEMICALS: A CYCLIC INTRICATE RELATIONSHIP

The intricate relationship between the products of crude oil refining, the production of plastics, and the subsequent generation of plastic waste, which is then converted back into refinery fuel products through the pyrolysis process, is shown in Figure 1.1. Initially, crude oil undergoes fractional distillation to yield various products like liquefied petroleum gas (LPG), light naphtha, petrol, aviation turbine fuel (ATF), diesel, and heavy fuel oil. The top distillation products, such as LPG and light

* Corresponding author's email: rohidas.chem@mnit.ac.in

DOI: 10.1201/9781003581000-2

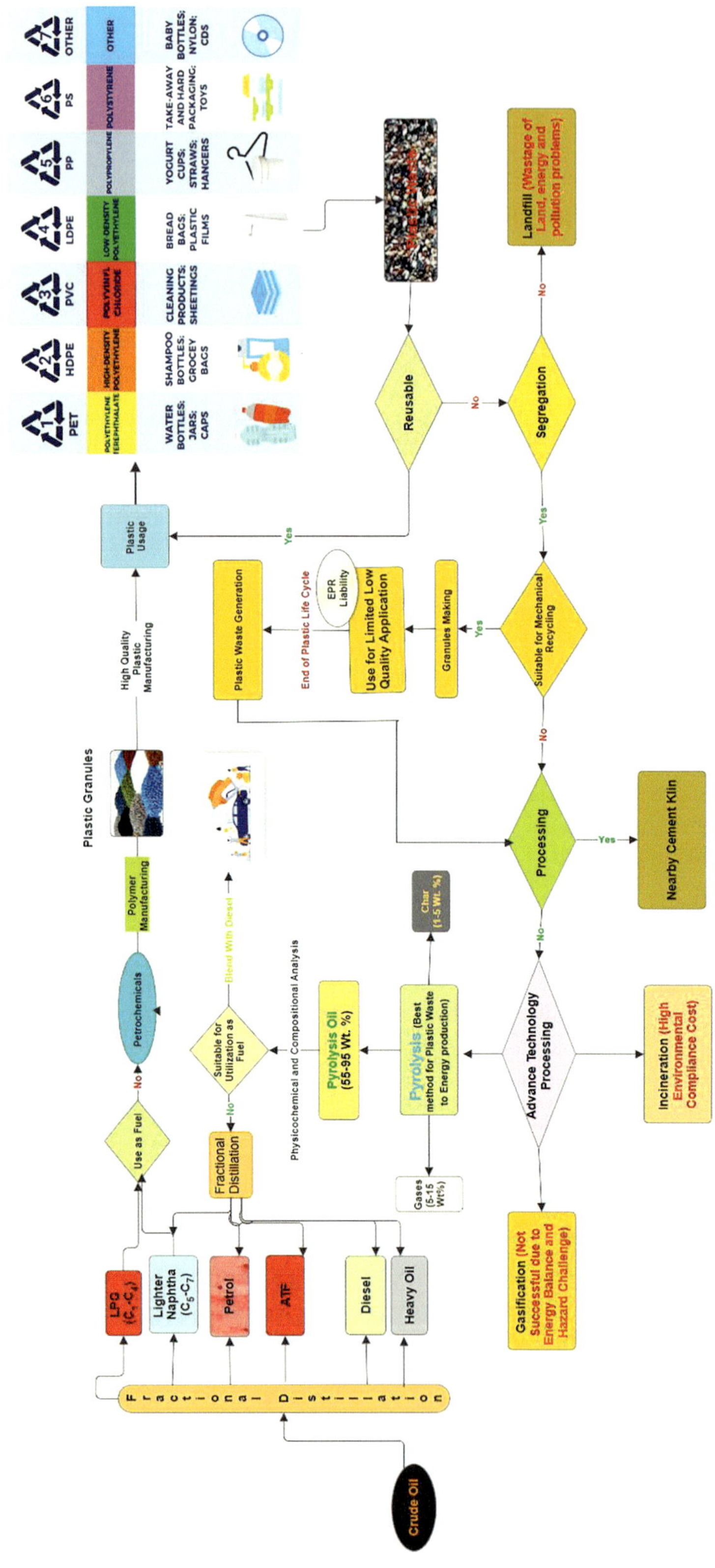

FIGURE 1.1 Process flow diagram of plastic production, plastic waste management, and generation of petrochemicals by waste plastic pyrolysis.

TABLE 1.1
Advantages and Disadvantages of Various Plastic Waste Management Methods

Serial Number	PW Management Strategies	Advantages	Challenges
1	Mechanical recycling: • Sorting/shredding/ extrusion/pelletizing	• Highly effective • Effective in managing large volume	• Elevated energy consumption • High operational cost
2	Landfilling	• Simple and cost-effective operation • No need for complex sorting • Suitable for all types of plastic waste	• Reduced soil fertility • Requiring extensive land usage
3	Incineration	• Production of heat and electricity • An effective solution for energy recovery from highly mixed plastic waste	• Release of harmful and toxic products • Increased CO_2 emissions
4	Gasification	• Capability to process mixed feedstock • Production of combustible gas products	• Requires raw material pretreatment • Low calorific value gases
5	Pyrolysis	• Adaptable operating conditions • Reduced greenhouse gas emissions • Valuable products such as oil, char and gas	• Significant energy demands • Challenges in scaling up and commercial viability

naphtha, are directly used as heating fuel, cooking gas, or polymer manufacturing. Petrochemicals such as ethylene, propylene, and styrene are used to manufacture polyethylene, polypropylene, and polystyrene granules, respectively. However, some polymers are manufactured using two or more types of monomers, such as polyethylene terephthalate (PET), which is manufactured using ethylene glycol and dimethyl terephthalate. Similarly, polyvinyl chloride (PVC) is manufactured by combining chlorine with ethylene. These manufactured granules are utilized to produce desired products as per the demand in the market by incorporating various additives such as plasticizers, antioxidants, stabilizers, colorants, and fillers to enhance the properties of the desired plastic. These additives are crucial for plastics to achieve desired characteristics such as strength, flexibility, and UV resistance. The seven common plastic types and their application and recycling codes are mentioned in Figure 1.1.

Global plastic production was 450 million tons in 2022, out of which 24.1 million tons was produced in India (Global Plastics Outlook, 2022). After utilizing these plastics in various household or industrial applications, they are treated as waste plastic and classified as single-use and multi-use plastics. Single-use plastics include plastic bags, straws, cutlery, Styrofoam cups, coffee stirrers, and so forth. Multi-use plastics are more durable and can be used repeatedly, such as water bottles, food containers, reusable shopping bags, durable kitchen utensils, and buckets and mugs. The advantages and disadvantages of various techniques used to treat waste plastics are summarized in Table 1.1.

1.3 PYROLYSIS OF PLASTIC WASTE FOR FUEL OIL PRODUCTION

Pyrolysis is a thermochemical process conducted without oxygen, wherein long-chain carbonaceous materials are broken down at high temperatures in an inert gas environment, either with or without a catalyst. Pyrolysis conducted without a catalyst is termed thermal pyrolysis, while with a catalyst it is known as catalytic pyrolysis. Plastic waste undergoes pyrolysis and typically yields three main products: pyrolysis oil, non-condensable gases (including CH_4, H_2, CO, and CO_2), and carbon-rich solid char. Pyrolysis oil mainly comprises aliphatic, naphthenic, and mono- and polyaromatic compounds. The oil can be used as a heat source in steam boilers or as a basis for monomer recovery, carbon, transportation fuels, and more (Meena et al., 2024). Alternatively, the waxy plastic oil, consisting of long-chain alkanes and alkenes with high boiling points (>500°C), requires further processing, such as fluid catalytic cracking, to convert it into liquid fuels or other petrochemical products. The non-condensable gases generated during pyrolysis can be recycled back into the process to serve as a heating or energy source, thus enabling a self-sustaining process without external energy sources. Additionally, the char produced can find utility as solid fuel, such as briquettes, and can be used to manufacture adsorbents, activated carbons, graphene, and its derivatives, among other applications.

The products obtained from plastic waste pyrolysis are primarily influenced by factors such as temperature, heating rate, and residence time. Other variables affecting the distribution of pyrolysis products include the type of plastic waste, reactor design, heating method, pressure, particle size, catalyst type, operation mode, and reaction atmosphere. The suitable parameters for oil production by pyrolysis of waste plastic are discussed in the following sections. Tables 1.3 and 1.4 provide an overview of pyrolysis systems (including reactor type) and the associated conditions (temperature, heating rate, residence time, particle size, reaction atmosphere, catalyst type, amount, and operation mode) for both thermal and catalytic pyrolysis of various plastic wastes (including single and mixed LDPE, HDPE, PP, and PS). The corresponding plastic oil yields for thermal and catalytic pyrolysis are also reported.

1.3.1 Common Plastics and Their Potential as Pyrolysis Feedstock

Plastic is formed primarily from fossil fuels and petrochemicals through polymerization or polycondensation. The six prevalent plastics find widespread use across various industries and everyday life, namely, LDPE, HDPE, PP, PET, PS, and PVC, as illustrated in Figure 1.1.

The composition and structure of plastic waste (PW) significantly influence its pyrolysis behavior. Key molecular properties such as bond energies, polymeric structure, composition, weak link positions in the polymer chain, and crystallinity are crucial factors (Maqsood et al., 2021). The proximate and ultimate analysis, along with the H/C ratio and calorific value of six common plastics, are mentioned in Table 1.2. The proximate analysis suggests that all plastic wastes have significantly higher volatile matter content (84.1–99.8 wt%) compared to moisture (0–0.67 wt%), fixed carbon (0 –13.9 wt%), and ash (0–4.45 wt%) content. This high volatile matter content makes them easy to ignite and highly responsive to pyrolysis. The high volatile matter and low ash content of plastic wastes suggest that they are likely to produce plastic oil and gas as primary products during pyrolysis, with a low char yield due to their low fixed carbon content. Generally, the thermal decomposition resistance of these materials follows this sequence: PS < PP ~ LDPE < HDPE (Diaz-Silvarrey et al., 2018). Compared to proximate analysis, ultimate analysis offers a more detailed assessment by determining the elemental composition of plastic waste.

As shown in Table 1.2, most plastic wastes, particularly LDPE, HDPE, PP, and PS, are primarily composed of carbon (83.7–91.6 wt%) and hydrogen (7.8–15.3 wt%), with minimal amounts of oxygen (0–0.45 wt%), nitrogen (0–0.67 wt%), sulfur (0–2.1 wt%), and chlorine (0–1.16 wt%). This composition suggests that they are likely to produce plastic oil with a higher heating value. On the other hand, PET and PVC wastes, which have lower carbon (38.3–64.1 wt%) and hydrogen (3.7–4.8 wt%) content and higher oxygen (33.0–34.2 wt%) content for PET or higher chlorine (56.4–56.5 wt%) content for PVC compared to LDPE, HDPE, PP, and PS, tend to produce plastic oil with a low HHV. Significant CO and CO_2 emissions were observed during pyrolysis of PET and HCl and chlorinated organic emissions for PVC (Miandad et al., 2019). Plastic waste's higher carbon and

TABLE 1.2
Physicochemical Properties of Common Plastics

Serial Number	Plastic Waste	Proximate Analysis (wt%)				Ultimate Analysis (wt%)						H/C Ratio	HHV (MJ/kg)	Temperature Onset-End-Set, °C	Reference
		Moisture	Volatile Matter	Fixed Carbon	Ash	C	H	O	N	S	Cl				
1	PET	0.0	84.1 to 94.4	5.6 to 13.9	0.0	62.6 to 64.1	3.7 to 4.4	33.0 to 34.2	0.0	0.0	0.0	0.7 to 0.82	21.6 to 24.15	380 to 490	Chattopadhyay et al., 2016
2	PS	0.0	99.6 to 99.8	0.2 to 0.4	0.0	90.4 to 91.6	7.8 to 8.6	0.18 to 0.45	0.15 to 0.18	0.04 to 0.51	0.0	1.0 to 1.12	38.5 to 43.58	360 to 460	van der Westhuizen et al., 2022
3	LDPE	0.0	99.1 to 99.8	0.0	0.2 to 0.9	85.8 to 86.4	13.6 to 14.0	0.0	0.0	0 to 0.07	0.0	1.8 to 1.93	44.2 to 47.23	330 to 430	Zheng et al., 2022
4	PP	0.0 to 0.16	93.8 to 96.9	0 .0 to 1.6	1.0 to 4.5	83.7 to 84.7	14.3 to 15.3	0.0 to 0.15	0.0 to 0.67	0.1 to 2.1	0.0 to 1.16	2.0 to 2.15	45.2 to 46.05	300 to 475	Hakeem et al., 2018
5	HDPE	0.0 to 0.20	97.2 to 99.6	0.0	0.2 to 0.8	85.5 to 86.5	14.2 to 15.1	0.0 to 0.2	0.0	0.0 to 0.25	0.0	1.9 to 2.08	43.9 to 45.38	340 to 435	Baloch et al., 2020
6	PVC	0.0	94.9 to 95.8	4.2 to 5.1	0.0	38.3 to 38.7	4.5 to 4.8	0.0	0.0 to 0.23	0.0 to 0.61	56.4 to 56.5	1.4 to 1.48	15.0 to 24.57	250 to 525	Ephraim et al., 2023

hydrogen content render them highly suitable for pyrolysis as feedstock. However, the emissions of CO and CO_2 during pyrolysis are primarily contingent on the plastic's hydrogen-to-carbon (H/C) ratios. Plastics with higher H/C ratios generally exhibit greater energy efficiency and lower CO and CO_2 emissions during pyrolysis. Consequently, polyolefin wastes like PP (with H/C ratios ranging from 2.04 to 2.15), HDPE (with H/C ratios ranging from 1.98 to 2.08), and LDPE (with H/C ratios ranging from 1.89 to 1.93), as well as PVC (with H/C ratios ranging from 1.40 to 1.48) and PS (with H/C ratios ranging from 1.03 to 1.12), are expected to yield higher energy efficiency and emit lower levels of CO and CO_2 in their exhaust compared to PET (with H/C ratios ranging from 0.70 to 0.82) (Pal et al., 2023).

HHVs of plastic wastes are primarily determined by their carbon, hydrogen, and oxygen contents, with increased carbon and hydrogen enhancing HHVs. In contrast, higher oxygen content decreases the HHV. Consequently, plastic wastes such as LDPE, HDPE, PP, and PS, characterized by higher carbon (ranging from 83.7 wt% to 91.6 wt%) and hydrogen (ranging from 7.8 wt% to 15.3 wt%) and lower oxygen (0 wt% to 0.45 wt%) contents, exhibit higher HHVs (38.51 to 47.23 MJ/kg) compared to PET and PVC wastes (15.06 to 24.57 MJ/kg). Plastic wastes with higher HHVs, such as LDPE, HDPE, PP, and PS, are anticipated to yield plastic oil with higher HHVs than those with lower HHVs, such as PET and PVC. Additionally, PS waste, containing aromatic compounds in its structure alongside its high HHV, emerges as a promising option for pyrolysis feedstock. However, precautions must be taken to mitigate the emission of polycyclic aromatic hydrocarbons (PAHs) during the pyrolysis of PS waste, including the use of catalysts (Verma et al., 2021). However, considering factors such as H/C ratios, energy efficiency, greenhouse gas emissions, and HHVs of various plastic wastes, it is evident that waste polyolefins like LDPE, HDPE, and PP wastes, along with PS waste, hold more promise than PVC and PET wastes as pyrolysis feedstock. Additionally, PVC waste tends to produce a significant amount of HCl. In contrast, PET waste generates benzoic and terephthalic acids, which are corrosive to reactors, leading to their exclusion from pyrolysis processes (Peng et al., 2022). This observation is supported by the extensive use of LDPE, HDPE, PP, and PS wastes as pyrolysis feedstock in numerous recent studies, compared to PVC and PET wastes, as documented in Tables 1.3 and 1.4.

1.4 PYROLYSIS SYSTEMS

The pyrolysis system consists of the feed hopper, plastic melter, reactor with heater, gas condensing system, product collector, and so forth. The reactor is at the core of the pyrolysis system and its choice depends on the desired processing capacity, feed quality, and desired products.

1.4.1 Reactor Types

Selecting the appropriate reactor is crucial to enhance the reaction rate and overall performance of plastic oil production in both thermal and catalytic pyrolysis of plastic waste. As shown in Tables 1.3 and 1.4, a variety of reactor types, ranging from laboratory to pilot scale, are utilized for plastic waste pyrolysis. These reactors predominantly include batch or semi-batch reactors operating under static or mixed conditions, such as round-bottom flask, tubular, retort, autoclave, and chemical vapor deposition reactors. Additionally, continuous-flow reactors like fixed bed and fluidized bed reactors are used. The prevalence of batch or semi-batch reactors in laboratory-scale thermal and catalytic pyrolysis is attributed to their straightforward and compact design, ease of operation, ability to assess multiple parameters simultaneously, and capacity to produce a variety of products using different reactants and operating conditions. However, they are associated with high labor costs and low production rates (Chang, 2023).

The conventional one-stage batch or semi-batch reactors are commonly employed in thermal pyrolysis or in situ catalytic pyrolysis. However, the challenges in catalyst separation from char residue and potential coke formation on the catalyst surface influence the pyrolysis process.

TABLE 1.3
Pyrolysis Systems and Process Parameters of Thermal Pyrolysis for Various Plastic Wastes and Oil Yields

		Pyrolysis Conditions					
Plastic Waste	Reactor Type	Reaction Temperature, T (°C)	Heating rate, β (°C/min)	Inert gas	Diameter of Feedstock, Dp (mm)	Oil Yield	Reference
LDPE	Batch	250 to 500	10 to 20	N_2	3 to 4	73.9 to 99.8	Hariadi et al., 2021
LDPE	FXR	350 to 800	5	N_2	0.04	62.5 to 85	Mariappan et al., 2021
HDPE	FXR	600	10	–	5	69.0	Akubo et al., 2019
PP	Batch	400 to 525	10 to 180	N_2	1 to 50	67.5 to 98.5	Parku, 2019
PP	Semi-Batch	500	20	N_2	0.1 to 10	86.25	Meena et al., 2023
PP	Batch	450	10	N_2	16	42.0	Miandad et al., 2019
PP	CSTR	450	–	N_2	–	87.0	Kusenberg et al., 2022b
PS	Semi-Batch	500	20	N_2	0.1 to 10	84.0	Singh et al., 2020
LDPE/PP/PVC mix	FBR	720	20	N_2	–	20.6	Park et al., 2022
LDPE/HDPE/PP mix	CSTR	450	–	N_2	–	89.0	Kusenberg et al., 2022a
LDPE/HDPE/PP/PS/PET mix	Batch	450	10	N_2	–	40.0	Miandad et al., 2017
LDPE/HDPE/PP/PS mix	Batch	450	10	N_2	–	49.0	Miandad et al., 2017

TABLE 1.4
Pyrolysis Systems and Process Parameters of Catalytic Pyrolysis for Various Plastic Wastes and the Oil Yields

		Pyrolysis Conditions						
Plastic Waste	Reactor Type	Reaction Temperature, T (°C)	Heating Rate, β (°C/min)	Catalyst-to-Feed Ratio	Diameter of Feedstock, Dp (mm)	Catalyst	Oil Yield	Reference
LDPE	FBR	500	200	3:10	0.04	Fly ash	79.0	Mariappan et al., 2021
HDPE	Batch	350	20	2:10	1 to 2	$CuCO_3$	94.0	Singh et al., 2018
HDPE	FBR	600	10	1:10	5	Zeolite-Y	45.0	Akubo et al., 2019
PP	Batch	450	30	5:10	–	Kaolin	79.9	Hakeem et al., 2018
PP	Batch	400	15	1:10	–	Spent FCC	83.3	Aisien et al., 2021
PS	Batch	450	10	0.5–1:10	–	MgO	89	Inayat et al., 2022
PET	Batch	450	–	2:10	3	$NiCl_2$	78	Jia et al., 2020
LDPE/HDPE/PP mix	Batch	450–500	10	1:10	–	Zeolite	53.0	Sivagami et al., 2022
HDPE/PP/PS/PET mix	Batch	500	10	0.5:10	–	Fe-pillared clay	79.3	Li et al., 2017
LDPE/HDPE/PP mix	2-Step-FBR	350	10	1:10	0.1 to 1	HZSM-5	60	Dwivedi et al., 2021b
LDPE	Semi-Batch Reactor	450	10	1:10	1 to 2	FCC Catalyst	79.70	Kim et al., 2017

Nevertheless, these issues can be addressed by employing corresponding two-stage configurations in ex situ catalytic pyrolysis, where the catalyst is not directly blended with the feedstock, thus avoiding direct contact between catalysts and feedstock while enhancing the exposure of volatile products to catalyst surfaces. This two-stage ex situ catalytic pyrolysis process not only accelerates the secondary polymerization and isomerization reactions of the volatiles but also increases the gas yield as well as improves the hydrocarbon distribution of plastic oil. However, the drawback is that it decreases the oil yield compared to in situ catalytic pyrolysis due to higher gas production. For example, in the catalytic pyrolysis of LDPE over HZSM-5, a slightly lower plastic oil yield was achieved in a two-stage ex situ catalytic process compared to a one-stage in situ counterpart (42.3 wt% versus 44.3 wt%, as shown in Table 1.4). Moreover, the plastic oil from the two-stage process exhibited a significantly higher concentration of aromatics (>77%) with approximately 72% consisting of gasoline-range monoaromatic hydrocarbons such as BTX (benzene, toluene, and xylene) compounds, compared to the one-stage process, which contained around 55% aromatics with approximately 50% being gasoline-range monoaromatic hydrocarbons. Readers are advised to refer to an article by Dwivedi et al. (2021a) for more information about two-stage reactors.

Continuous-flow reactors offer advantages over batch or semi-batch reactors due to their lower labor costs and higher production rates, making them more suitable for large-scale thermal and catalytic pyrolysis. Among continuous-flow reactors, the fluidized bed reactor is the most efficient for plastic oil production in thermal and catalytic pyrolysis. Its efficiency stems from high heat and mass transfer rates and large surface areas for reaction, facilitated by intimate contact among granulated inert bed materials (such as sand), plastic waste particles, and catalyst (in the case of in situ catalytic pyrolysis) with short residence times. These reactors result in high plastic oil yields, reaching up to 99.7 wt% (Garcia-Nunez et al., 2017). Additionally, the fluidized bed reactor enhances catalyst reusability without requiring discharge, making it particularly favorable for catalytic pyrolysis. However, it poses operational challenges due to the need for a large volume of carrier gas and bed materials compared to other reactors like fixed bed, auger, and continuous stirred tank reactors, which are easier to operate due to their more straightforward construction and compact design. While most continuous-flow reactors for plastic waste pyrolysis are one-stage, some two-stage fixed bed and fluidized bed reactors have also been reported (Sekar et al., 2022).

1.4.2 Heating Source

Selecting an efficient heating method with effective heat transfer in the pyrolysis reactor is pivotal for boosting the reaction rate and, consequently, enhancing the yield of plastic oil. In Tables 1.3 and 1.4, mainly electric heating was reported as the most prevalent heating approach for pyrolysis reactors processing plastic waste. Alternative heating methods utilized for pyrolysis are pyrolysis gas, fossil fuels such as kerosene and propane, biomass, microwave heating and solar energy (Kaminsky, 2021; Maqsood et al., 2021; Singh et al., 2019). Pyrolysis gas generated during the pyrolysis process to heat the reactor tends to foster a self-sufficient pyrolysis system. It reduces utility costs by minimizing electricity expenses or fossil fuel usage. Nevertheless, pyrolysis gas's inherently low heating value has led to its limited fuel efficiency.

Microwave heating relies on high-frequency electromagnetic waves generated by magnetrons, enabling rapid, uniform, and selective volumetric heating within the plastic waste feedstock through convection rather than conduction, as seen in conventional methods. However, microwave heating may be ineffective for materials with low dielectric constants, such as plastic waste, necessitating the addition of microwave absorbers like activated carbon and silicon carbide to enhance energy absorption and conversion efficiency. Therefore, it is not widely suitable for plastic waste pyrolysis (Prathiba et al., 2018). Meanwhile, researchers obtained a substantial plastic oil yield from solar pyrolysis of LDPE waste using bifunctional photothermal catalysts. Also, a significantly reduced energy consumption was reported in the thermal pyrolysis of HDPE waste by utilizing solar heating while maintaining high plastic oil yields. Despite its advantages, solar heating faces challenges such as solar intermittency, low energy density, and the requirement for solar receivers and concentrators, leading to high initial capital costs (Habtewold et al., 2020; Luo et al., 2022).

1.5 PYROLYSIS PROCESS PARAMETERS

In addition to the selection of reactor type and heating method, various operational parameters during thermal and catalytic pyrolysis, including temperature, heating rate, residence time, pressure, particle size, reaction atmosphere, and catalyst type and its operational modes, as well as the composition of single or mixed plastic wastes, significantly influence both the quantity and quality of the resulting plastic oil yield.

1.5.1 Temperature

Temperature is a critical factor in pyrolysis processes. Initially, the thermal degradation of polymers involves the random breaking of C—C and C—H bonds when the supplied energy exceeds their respective bond enthalpies, forming free radicals. This process is followed by intermolecular and intramolecular hydrogen transfers and β-scission pathways, producing lower molecular weight products. The activation energies of different reactions determine the influence of pyrolysis temperature in plastic waste pyrolysis. For example, higher temperatures enhance monomer yields by overcoming the high energy barrier of β-scission reactions. The presence of van der Waals interactions between polymer molecules makes it challenging to break the non-polar C—C bonds, requiring significant energy input (Banu et al., 2020; Nanda and Berruti, 2021).

At lower temperatures (350–450°C), thermal degradation of plastic waste yields a waxy oil, primarily composed of C_{20+} hydrocarbons due to incomplete polymer decomposition (Mariappan et al., 2021). Increasing the pyrolysis temperature (450–600°C) promotes the production of gases containing $C_1 - C_4$ olefins and liquid fuel containing $C_5 - C_{20}$ paraffin and olefins, which further break down into shorter molecules. Another study investigated the extent of HDPE cracking during pyrolysis at varying temperatures ranging from 500°C to 800°C. The maximum yield of liquid/wax product (70 wt%) was achieved at 550°C, predominantly featuring $C_8 - C_{12}$ gasoline-range hydrocarbons. Conversely, at temperatures exceeding 750°C, the pyrolysis process was more selective toward gaseous products, mainly $C_2 - C_4$ hydrocarbons (>70%). However, higher gas yield is obtained at the expense of liquid products due to intensified cracking and secondary reactions. Moreover, extremely high temperatures induce aromatics hydrogenation and enhance PAH formation (Al-Salem, 2019).

The overall effect of temperature on pyrolysis oil and its quality is depicted in Figure 1.2. The yield of plastic oil increases with temperature but begins to decline once the temperature surpasses the plastic waste's end-set degradation temperature. The end-set temperature can be determined by TGA analysis of waste plastic. The temperature above end-set enhances the PAH content in the pyrolysis oil. Thus, the optimal temperature range for specific plastics for maximum oil production lies between the maximum and end-set degradation temperatures.

1.5.2 Heating Rate and Residence Time

Both heating rate and residence time are critical process parameters. The heating to elevated temperatures is the initial step in the pyrolysis reaction, and the duration of time that feedstock or vapor products spend in the reactor influences the extent of the pyrolysis reaction. These factors play a crucial role in modulating the degrees of primary and secondary pyrolysis reactions by affecting mass and heat transfer within the reactor. Consequently, they exert a significant influence on the composition of the final pyrolysis products. Pyrolysis can be classified into two types based on the heating rate: slow pyrolysis and fast pyrolysis. In slow pyrolysis, the heating rate is low to moderate (5–200°C/min). Due to this low heating rate, the residence time of solids and gases in the reactor

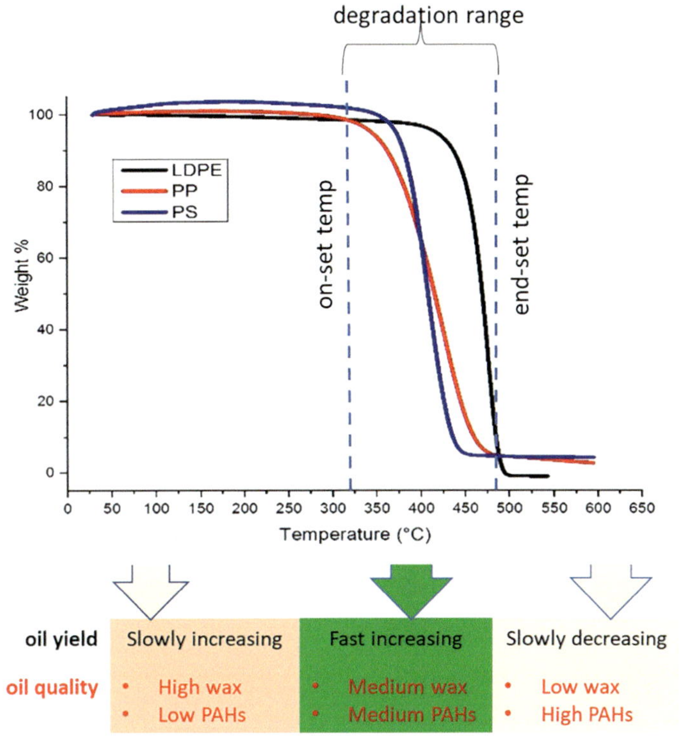

FIGURE 1.2 TGA plot and variation of pyrolysis oil yield and quality with regard to PAHs and wax content with temperature.

is high (30–300 min), consequently promoting the production of pyrolysis oil with relatively lower yields of gas and char. According to Das and Tiwari (2018), slow pyrolysis of PP at a heating rate of 10°C/min produced oil yields of 83%, which shows a high research octane number (RON) of 92% due to a higher fraction of $C_6 - C_{11}$ hydrocarbons. In another study, pyrolyzing municipal plastic waste (mixed plastic) at 500°C with a heating rate of 10°C/min resulted in a product yield of 75.8% oil, 10.0% solid char, and 14.2% gases (Singh et al., 2019).

In fast pyrolysis, a higher heating rate (200–1000°C/min) results in shorter vapor residence times at higher temperatures. The heating enables efficient heat transfer and rapid decomposition of the plastic waste, leading to extensive cracking of vapors, reduced secondary cracking reactions, increased production of gaseous products, and decreased liquid products. During fast heating, the degradation of plastic waste initiated with β-scission, followed by both intermolecular and intramolecular hydrogen transfer, results in the formation of a higher quantity of monomers and short-chain hydrocarbons alongside significant hydrogen, leading to increased gas production with relatively lower amounts of plastic oil and char.

1.5.3 Catalytic Pyrolysis

The primary distinction between catalytic and thermal pyrolysis lies in the utilization of a catalyst. Various catalysts, predominantly solid acid and base catalysts, have been employed for catalytic pyrolysis of plastic waste. These catalysts include zeolites and their derivatives (such as ZSM-5, H-ZSM-5, Zn-ZSM-5, zeolite-Y, natural zeolites, and fluid cracking catalyst), metal catalysts (such as magnesium oxide, copper carbonate, nickel-aluminum, nickel chloride, zinc, and sulfated zirconia), minerals (such as kaolin, iron-pillared clay, and dolomite), and fly ash, as indicated in Table 1.4.

Generally, polyolefins like LDPE, HDPE, and PP, as well as polyesters like PET and aromatic polymers like PS, are readily cracked with solid acid catalysts like zeolites and their derivatives, clays, and fly ash into lighter hydrocarbons due to their abundant Brønsted and Lewis acid sites (Peng et al., 2022). Researchers suggest that solid acid catalysts like zeolites are more effective than solid base catalysts like magnesium oxide in depolymerizing polymeric macromolecules like styrene dimers into monomers and other valuable chemicals like styrene, toluene, and ethylbenzene. Studies on the catalytic degradation and dechlorination of chloropolymers such as PVC have indicated that metal oxide–based solid base catalysts like zinc oxide and lanthanum oxide tend to adsorb the evolved HCl or inhibit the formation of HCl depending on their basicity, while solid acid catalysts like zeolites and noble metal catalysts can produce lighter plastic oil based on the total number of acid sites. The reaction mechanisms for solid acid and solid base catalysis of various plastic wastes have been extensively discussed in Al-Salem et al. (2017).

A variety of heterogeneous catalysts, including microporous zeolites such as HZSM-5, USY, HY, Hβ, and HMOR, mesoporous materials like SBA-15 and MCM-41, Fluid Catalytic Cracking (FCC) catalysts and spent FCC catalysts, clays such as kaolin, conventional base salts, and oxides such as ZnO, $CaCO_3$, MgO, CaO, Fe_2O_3, $CuCO_3$, and SiO_2, as well as metal or metal oxide–incorporated catalysts, have been utilized in the conversion of plastic waste to fuels. Recent studies on the thermo-catalytic pyrolysis of plastic waste have been compiled in Table 1.4, outlining the effects of different catalysts on pyrolysis product distribution. It is also discussed in the following sections.

1.5.4 Zeolites

Zeolites possessing varied acidities and structural features have demonstrated significant influence on both the reaction kinetics and the product profile of waste plastic pyrolysis. Specifically, zeolite catalysts offer favorable access for reactant molecules and transition states to Brønsted/Lewis acid sites. Additionally, their distinctive pore structures facilitate desired reactions, leading to enhanced selectivity toward desired products. For example, HZSM-5 is highly effective in promoting cracking reactions, such as chain-end scission and β-scission, due to its strong acidity, which boosts the production of $C_1 - C_4$ gaseous hydrocarbons, predominantly light olefins. HZSM-5 is highly effective in promoting cracking reactions, such as chain-end scission and β-scission, due to its strong acidity, which boosts the production of $C_1 - C_4$ gaseous hydrocarbons, predominantly light olefins (Peng et al., 2022). A study reported catalytic pyrolysis of LDPE at 380°C for 2 hours with HZSM-5, achieving a 98 wt% conversion rate and producing 40 wt% light hydrocarbons ($C_1 - C_4$), 55 wt% medium hydrocarbons ($C_5 - C_{12}$), and 5 wt% heavy hydrocarbons ($>C_{12}$). These results contrasted with those of a non-catalytic test, which showed a 31 wt% conversion and yielded 36 wt% light hydrocarbons, 52 wt% medium hydrocarbons, and 12 wt% heavy hydrocarbons (Figueiredo et al., 2016).

Overall, zeolite catalysts enhance the oil yield as well as improve the quality of oil by shifting the oil composition toward gasoline and diesel range hydrocarbons. However, the higher acidity of the zeolite catalyst promotes the production of gaseous compounds ($C_1 - C_5$)due to extensive cracking.

1.5.5 Fluid Catalytic Cracking

The FCC catalyst, comprising an acidic silica–alumina matrix, a binder, usually clay, and zeolite-Y crystals, is widely acknowledged for converting heavy oils into lighter hydrocarbons (gasoline) for industrial purposes. It has also been recognized as an effective catalyst for increasing the yield of liquid products up to 90 wt% and preventing wax formation in plastic waste pyrolysis (Banu et al., 2020). The FCC catalyst is extensively used in the petrochemical industry and remains in use for multiple cycles. However, after multiple uses, it could lose its activity amid the refinery process due to deposition of coke and metals on its surface; this waste catalyst is referred to as "spent FCC" catalyst. Worldwide, roughly 200,000 tons of spent FCC catalysts are generated annually that were ultimately disposed of in landfill. This spent FCC catalyst, typically discarded as waste from FCC operations, has proven to be effective in converting plastic waste into valuable products, enhancing the economic viability of the process. The enhanced pre-cracking activity in the spent FCC catalyst is attributed to the presence of metal deposits such as Ni, Fe, and V (Chiranjeevi et al., 2016; Meena et al., 2024). A study reported the pyrolysis of the mixture of LDPE, HDPE, and PP using spent FCC catalyst. This study propounds that spent FCC catalyst suppressed the formation of aromatic compounds and facilitated the production of diesel range hydrocarbon. The similarity in properties enables the plastic pyrolysis oil to serve as a viable alternative or additive to conventional gasoline and diesel, offering potential benefits in terms of fuel versatility and reducing reliance on traditional fossil fuels (Kongngoen et al., 2023).

1.5.6 Clays

In recent years, a significant focus has been seen on utilizing low-cost clays such as montmorillonite, saponite, beidellite, and hectorite for green energy production, primarily due to the reduction in processing costs. The layered structure of clays allows for the creation of a customizable porous network by inserting pillars between the sheets, resulting in the formation of a two-dimensional network with interconnected micropores typically more extensive than those found in zeolites. Due to their moderate acidity, clays discourage over-cracking and hydrogen transfer reactions, promoting higher liquid yields with a predominance of heavier hydrocarbons. Clay catalysts produce less coke than zeolites and can be recycled for multiple cycles (Moorthy et al., 2020).

In another study, Budsaereechai et al. (2019) conducted pyrolysis of HDPE, LDPE, PP, and PS using binder-free pelletized bentonite clay at 700°C, yielding 86–90 wt% liquid product for all plastics. The liquid product primarily comprised $C_5 - C_{24}$ hydrocarbons for PS pyrolysis and aliphatic diesel-range hydrocarbons for other plastics. Furthermore, Hakeem et al. (2018) utilized Kaolin clay in the pyrolysis of PP at 350°C, resulting in a maximum of 79.9 wt% liquid product mainly consisting of gasoline- and diesel-range hydrocarbons. A study found that clays, specifically saponite and montmorillonite, exhibited lower activity than USY at temperatures below 327°C, but their performance surpassed that of USY at higher temperatures. Clays yielded approximately 70 wt% liquid oil compared to 20 wt% with USY. Moreover, using clay catalysts resulted in a higher fraction of olefins due to reduced hydrogen-transfer secondary reactions resulting from lower acidity. Additionally, clays exhibited a 10 wt% lower coke formation compared to the USY catalyst (Li et al., 2017).

1.5.7 Base Salts and Oxides

Base catalysts and oxides are not as widely utilized as acid catalysts in waste plastic pyrolysis. Unlike acid catalysts, the liquid oil obtained with base catalysts is characterized by a higher proportion of paraffin and olefins, but typically contains fewer branched isomers and aromatics. Obeid et al. (2014) investigated the catalytic performance of NaOH, HUSY, and Hβ on HDPE pyrolysis at 450°C. NaOH yielded the highest liquid product yield (81 wt%) and the lowest gas yield (19 wt%). The liquid oil produced with NaOH consisted of 41% paraffins, 47% olefins, 10% alcohols, and minimal amounts of aromatics and other hydrocarbons. In one such study, HDPE was pyrolyzed thermally and catalytically using $BaCO_3$ catalyst to optimize temperature, catalyst/plastic ratio, and reaction time. The oil obtained at optimum conditions (450°C, 0.1 cat/plastic ratio, and 2 h reaction time) was analyzed using GC–MS; a significant fraction of hydrocarbon was in the range of $C_7 - C_{18}$. In another study, $BaCO_3$ nanomaterial catalyst was used to produce aromatic derivatives by the pyrolysis of waste polystyrene thermocol and virgin polystyrene. The results revealed that virgin polystyrene produced 99.29% of aromatic derivatives, while waste polystyrene thermocol produced 88.77% (Singh, 2023).

1.5.8 Metal or Metal Oxide–Incorporated Catalysts

Metal or metal oxides can be added to various catalysts to enhance the effectiveness of plastic waste pyrolysis, depending on the desired products, largely by modifying the distribution of acid sites. In highly acidic catalysts, the combination of Lewis acid sites with Brønsted acid sites on the support creates bifunctional catalysts that enhance the commercial viability of plastic waste pyrolysis due to versatile functionality (Akubo et al., 2019). While the metal oxide sites catalyze H-abstraction and hydrogenation/dehydrogenation reactions, the support's acid sites generate carbonium ions and facilitate cracking, isomerization, and cyclization reactions. This combination enables various reforming reactions, such as isomerization of straight-chain paraffin into branched paraffin, dehydrocyclization of straight-chain paraffin into naphthene, and dehydrogenation of naphthenes into aromatics, thereby improving the octane numbers of the resulting oil across different carbon numbers. Additionally, maintaining a balance between acid and metal sites promotes self-hydrogen generation, facilitating hydrogen transfer reactions and suppressing coke formation due to the catalyst's high activity in hydrogenation reactions.

Ni nanoparticles incorporated into ZSM-5 (Ni-ZSM-5) exhibited reduced activity compared to Ni^{2+} exchanged zeolite in the catalytic pyrolysis of HDPE, highlighting the importance of NiO as Lewis acid sites for aromatization reactions. Moreover, adding NiO into HY-enhanced olefin and aromatic contents in LDPE pyrolysis promotes H-transfer and dehydrogenation reactions. In addition to NiO, other metals with diverse properties have been incorporated into acidic supports to enhance the efficiency of plastic waste pyrolysis (Vichaphund et al., 2017). For instance, Akubo et al. (2019) observed that while thermal pyrolysis of HDPE at 600°C produced liquid, rich in aliphatic products, the introduction of transition metals (Ni, Fe, and Co) into zeolite-Y shifted the product distribution toward aromatics as the major components. Incorporating metals such as Fe, Co, and Zr into HY zeolite decreased the fraction of Brønsted acid sites and promoted the formation of aromatics (Akubo et al., 2019). In another study, it was noted that impregnating Pd into Hβ significantly improved wax conversion, dehydrogenation of paraffin, and hydrogenation of olefins, attributed to enhanced Lewis acidity. The study highlighted that while Hβ catalyzed the dehydrogenation of heavy wax to light olefins over its acidic surface, the bifunctional Pd/Hβ catalyst facilitated wax dehydrogenation on its Pd metal sites and subsequently degraded or isomerized the dehydrogenated intermediates over its acid sites (Choi et al., 2022).

In brief, the introduction of metal and metal oxides as Lewis acid sites into strongly acidic catalysts leads to increased aromatic compounds and decreased aliphatic compounds. Conversely, integrating metals and metal oxides as Brønsted acid sites into catalysts with lower acidity levels enhances the aliphatic fraction.

1.6 IN SITU VERSUS EX SITU APPROACH

In catalytic pyrolysis, the catalyst can be employed either within the same reactor as the plastic waste feedstock, termed in situ catalytic pyrolysis, or in a separate reactor, known as ex situ catalytic pyrolysis. In the case of in situ catalytic pyrolysis, the catalyst and plastic waste feedstock are physically mixed, allowing for catalytic upgrading of pyrolysis products within the same reactor. Conversely, ex situ catalytic pyrolysis involves thermally pyrolyzing the plastic waste feedstock and upgrading the pyrolysis vapor in a downstream catalytic bed before exiting the system. Each mode of catalyst operation, whether in situ or ex situ catalysis, comes with its own set of advantages and disadvantages. In situ catalysis is typically simpler, as it does not require retrofitting the existing pyrolysis system, and it consumes less energy for pyrolysis due to the lower decomposition temperature. However, it faces challenges such as difficult catalyst recovery from its mixture with solid residue and catalyst deactivation by the solid residue. On the other hand, with ex situ catalysis, it is easier to regenerate the catalyst, but it often necessitates an external catalytic bed and higher temperatures in the pyrolysis reactor to ensure effective upgrades of the pyrolysis vapor.

Ex situ catalysis tends to reduce the amount of wax produced compared to in situ catalysis, as the external downstream catalyst bed is preheated to the target temperature before contacting the catalyst particles with the pyrolysis vapor. In in situ catalysis, the simultaneous heating of feedstock and catalysts in the pyrolysis reactor may favor cross-linking reactions over cracking, leading to more wax formation (Inayat et al., 2022). Moreover, the choice between in situ and ex situ catalysis can influence the overall product distribution of pyrolysis due to differences in feedstock–catalyst contacts and interactions. Irrespective of whether in situ or ex situ catalytic pyrolysis is employed, either offers several advantages over thermal pyrolysis when dealing with plastic waste. These benefits include lower pyrolysis temperature, shorter residence time, higher reaction rate, reduced energy requirement, lower ash and wax yields, and improved quality of plastic oil. This superiority is primarily attributed to the catalyst's ability to lower the pyrolysis reaction's energy barrier or activation energy, thereby reducing the reaction temperature and energy requirement while enhancing reaction efficiency. Additionally, catalysts can alter the pyrolysis pathways and product selectivity, favoring the formation of desired products such as gasoline ($C_3 - C_{12}$) and diesel ($C_{13} - C_{20}$), and reduced the wax (>C_{203}) formation in plastic oil (Peng et al., 2022).

1.7 SUSTAINABLE PYROLYSIS PROCESS

The global plastic waste crisis requires innovative solutions, and pyrolysis has emerged as a promising technology. Plastic pyrolysis can address the twin problems of plastic waste disposal and depletion of fossil fuel reserves. Pyrolysis can convert 60–80 % of plastic waste into liquid fuels, with yields of up to 8 % in fast pyrolysis processes conducted at temperatures between 450°C and 600°C. The process also reduces greenhouse gas emissions by 40 %, mitigating 3.5 tons of CO_2-equivalent per ton of plastic waste processed. The pyrolysis gas, a hydrocarbon-rich gas with a calorific value of 25–24 MJ/kg, can be fed back to the process to extract energy for heating. This can reduce the dependence on external energy substantially. The hazardous gas emission from the pyrolysis process is less than that from incineration. Also, the reaction temperature in pyrolysis is less than that of incineration.

1.8 CONCLUSIONS AND PERSPECTIVES

Population growth, economic growth, and global industrialization have led to the generation of vast quantities of waste, including plastic waste, particularly single-use plastic. Pyrolysis can be conducted in the absence or presence of the catalyst. The performance of the pyrolysis process depends on reactor type, heating source, residence time, heating rate, and so forth. The primary products of plastic pyrolysis are oil, gas, and char. Further, slow or fast pyrolysis greatly affects the final product distribution. Various catalysts, predominantly solid acid and base catalysts, have been employed for catalytic pyrolysis of plastic waste. These catalysts include zeolites and their derivatives (such as ZSM-5, H-ZSM-5, Zn-ZSM-5, zeolite-Y, natural zeolites, and fluid cracking catalyst), metal catalysts (such as magnesium oxide, copper carbonate, nickel-aluminum, nickel chloride, zinc, and sulfated zirconia), minerals (such as kaolin, iron-pillared clay, and dolomite), and fly ash.

Challenges include high energy requirements and more efficient catalysts, which could improve liquid fuel yields. Future research should prioritize developing cost-effective and durable catalysts, improving energy efficiency in large-scale reactors, and integrating renewable energy sources to enhance sustainability further. Notably, the pyrolysis process can be feasible only if there is a consistent supply of waste materials. In the absence of a steady supply, the full capacity of the pyrolysis process cannot be utilized, leading to uneconomical operations. In situ pyrolysis and ex situ pyrolysis offer the choice between maximizing oil yield or gas yield. However, the in situ operation is similar, requiring lower investment and operating costs. But ex-situ operation, though expensive, provides control over final products such as gasoline and diesel. Numerous catalytic works are referred to in this chapter, giving detailed information on catalyst effectiveness, catalyst recovery, and product distribution. The choice of the catalyst and other operating conditions depends on the yield of the desired product.

Plastic pyrolysis can address the twin problems of plastic waste disposal and depletion of fossil fuel reserves. Further, the hazardous gas emission from the pyrolysis process is less than that from incineration. Also, the reaction temperature in pyrolysis is less than that of incineration leading, to a sustainable way of waste management.

REFERENCES

Aisien, E.T., Otuya, I.C., Aisien, F.A., 2021. Thermal and catalytic pyrolysis of waste polypropylene plastic using spent FCC catalyst. *Environ. Technol. Innov.* 22, 101455. https://doi.org/10.1016/j.eti.2021.101455

Akubo, K., Nahil, M.A., Williams, P.T., 2019. Aromatic fuel oils produced from the pyrolysis-catalysis of polyethylene plastic with metal-impregnated zeolite catalysts. *J. Energy Inst.* 92, 195–202. https://doi.org/10.1016/j.joei.2017.10.009

Al-Salem, S.M., 2019. Thermal pyrolysis of high density polyethylene (HDPE) in a novel fixed bed reactor system for the production of high value gasoline range hydrocarbons (HC). *Process Saf. Environ. Prot.* 127, 171–179. https://doi.org/10.1016/j.psep.2019.05.008

Al-Salem, S.M., Antelava, A., Constantinou, A., Manos, G., Dutta, A., 2017. A review on thermal and catalytic pyrolysis of plastic solid waste (PSW). *J. Environ. Manage.* 197, 177–198. https://doi.org/10.1016/j.jenvman.2017.03.084

Baloch, H.A., Siddiqui, M.T.H., Nizamuddin, S., Mubarak, N.M., Khalid, M., Srinivasan, M.P., Griffin, G.J., 2020. Solvothermal co-liquefaction of sugarcane bagasse and polyethylene under sub-supercritical conditions: Optimization of process parameters. *Process Saf. Environ. Prot.* 137, 300–311. https://doi.org/10.1016/j.psep.2020.01.018

Banu, J.R., Sharmila, V.G., Ushani, U., Amudha, V., Kumar, G., 2020. Impervious and influence in the liquid fuel production from municipal plastic waste through thermo-chemical biomass conversion technologies—A review. *Sci. Total Environ.* 718, 137287. https://doi.org/10.1016/j.scitotenv.2020.137287

Budsaereechai, S., Hunt, A.J., Ngernyen, Y., 2019. Catalytic pyrolysis of plastic waste for the production of liquid fuels for engines. *RSC Adv.* 9, 5844–5857. https://doi.org/10.1039/c8ra10058f

Chang, S.H., 2023. Plastic waste as pyrolysis feedstock for plastic oil production: A review. *Sci. Total Environ.* 877, 162719. https://doi.org/10.1016/j.scitotenv.2023.162719

Chattopadhyay, J., Pathak, T.S., Srivastava, R., Singh, A.C., 2016. Catalytic co-pyrolysis of paper biomass and plastic mixtures (HDPE (high density polyethylene), PP (polypropylene) and PET (polyethylene terephthalate)) and product analysis. *Energy* 103, 513–521. https://doi.org/10.1016/j.energy.2016.03.015

Chiranjeevi, T., Pragya, R., Gupta, S., Gokak, D.T., Bhargava, S., 2016. Minimization of waste spent catalyst in refineries. *Procedia Environ. Sci.* 35, 610–617. https://doi.org/10.1016/j.proenv.2016.07.047

Choi, I.H., Lee, H.J., Rhim, G.B., Chun, D.H., Lee, K.H., Hwang, K.R., 2022. Catalytic hydrocracking of heavy wax from pyrolysis of plastic wastes using Pd/Hβ for naphtha-ranged hydrocarbon production. *J. Anal. Appl. Pyrolysis* 161, 105424. https://doi.org/10.1016/j.jaap.2021.105424

Das, P., Tiwari, P., 2018. The effect of slow pyrolysis on the conversion of packaging waste plastics (PE and PP) into fuel. *Waste Manag.* 79, 615–624. https://doi.org/10.1016/j.wasman.2018.08.021

Diaz-Silvarrey, L.S., McMahon, A., Phan, A.N., 2018. Benzoic acid recovery via waste poly(ethylene terephthalate) (PET) catalytic pyrolysis using sulphated zirconia catalyst. *J. Anal. Appl. Pyrolysis* 134, 621–631. https://doi.org/10.1016/j.jaap.2018.08.014

Dwivedi, U., Naik, S.N., Pant, K.K., 2021a. High quality liquid fuel production from waste plastics via two-step cracking route in a bottom-up approach using bi-functional Fe/HZSM-5 catalyst. *Waste Manag.* 132, 151–161. https://doi.org/10.1016/j.wasman.2021.07.024

Dwivedi, U., Pant, K.K., Naik, S.N., 2021b. Controlling liquid hydrocarbon composition in valorization of plastic waste via tuning zeolite framework and SiO2/Al_2O_3 ratio. *J. Environ. Manage.* 297, 113288. https://doi.org/10.1016/j.jenvman.2021.113288

Ephraim, A., Pozzobon, V., Lebonnois, D., Peregrina, C., Sharrock, P., Nzihou, A., Pham Minh, D., 2023. Pyrolysis of wood and PVC mixtures: Thermal behaviour and kinetic modelling. *Biomass Convers. Biorefinery* 13, 8669–8683. https://doi.org/10.1007/s13399-020-00952-2

Figueiredo, A.L., Araujo, A.S., Linares, M., Peral, Á., García, R.A., Serrano, D.P., Fernandes, V.J., 2016. Catalytic cracking of LDPE over nanocrystalline HZSM-5 zeolite prepared by seed-assisted synthesis from an organic-template-free system. *J. Anal. Appl. Pyrolysis* 117, 132–140. https://doi.org/10.1016/j.jaap.2015.12.005

Garcia-Nunez, J.A., Pelaez-Samaniego, M.R., Garcia-Perez, M.E., Fonts, I., Abrego, J., Westerhof, R.J.M., Garcia-Perez, M., 2017. Historical developments of pyrolysis reactors: A review, *Energy and Fuels*. https://doi.org/10.1021/acs.energyfuels.7b00641

Global Plastics Outlook, 2022. *Global Plastics Outlook*. https://doi.org/10.1787/aa1edf33-en

Habtewold, A.T., Ambie, D.A., Eremed, W.B., 2020. Solar assisted pyrolysis system for High-Density polyethylene plastic waste to fuel conversion. *AIMS Energy* 8, 455–473. https://doi.org/10.3934/energy.2020.3.455

Hakeem, I.G., Aberuagba, F., Musa, U., 2018. Catalytic pyrolysis of waste polypropylene using Ahoko kaolin from Nigeria. *Appl. Petrochemical Res.* 8, 203–210. https://doi.org/10.1007/s13203-018-0207-8

Hariadi, D., Saleh, S.M., Anwar Yamin, R., Aprilia, S., 2021. Utilization of LDPE plastic waste on the quality of pyrolysis oil as an asphalt solvent alternative. *Therm. Sci. Eng. Prog.* 23, 100872. https://doi.org/10.1016/j.tsep.2021.100872

Inayat, Amer, Inayat, Alexandra, Schwieger, W., Sokolova, B., Lestinsky, P., 2022. Enhancing aromatics and olefins yields in thermo-catalytic pyrolysis of LDPE over zeolites: Role of staged catalysis and acid site density of HZSM–5. *Fuel* 314, 123071. https://doi.org/10.1016/j.fuel.2021.123071

Jia, H., Ben, H., Luo, Y., Wang, R., 2020. Catalytic fast pyrolysis of poly (ethylene terephthalate) (PET) with zeolite and nickel chloride. *Polymers (Basel)* 12(3), 705. https://doi.org/10.3390/polym12030705

Kaminsky, W., 2021. Chemical recycling of plastics by fluidized bed pyrolysis. *Fuel Commun.* 8, 100023. https://doi.org/10.1016/j.jfueco.2021.100023

Kim, Y.M., Lee, H.W., Choi, S.J., Jeon, J.K., Park, S.H., Jung, S.C., Kim, S.C., Park, Y.K., 2017. Catalytic copyrolysis of polypropylene and Laminaria japonica over zeolitic materials. *Int. J. Hydrogen Energy* 42, 18434–18441. https://doi.org/10.1016/j.ijhydene.2017.04.139

Kongngoen, P., Phetwarotai, W., Assabumrungrat, S., Phusunti, N., 2023. Possible use of spent FCC catalyst for upgrading of wax from the pyrolysis of plastics to liquid fuel. *J. Anal. Appl. Pyrolysis* 173, 106076. https://doi.org/10.1016/j.jaap.2023.106076

Kusenberg, M., Eschenbacher, A., Djokic, M.R., Zayoud, A., Ragaert, K., De Meester, S., Van Geem, K.M., 2022a. Opportunities and challenges for the application of post-consumer plastic waste pyrolysis oils as steam cracker feedstocks: To decontaminate or not to decontaminate? *Waste Manag.* 138, 83–115. https://doi.org/10.1016/j.wasman.2021.11.009

Kusenberg, M., Zayoud, A., Roosen, M., Thi, H.D., Abbas-Abadi, M.S., Eschenbacher, A., Kresovic, U., De Meester, S., Van Geem, K.M., 2022b. A comprehensive experimental investigation of plastic waste pyrolysis oil quality and its dependence on the plastic waste composition. *Fuel Process. Technol.* 227, 107090. https://doi.org/10.1016/j.fuproc.2021.107090

Li, K., Lei, J., Yuan, G., Weerachanchai, P., Wang, J.Y., Zhao, J., Yang, Y., 2017. Fe-, Ti-, Zr- and Al-pillared clays for efficient catalytic pyrolysis of mixed plastics. *Chem. Eng. J.* 317, 800–809. https://doi.org/10.1016/j.cej.2017.02.113

Luo, H., Yao, D., Zeng, K., Li, J., Yan, S., Zhong, D., Hu, J., Yang, H., Chen, H., 2022. Solar pyrolysis of waste plastics with photothermal catalysts for high-value products. *Fuel Process. Technol.* 230, 107205. https://doi.org/10.1016/j.fuproc.2022.107205

Maqsood, T., Dai, J., Zhang, Y., Guang, M., Li, B., 2021. Pyrolysis of plastic species: A review of resources and products. *J. Anal. Appl. Pyrolysis* 159, 105295. https://doi.org/10.1016/j.jaap.2021.105295

Mariappan, M., Panithasan, M.S., Venkadesan, G., 2021. Pyrolysis plastic oil production and optimisation followed by maximum possible replacement of diesel with bio-oil/methanol blends in a CRDI engine. *J. Clean. Prod.* 312, 127687. https://doi.org/10.1016/j.jclepro.2021.127687

Meena, P., Singh, S., Sharma, N., Saharan, V.K., George, S., Bhoi, R., 2023. Performance, combustion, and emission characteristics of bio-oil produced by in situ catalytic pyrolysis of polypropylene using spent FCC. *Environ. Sci. Pollut. Res.* https://doi.org/10.1007/s11356-023-30786-0

Meena, P., Singh, S., Sharma, M., Saharan, V.K., George, S., Bhoi, R., 2024. Enhanced oil yield by catalytic pyrolysis of thermoplastics using cost-effective spent FCC and BaCO3 and its valorization to gasoline and diesel grade Fuel via Fractionation. *Korean J. Chem. Eng.* https://doi.org/10.1007/s11814-024-00041-4

Miandad, R., Barakat, M.A., Rehan, M., Aburiazaiza, A.S., Ismail, I.M.I., Nizami, A.S., 2017. Plastic waste to liquid oil through catalytic pyrolysis using natural and synthetic zeolite catalysts. *Waste Manag.* 69, 66–78. https://doi.org/10.1016/j.wasman.2017.08.032

Miandad, R., Rehan, M., Barakat, M.A., Aburiazaiza, A.S., Khan, H., Ismail, I.M.I., Dhavamani, J., Gardy, J., Hassanpour, A., Nizami, A.S., 2019. Catalytic pyrolysis of plastic waste: Moving toward pyrolysis based biorefineries. *Front. Energy Res.* 7. https://doi.org/10.3389/fenrg.2019.00027

Moorthy Rajendran, K., Chintala, V., Sharma, A., Pal, S., Pandey, J.K., Ghodke, P., 2020. Review of catalyst materials in achieving the liquid hydrocarbon fuels from municipal mixed plastic waste (MMPW). *Mater. Today Commun.* 24, 100982. https://doi.org/10.1016/j.mtcomm.2020.100982

Nanda, S., Berruti, F., 2021. Thermochemical conversion of plastic waste to fuels: A review. *Environ. Chem. Lett.* 19, 123–148. https://doi.org/10.1007/s10311-020-01094-7

Nizzetto, L., Futter, M., Langaas, S., 2016. Are agricultural soils dumps for microplastics of urban origin? *Environ. Sci. Technol.* 50, 10777–10779. https://doi.org/10.1021/acs.est.6b04140

Obeid, F., Zeaiter, J., Al-Muhtaseb, A.H., Bouhadir, K., 2014. Thermo-catalytic pyrolysis of waste polyethylene bottles in a packed bed reactor with different bed materials and catalysts. *Energy Convers. Manag.* 85, 1–6. https://doi.org/10.1016/j.enconman.2014.05.075

Pal, S., Kumar, A., Ali, M.A., Gupta, N.K., Pandey, S., Ghodkhe, P.K., Bhurat, S., Alam, T., Eldin, S.M., Dobrota, D., 2023. Experimental evaluation of diesel blends mixed with municipal plastic waste pyrolysis oil on performance and emission characteristics of CI engine. *Case Stud. Therm. Eng.* 47, 103074. https://doi.org/10.1016/j.csite.2023.103074

Park, K.B., Choi, M.J., Chae, D.Y., Jung, J., Kim, J.S., 2022. Separate two-step and continuous two-stage pyrolysis of a waste plastic mixture to produce a chlorine-depleted oil. *Energy*. https://doi.org/10.1016/j.energy.2021.122583

Parku, G.K., 2019. *Pyrolysis of waste polypropylene plastics for energy recovery: Investigation of operating parameters and process development at pilot scale*, Master's Thesis, Faculty of Engineering at Stellenbosch University, 2019. https://scholar.sun.ac.za/server/api/core/bitstreams/57b22368-d606-4f54-8f3b-90141899881c/content

Peng, Y., Wang, Y., Ke, L., Dai, L., Wu, Q., Cobb, K., Zeng, Y., Zou, R., Liu, Y., Ruan, R., 2022. A review on catalytic pyrolysis of plastic wastes to high-value products. *Energy Convers. Manag.* 254, 115243. https://doi.org/10.1016/j.enconman.2022.115243

Prajapati, R., Kohli, K., Maity, S.K., Sharma, B.K., 2021. Potential chemicals from plastic wastes. *Molecules* 26, 1–22. https://doi.org/10.3390/molecules26113175

Prathiba, R., Shruthi, M., Miranda, L.R., 2018. Pyrolysis of polystyrene waste in the presence of activated carbon in conventional and microwave heating using modified thermocouple. *Waste Manag.* 76, 528–536. https://doi.org/10.1016/j.wasman.2018.03.029

Sekar, M., Ponnusamy, V.K., Pugazhendhi, A., Nižetić, S., Praveenkumar, T.R., 2022. Production and utilization of pyrolysis oil from solidplastic wastes: A review on pyrolysis process and influence of reactors design. *J. Environ. Manage.* 302. https://doi.org/10.1016/j.jenvman.2021.114046

Singh, M.V., 2023. A BaCO3 nanomaterial for pyrolysis of sustainable waste and virgin polystyrene into green aromatic derivatives. *ChemistrySelect* 8, 2–11. https://doi.org/10.1002/slct.202204400

Singh, M.V., Kumar, S., Sarker, M., 2018. Waste HD-PE plastic, deformation into liquid hydrocarbon fuel using pyrolysis-catalytic cracking with a CuCO 3 catalyst. *Sustain. Energy Fuels* 2, 1057–1068. https://doi.org/10.1039/c8se00040a

Singh, R.K., Ruj, B., Sadhukhan, A.K., Gupta, P., 2019. Impact of fast and slow pyrolysis on the degradation of mixed plastic waste: Product yield analysis and their characterization. *J. Energy Inst.* 92, 1647–1657. https://doi.org/10.1016/j.joei.2019.01.009

Singh, R.K., Ruj, B., Sadhukhan, A.K., Gupta, P., Tigga, V.P., 2020. Waste plastic to pyrolytic oil and its utilization in CI engine: Performance analysis and combustion characteristics. *Fuel* 262, 116539. https://doi.org/10.1016/j.fuel.2019.116539

Sivagami, K., Kumar, K. V., Tamizhdurai, P., Govindarajan, D., Kumar, M., Nambi, I., 2022. Conversion of plastic waste into fuel oil using zeolite catalysts in a bench-scale pyrolysis reactor. *RSC Adv.* 12, 7612–7620. https://doi.org/10.1039/d1ra08673a

van der Westhuizen, S., Collard, F.X., Görgens, J., 2022. Pyrolysis of waste polystyrene into transportation fuel: Effect of contamination on oil yield and production at pilot scale. *J. Anal. Appl. Pyrolysis* 161, 1–9. https://doi.org/10.1016/j.jaap.2021.105407

Verma, A., Sharma, S., Pramanik, H., 2021. Pyrolysis of waste expanded polystyrene and reduction of styrene via in-situ multiphase pyrolysis of product oil for the production of fuel range hydrocarbons. *Waste Manag.* 120, 330–339. https://doi.org/10.1016/j.wasman.2020.11.035

Vichaphund, S., Aht-ong, D., Sricharoenchaikul, V., Atong, D., 2017. Effect of CV-ZSM-5, Ni-ZSM-5 and FA-ZSM-5 catalysts for selective aromatic formation from pyrolytic vapors of rubber wastes. *J. Anal. Appl. Pyrolysis* 124, 733–741. https://doi.org/10.1016/j.jaap.2016.11.011

Zheng, K., Wu, Y., Hu, Z., Wang, S., Jiao, X., Zhu, J., Sun, Y., Xie, Y., 2022. Progress and perspective for conversion of plastic wastes into valuable chemicals. *Chem. Soc. Rev.* 52, 8–29. https://doi.org/10.1039/d2cs00688j

2 Hydrogen Production from Plastic Waste

Nishika Sharma[1], *Smita Biswas*[1], *Arpita Kitchloo*[1], *Surbhi Sharma*[1], *Manju M. Gupta*[2], *Ashok Pandey*[3], *and Abha Kumari*[1*]
[1]Amity Institute of Biotechnology, Amity University, Noida, India
[2]School of Earth and Sustainability, Northern Arizona University, USA
[3]Centre for Energy and Environmental Sustainability, Lucknow, India

2.1 INTRODUCTION

Production of hydrogen from plastic waste is one of the emerging ways to tackle the two major problems of waste management and sustainable production of energy. Since plastic wastes increase in volume rapidly and their harmful impacts on the environment are hazardous, the present initiative should be taken to conduct profound research into advanced thermochemical and photochemical processes for hydrogen production, which has become an innocuous carrier of energy. Pyrolysis is a thermal decomposition process that takes place in the absence of oxygen. It breaks down plastic waste into syngas, oil, and char, which can recover hydrogen-rich gases. However, the temperature and catalyst conditions must be optimized for the effective conversion of the plastic waste (Li et al., 2023a). Gasification is another high-temperature process that involves limited oxygen or steam to convert plastics into syngas, a mixture of hydrogen and carbon monoxide. This method is effective but requires advanced gas-cleaning technologies to remove impurities (Kaydouh & El Hassan, 2022). Photo-reforming employs solar energy and photocatalysts to degrade plastics and water, producing hydrogen with minimal environmental emissions. Although in its experimental stages, this method holds promise for sustainable hydrogen production with improvements in catalyst efficiency and durability (Zhang et al., 2023a). The catalytic system uses electric currents to electro-reform, decomposing the plastics into hydrogen and oxygen gases.

While offering controlled reaction rates, its high energy demands can be mitigated using renewable energy sources (Wang et al., 2022b). The methods have unique technical and economic implications. All these routes of hydrogen production from plastic waste entail specific technical and economic impacts that make new research a necessity to optimize the processes to yield hydrogen as an important clean energy source from otherwise environmentally detrimental waste products. In the case of specific achievements in these technologies, HGPW will play a more important role in reducing accumulation and waste, promoting a circular economy, and globalization toward energy sustainability and protection of the environment. Research is imperative to optimize these technologies to sustainably produce hydrogen while addressing plastic waste accumulation, promoting circular economies, and advancing global energy sustainability. Integrating waste management with the production of hydrogen addresses the plastic pollution problem, but it is also in tandem with the circular economy principles. Co-pyrolysis strategy, which has shown promise—co-pyrolyzing plastic wastes with other material streams such as biomass—would enhance thermal stability and yields for hydrogen. The results from some studies on biomass-plastic co-conversion are found in Adeniyi

* Corresponding author's email: akumari@amity.edu

DOI: 10.1201/9781003581000-3

et al. (2024). Additionally, emerging technologies such as photocatalytic hydrogen production are gaining traction for their ability to utilize sunlight, further reducing carbon emissions (Zhang et al., 2023b). However, challenges like feedstock variability and the high initial costs of reactor systems remain barriers to widespread adoption (Kibria et al., 2023).

The environmental and economic impacts of using plastic waste for hydrogen production are significant. It has the added advantage of countering the adverse effects of plastic waste on ecosystems and developing a clean energy carrier. Ways to scale this technology include upgrading collection systems for plastic waste, installing decentralized pyrolysis units, and incorporating hydrogen production into existing waste management infrastructure (Jambeck et al., 2015). In addition, policies for investment in waste-to-energy technologies are expected to further boost the move toward a low-carbon economy. As more research is developed, cost optimization and process minimization would be expected to make hydrogen from plastic waste a viable and sustainable source of energy (Ren et al., 2023).

The economic potential in the production of hydrogen from plastic waste is extensive, which could lead to excellent profitability for its stakeholders. There is a possibility that the market for hydrogen is going to surge exponentially, as people increase their demand for clean energy, coupled with policies encouraging such consumption. Recycling plastic waste into hydrogen generates revenue from hydrogen sales while saving money on waste disposal and the cost of operating landfills (Kibria et al., 2023). Moreover, decentralized hydrogen production facilities will create employment opportunities in engineering, operations, and maintenance, hence boosting the local economy. For example, small-scale pyrolysis units can provide sustainable business models for communities, allowing them to monetize waste while contributing to energy self-sufficiency (Adeniyi et al., 2024).

The integration of plastic-to-hydrogen technologies into existing waste management systems, viewed from a cost–benefit perspective, will increase resource efficiency and bring about long-term savings. In turn, transforming a problematic waste stream into a lucrative energy resource brings considerable returns on investment to the stakeholders. The advanced catalysts along with optimized reactor designs can significantly improve the hydrogen yields, making the process economically viable even at a smaller scale (Zhang et al., 2023a). In addition, the reduction of carbon emissions due to the replacement of fossil fuels with hydrogen can attract green subsidies and tax benefits, further working in favor of profitability. Governments and private investors alike recognize the potential of this approach, with increased funding for waste-to-energy technologies and the establishment of hydrogen-focused infrastructure (Ren et al., 2023).

Apart from financial benefits, the production of hydrogen from plastic waste provides considerable environmental and policy-driven benefits. With countries across the globe setting net-zero carbon targets, hydrogen is turning out to be a critical component in decarbonizing transportation, power, and manufacturing sectors (Zhang et al., 2023a). Furthermore, these technologies can attract green subsidies, tax benefits, and private investment, thus further increasing profitability. Recent studies highlight the scaling potential of hydrogen production technologies, especially in cities where plastic waste is abundant (Pandey et al., 2023). With proper alignment of economic incentives with environmental objectives, plastic-to-hydrogen processes will be the foundation of sustainable development. Figure 2.1 shows data on the generation of different kinds of plastic waste globally.

2.2 GLOBAL PLASTIC WASTE GENERATION

The most pressing environmental issue of the 21st century is the proliferation of plastic generation at a rate of 300 million metric tons every year. A significant portion of the total becomes waste (Borrelle et al., 2020). It's durable, affordable, and versatile in its use yet also persists in the environment, posing severe risks. Studies have also shown that, so far, only 9% of the plastic waste that has ever been produced has been recycled, whereas 79% has accumulated in landfills or the natural environment (MacLeod et al., 2021). Further, plastic waste affects marine ecosystems in a way that

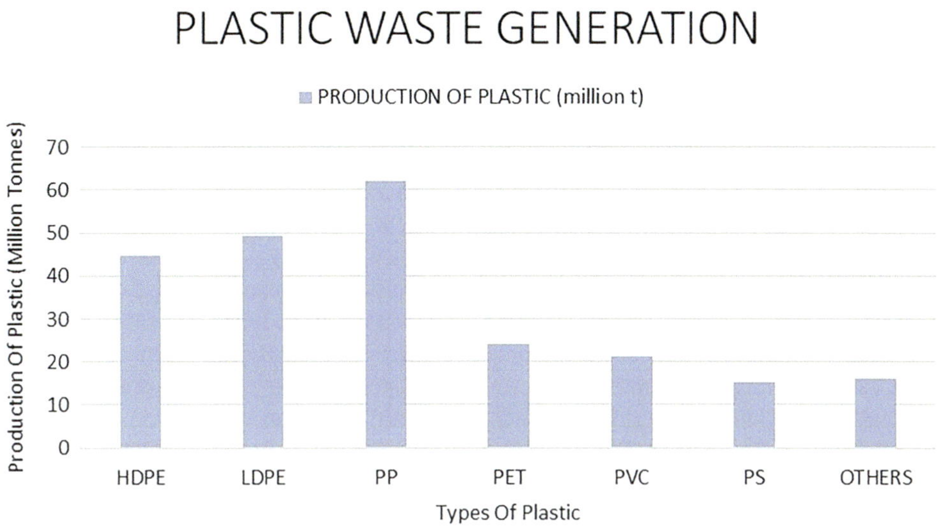

FIGURE 2.1 The generation of different kinds of plastic waste globally.

it becomes entangled, ingested, and causes habitat destruction (Eriksen et al., 2014). Projections indicate that plastic production and associated waste will triple by 2060 without significant intervention, further exacerbating its ecological footprint (Borrelle et al., 2020).

Plastic pollution also directly and indirectly impacts human health. Microplastics, plastic particles smaller than 5.0 mm, are particularly worrying because they are widespread and can bioaccumulate (Campanale et al., 2020). The presence of these particles has been documented in drinking water, seafood, and even human blood (Leslie et al., 2022). Exposure to microplastics has been reported to induce oxidative stress, cytotoxicity, and inflammation (Bhuyan, 2022). In addition, additives to plastics like phthalates and bisphenol A are known endocrine disruptors, which may lead to developmental, reproductive, or metabolic disorders (Li et al., 2023b). The presence of microplastics in the food chain underlines the interconnection between environmental degradation and human health impacts (Prata et al., 2019).

The environmental impacts of plastic waste are most prominent in aquatic environments. Oceans have the highest concentrations of plastic waste, with over 5 trillion plastic particles afloat (Eriksen et al., 2014). The plastics break down into microplastics, which interfere with feeding channels and allow toxic chemicals to be released into marine organisms (Gewert et al., 2021). These also include coral reefs, which play a significant role in biodiversity and coastal protection. Plastic debris further promotes the spread of coral diseases (Pinheiro et al., 2023). The economic impact is also enormous, with the estimated costs for marine industries, such as fisheries and tourism, running into billions of dollars every year (Napier & Thompson, 2020). Such environmental and economic damages highlight the importance of reducing plastic waste generation.

The need to address the crisis calls for a systemic change in waste management and plastic use. The circular economy framework supports the reduction, reuse, and recycling of plastics to decrease environmental impact (Walker & Fequet, 2023). However, efforts such as biodegradable plastics have yet to be confirmed due to their incomplete degradation in natural conditions (Royer et al., 2023). Policies on single-use plastics, upgraded recycling infrastructures, and international cooperation are all essential to plastic pollution control. Stakeholder partnerships, especially between governments, industries, and consumers, are important for the adoption of sustainable practices and the reduction of dependence on virgin plastic production (Lampitt et al., 2023).

2.3 SOURCES OF PLASTIC

Plastic waste has become a major global concern due to its high usage in all industries and failure of waste management systems. The use of plastic is highly widespread because it is versatile and cheap to produce, thus finding its way into the packaging, construction, and electronics sectors. However, this has led to the generation of the most massive volumes of waste produced, estimated to be 275 million metric tons of plastic waste in 2010, where 4.8 to 12.7 million entered marine environments (Jambeck et al., 2015). The single-use packaging material, which majorly comprises bags, bottles, and wrappers, accounts for about half of this waste since society relies greatly on disposable items (Ncube et al., 2021). The linear consumption pattern of producing, using, and discarding intensifies the problem, especially in countries that do not have a well-developed infrastructure for waste management. For instance, low- and middle-income countries significantly contribute to marine plastic pollution due to inefficient systems for capturing and recycling waste (Borrelle et al., 2020). In this regard, plastic pollution necessitates systemic change in the patterns of production, consumption, and waste management through innovative recycling methods and strict regulatory frameworks (Table 2.1).

TABLE 2.1
Applications of Different Plastic Types

Type of Plastic	Common Uses	Chemical Composition	Characteristics	Reference
Polyethylene (PE)	Packaging films, grocery bags, detergent bottles, piping, geomembranes	$(C_2H_4)_n$	Tough, near-zero moisture absorption, excellent chemical inertness, low coefficient of friction, ease of processing, and electrical conductivity	Mahmood et al. (2017)
Polypropylene (PP)	Trays, funnels, pails, bottles, carboys and instrument jars (sterilized) used in clinics	$(C_3H_6)_n$	Low-density, high-melting temperature, and chemical inertness with low cost in making PP	Kim et al. (2023)
Polystyrene (PS)	Construction materials (insulation), packaging foam, food containers, disposable cups, plates, cutleries, cassette boxes, and compact disks	$(C_8H_8)_n$	Low cost, lightweight, ease of manufacture, versatility, thermal efficiency, durability, and moisture resistant	Turner (2020)
Polyethylene Terephthalate (PET)	Beverage bottles, food packaging, synthetic fibers, thermally stabilized films, and electrical components	$(C_{10}H_8O_4)_n$	Strong, lightweight, thermally stable, excellent moisture and gas barrier properties, transparent polymer, with good mechanical properties and good stability under variable load	Wei et al. (2019)
Polyvinyl Chloride (PVC)	Pipes, cable insulation, clothing, flooring, foil, blisters or light-transmitting panels, food wrap	$(C_2H_3Cl)_n$	Flexible, high chemical resistance and favorable mechanical properties, resistance to water and weather conditions	Fakhri et al. (2022)

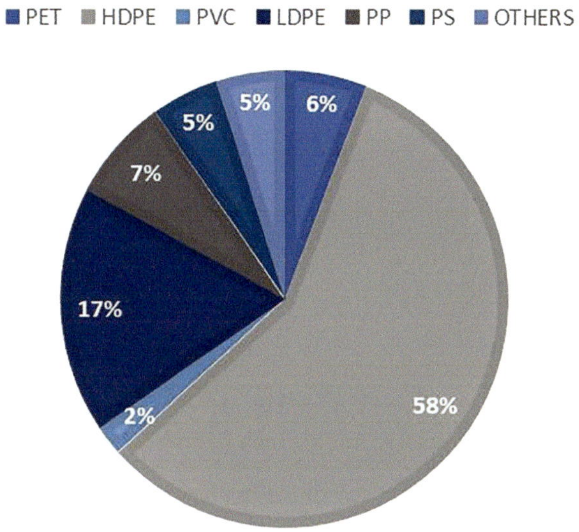

FIGURE 2.2 The composition of municipal plastic waste by weight.

The primary source of plastic pollution mainly induced by urbanization and the increase in population is municipal waste. Figure 2.2 shows the composition of municipal plastic waste. There are immense amounts of single-use plastics, such as food wrapping material, water bottles, and disposable cutlery items, in municipal solid waste, which are discarded after a very short period. Municipal systems worldwide gather and treat nearly 2 billion tons of waste every year, but most of it remains unmanaged, especially in developing areas, as indicated by Sousa in 2021. Urban centers in India and Indonesia, for example, have challenges in handling municipal plastic waste since segregation, collection, and recycling capabilities are very limited (Kibria et al., 2023). In the worst-case scenario, the wastes are disposed of in landfills and contaminate the soil and groundwater, or leak into rivers and reach the oceans. Initiatives led by communities, such as improved waste segregation programs and plastic buy-back schemes, have the potential to reduce the environmental footprint of municipal waste. For example, decentralized waste-to-energy programs in urban India have shown that community involvement in waste management significantly reduces plastic pollution while offering economic benefits (Pandey et al., 2023).

Another major source of plastic pollution is e-waste, which consists of discarded electrical and electronic devices. Modern electronics are highly dependent on plastics for casings, insulation, and structural components, making them a critical contributor to the increasing volume of e-waste. E-waste generation around the world increased to 53.6 million metric tons in 2019 and is projected to continue rising because of fast technological change and shorter lifespans of devices (Ren et al., 2023). It is quite difficult to recycle plastics due to their mixtures with hazardous materials, such as lead, cadmium, and flame retardants, in e-waste. Such substances cause complications in recycling and create severe health risks when not handled properly. In countries like Ghana and Nigeria, where recycling infrastructure is less developed, informal forms of recycling often take place, and they involve open burning of e-waste in open-air areas and subsequently emit toxic fumes and microplastics into the environment (Gaur et al., 2022). Innovations in recycling technologies, including advanced polymer recovery and chemical recycling, seek to extract high-quality materials from e-waste plastics, but large-scale adoption is limited. Industrial processes also generate a lot of plastic waste, both directly as manufacturing by-products and indirectly through packaging.

Innovative solutions are being sought in the construction industry, where plastic waste is repurposed into sustainable materials such as plastic-enhanced bricks and concrete (Haq et al., 2023).

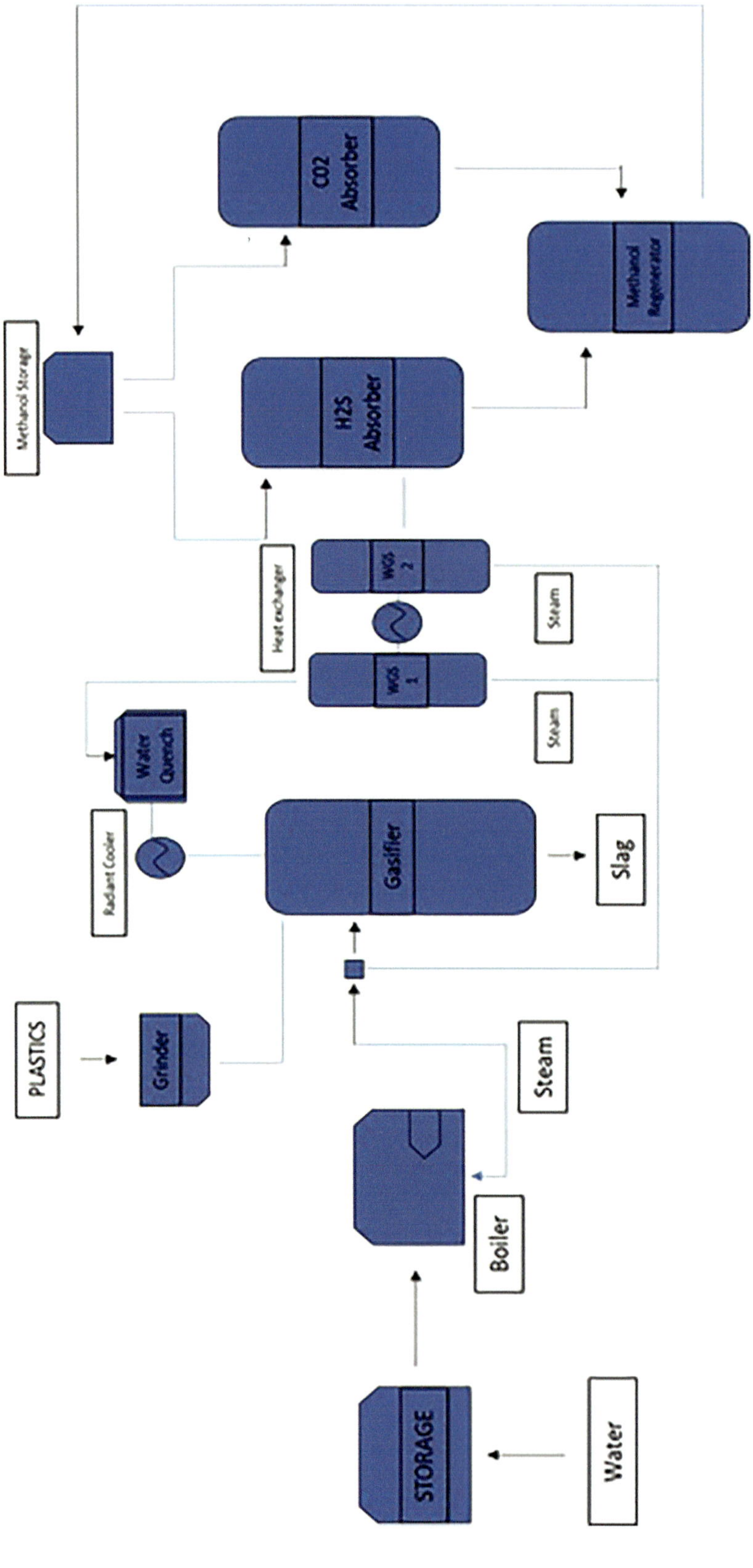

FIGURE 2.3 The production of hydrogen from polyethylene and polypropylene waste.

Such materials have the potential to reduce the environmental footprint of construction while enhancing material properties, such as durability and thermal resistance. Challenges in scaling up these initiatives, however, are encountered due to variability in the availability and quality of plastic waste. More and more, industries around the world are adopting circular economy principles to reduce the generation of industrial waste by designing products for recyclability and improving material reuse. However, many industrial waste management systems face problems such as high operation costs, bureaucratic barriers, and a lack of demand from consumers for recycled materials (Zhang et al., 2023b). To overcome these issues, there must be a cooperative effort between policymakers, industry players, and consumers to develop sustainable industrial practices. Agricultural activities are also responsible for plastic waste mainly through the usage of synthetic materials such as mulch films, irrigation pipes, and fertilizer bags. Although plastics are important to enhance agricultural productivity by reducing the evaporation of water and enhancing weed control, their improper disposal leads to serious environmental issues.

Plastics used in agriculture often degrade into microplastics, contaminating soil and water systems and affecting crop health and ecosystem stability (Islam et al., 2022). For instance, studies in Bangladesh have documented the accumulation of microplastics in farmlands due to improper disposal of plastic materials, threatening soil fertility and food safety. The use of biodegradable alternatives, such as compostable mulch films, has been promising in reducing the environmental impact of agricultural plastics. However, the high costs and limited availability of such alternatives are major barriers to their widespread implementation (Sahani et al., 2022). Improved regulations, education, and incentivization programs for farmers may be effective in promoting sustainable agricultural practices to reduce plastic pollution.

2.4 PLASTIC WASTE AS A FEEDSTOCK FOR HYDROGEN PRODUCTION

Plastic waste is a rich and relatively underutilized feedstock for hydrogen production due to its unique chemical composition. Most plastics, including polyethylene (PE), polypropylene (PP), polystyrene (PS), and polyethylene terephthalate (PET), are primarily composed of hydrocarbons, making them a high hydrogen-to-carbon ratio feedstock. Compared to biomass or fossil fuels, plastics are very hydrogen-rich feedstocks, hence holding a lot of energy potential during conversion processes (Yang et al., 2022). For instance, PE and PP, which constitute more than 50% of global plastic production, are polymers that contain simple hydrocarbon chains that make hydrogen recovery very efficient. Studies show that such polymers can yield a considerable amount of hydrogen per unit mass when processed thermochemically, providing a sustainable alternative to traditional hydrogen sources (Chen et al., 2023; Al-Fatesh et al., 2023).

The global prevalence of plastic waste further adds to its appeal as a source of hydrogen. Estimates indicate the generation of over 300 million metric tons of plastic waste annually, most of which is not recycled or disposed of properly (Jambeck et al., 2015). This waste is most often a burden on landfills and contributes to degrading the environment. By repurposing plastic waste as a feedstock for hydrogen production, industries can address two critical global challenges: waste management and the transition to clean energy (Lan & Yao, 2022). Additionally, plastic waste is readily available and geographically distributed, offering localized opportunities for hydrogen generation in regions with limited access to conventional energy resources. This can help reduce reliance on fossil fuels and promote energy independence in developing economies (Hren et al., 2023).

PET and PS are among the plastics with specific advantages in hydrogen feedstocks. PET, used in many beverage bottles and packaging, offers a combination of high density and hydrogen yield potential, making it an attractive option for industrial applications (Ren et al., 2023). Similarly, PS is commonly used in insulations and disposable food containers, which possess a high calorific value to be efficiently converted into hydrogen and other valuable by-products (Yang et al., 2022). The flexibility of processing plastic wastes, such as mixed or contaminated streams, further justifies the applicability of plastic waste in the production of hydrogen. This flexibility reduces the amount of sorting and preprocessing needed, decreasing operation costs and making the whole process more economically viable (Khan et al., 2023).

From an economic point of view, plastic waste is a relatively inexpensive and readily available feedstock as compared to natural gas or coal. The ever-increasing price of fossil fuels and the importance placed on decarbonization make plastic waste a viable substitute. Additionally, hydrogen production from plastic waste has economic benefits since it creates a market for waste that was once considered worthless. This would stimulate employment opportunities in waste collection, sorting, and processing industries, especially in areas facing unemployment and poor waste management infrastructure (Lan & Yao, 2022). In addition, the high calorific value of plastics is efficiently converted into energy recovery in the conversion processes, thus making the production of hydrogen more economically viable (Ren et al., 2023).

Scalability and resource availability also make plastic waste an advantage. Unlike solar or wind-based renewable hydrogen sources, plastic waste does not depend on weather conditions or geographical restrictions. It is consistently available, and this reliability is very valuable in regions that have a high generation of plastic waste but not much renewable energy capacity. Additionally, the use of plastic waste in hydrogen production aligns with global sustainability goals, reducing greenhouse gas emissions and promoting the responsible management of waste resources (Al-Qadri et al., 2022; Chen et al., 2023).

2.5 TECHNIQUES OF HYDROGEN PRODUCTION FROM PLASTIC WASTE

Hydrogen production from plastic waste employs various innovative techniques that transform this environmental challenge into a resource. The primary methods include thermochemical processes, photocatalytic approaches, and electrocatalytic routes, each with distinct operational parameters and efficiencies. Table 2.2 shows the key techniques for hydrogen production from plastic waste with some examples of cases where hydrogen is being successfully produced from plastic waste.

2.5.1 Thermochemical Processes

2.5.1.1 Pyrolysis

Pyrolysis is the thermal decomposition of plastic waste within a high-temperature range of 500–1000°C in an environment without oxygen (Figure 2.4). The degradation process converts plastics into hydrocarbons and hydrogen. For example, hydrogen can be obtained from the pyrolysis of polypropylene with yields that may reach 2360 Nm^3/ton (Hossain & Mahmud, 2024). It has already been found that Fe/Al_2O_3 can act as a catalyst for improving hydrogen yields. For example, pyrolysis combined with steam reforming and water gas shift reaction can reach a hydrogen production of 122 mmol/plastic for optimized conditions (Alshareef et al., 2023). More recently, reactors such as fluidized bed pyrolysis have been able to enhance heat transfer efficiency, which is necessary for higher rates of conversion. Shah et al. (2024) reported that a proper reactor configuration provided superior output of hydrogen.

2.5.1.2 Steam Reforming

Pyrolysis oils are subjected to steam reforming, where they react with steam in the presence of catalysts such as $Ni/Ce\text{-}Zr\text{-}Mg/Al_2O_3$ to produce hydrogen (Figure 2.5). This process achieves efficiencies as high as 91.3%, making it a promising method for large-scale hydrogen production (Patlolla et al., 2023). The addition of bimetallic catalysts such as Ni-Fe further enhances conversion efficiency (Hsu et al., 2023).

TABLE 2.2
Generation of Hydrogen from Plastic Wastes Using Different Methodologies

Plastic Type	Process Description	Result	Reference
Polyethylene (PE)	Pyrolysis at 550°C, catalytic steam reforming with Fe/Al_2O_3 at 350–550°C under optimized steam/carbon ratios.	122 mmol/g hydrogen	Alshareef et al. (2023)
Polyethylene (PE)	Pyrolysis of polyethylene at 500°C, followed by hydrogen separation using a proprietary hydrogen extraction unit.	90% hydrogen recovery	Ismail and Dincer (2023a)
Mixed plastics	Pyrolysis at 700°C, gasification with Ni-based catalysts under atmospheric pressure.	64 mmol/g hydrogen	Al-Fatesh et al. (2023)
Mixed plastics	Gasification of mixed plastics at 800°C, with in situ carbon capture using CaO.	70% hydrogen syngas	Lan and Yao (2022)
Mixed plastics	Catalytic pyrolysis at 600°C using Cu/Al_2O_3 catalyst under controlled pressure of 5 bar.	85% hydrogen yield	Mishra et al. (2023)
High-density PE (HDPE)	Hydrocracking of HDPE at 300°C, Ni/ZrO_2-SO_4 catalyst, under 20 mL/min hydrogen flow.	44% hydrogen yield	Amin et al. (2018)
Mixed plastics	Steam reforming at 850°C, Ni-based core-shell catalysts with ceria-zirconia shell.	90 mmol/g hydrogen	Wu et al. (2020)
High-density PE (HDPE)	Gasification of HDPE plastic at 800°C, catalytic activity enhanced by Ni-MgO/γ-Al_2O_3.	70% hydrogen	Ibrahim et al. (2021

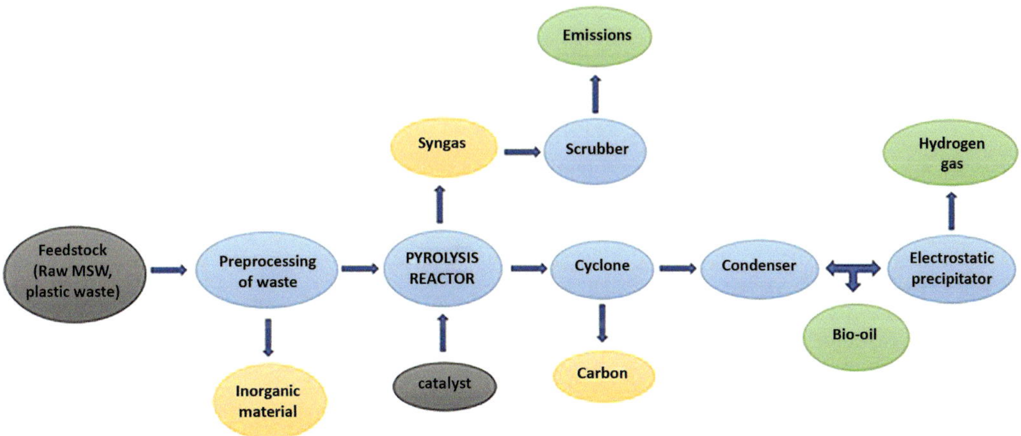

FIGURE 2.4 General flow sheet for the production of hydrogen from plastic waste by pyrolysis.

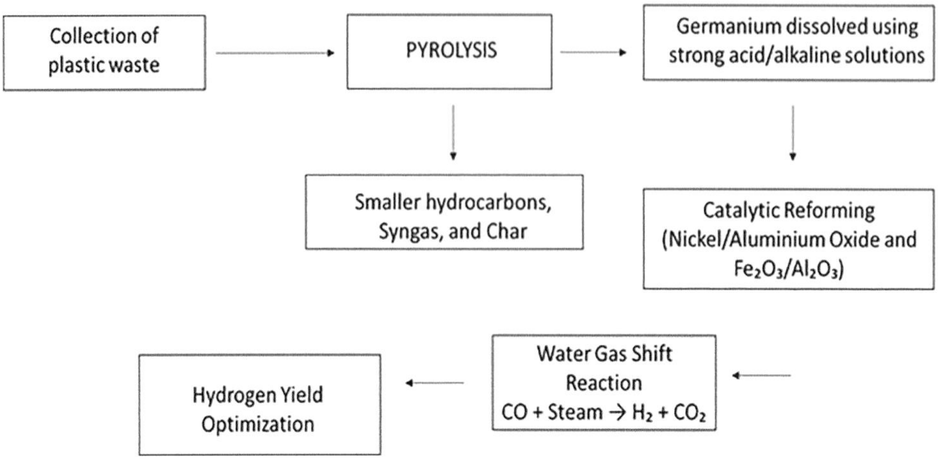

FIGURE 2.5 Production of hydrogen from plastic waste using a water gas shift reaction.

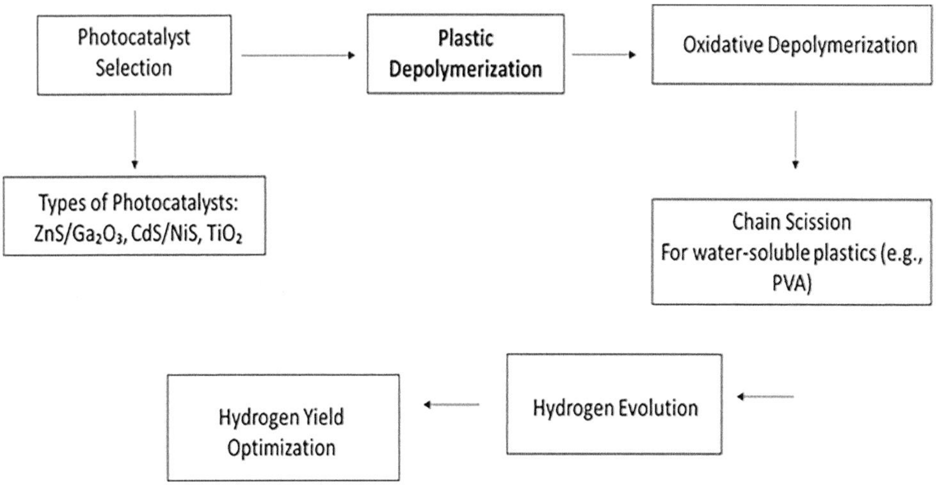

FIGURE 2.6 Photocatalytic degradation of plastics to generate hydrogen.

2.5.2 Photocatalytic Methods

Photocatalytic reforming of plastics is an environmentally friendly process where light degrades polymers such as PET and LDPE into hydrogen and valuable organic by-products (Figure 2.6). For example, this process may produce substantial amounts of hydrogen, with PET being more effective (Thiloka Edirisooriya et al., 2023). High efficiency in the evolution of hydrogen has been achieved through the use of advanced photocatalysts, including defect-rich $NiPS_3$ nanosheets. For instance, Zhang et al. (2023a) (photoreforming) illustrated an ultrahigh rate of hydrogen evolution at 40 mmol/gram/h accompanied by the production of organic acids at ambient conditions.

Integrated plasma-photocatalytic systems are another advanced upcycling method for polyolefin plastics. In such a system, plasma can be used to introduce oxygenated functional groups onto the plastic surface to facilitate the subsequent photocatalytic production of hydrogen with outstanding efficiency (Wang et al., 2025).

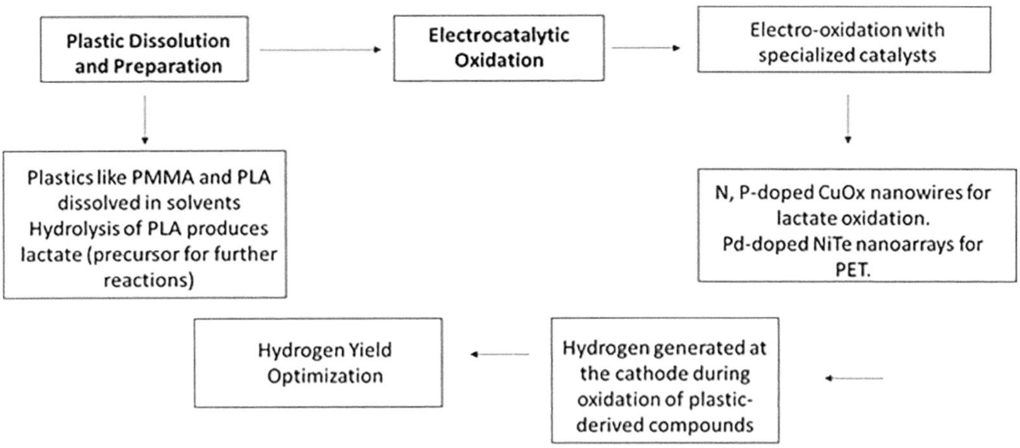

FIGURE 2.7 Hydrogen production from plastic waste using electrocatalytic oxidation.

2.5.3 Electrocatalytic Techniques

Electrocatalysis is a clean technology for the conversion of plastic waste into hydrogen by using electrochemical reactions (Figure 2.7). For example, the application of advanced catalysts like $Au/Ni(OH)_2$ has made it possible to produce biodegradable polymer monomers in situ with hydrogen. Yan et al. (2023) have shown that a selectivity greater than 90% for particular products and hydrogen production rates at industrially relevant current densities can potentially be achieved through this technique of hydrogen production. Recent developments include the synthesis of bimetallic catalysts, such as $NiO@CeO_2$, based on metal–organic frameworks, which show outstanding resistance to coking and high hydrogen yields (Hsu et al., 2023). New methods like tandem electrocatalysis have shown improved efficiencies in hydrogen production. For example, Zhang et al. (2023a) proposed MBenes catalysts for nitrate reduction, which showed outstanding hydrogen evolution rates.

2.6 HYDROGEN PRODUCTION FROM SPECIFIC TYPES OF PLASTIC WASTE

2.6.1 Polyethylene

Polyethylene (PE), one of the most widely produced and consumed plastics globally, has garnered significant attention as a feedstock for hydrogen production. With its high hydrogen-to-carbon ratio, PE serves as an abundant and energy-rich resource. PE waste, particularly from packaging and disposable products, contributes to the growing plastic pollution problem. By utilizing this waste for hydrogen production, not only can environmental challenges be mitigated, but a sustainable energy resource can also be developed. Techniques such as pyrolysis, gasification, photo-reforming, and electro-reforming have emerged as the leading methods to convert PE into hydrogen (Yang et al., 2022). Each of these approaches exploits the chemical structure of PE, which consists of long hydrocarbon chains, to release hydrogen efficiently. The scalability and versatility of these methods make PE a promising candidate for addressing both energy and waste management crises.

Pyrolysis, a thermochemical decomposition process, is one of the most explored techniques for extracting hydrogen from PE. This method involves heating PE in an oxygen-free environment and breaking down its polymeric chains into smaller hydrocarbons and hydrogen. The use of catalysts such as nickel or copper can enhance the efficiency of hydrogen recovery during pyrolysis. Studies indicate that PE pyrolysis can achieve conversion efficiencies of up to 90%, producing not only hydrogen but also valuable by-products like synthetic gases and liquid fuels (Singh et al., 2018). Catalytic pyrolysis further reduces energy requirements and facilitates selective hydrogen production. This method's adaptability to various feedstock conditions, including contaminated or mixed PE waste, enhances its practicality for industrial applications (Xu et al., 2023).

Gasification is another thermochemical method where PE is converted into hydrogen-rich syngas by reacting it with controlled amounts of oxygen or steam at high temperatures. This method efficiently breaks down complex hydrocarbon structures in PE, producing hydrogen alongside carbon monoxide and methane (Jeong et al., 2020). Gasification has been demonstrated to work effectively with various waste types, including low-density and high-density PE. The introduction of plasma or supercritical water gasification technologies has further optimized hydrogen yields and reduced operational emissions. Additionally, integrating gasification with carbon capture technologies can significantly lower the environmental footprint of the process, making it a sustainable alternative for hydrogen generation (Peng et al., 2023)

Photoreforming, a relatively novel approach, utilizes photocatalysts under light irradiation to produce hydrogen from PE. This method leverages advanced photocatalytic materials like titanium dioxide (TiO_2) or graphitic carbon nitride to facilitate the oxidation of PE polymers, simultaneously releasing hydrogen. The process operates under mild conditions, reducing energy input and environmental impact compared to conventional thermal methods. Research indicates that combining photo-reforming with solar energy can transform PE waste into hydrogen while generating additional value-added chemicals like formic acid or acetate (Du et al., 2022). This dual benefit of energy recovery and chemical synthesis underscores the potential of photoreforming in circular economy models.

Electro-reforming combines electrochemical techniques with catalytic processes to convert PE into hydrogen. This method often involves the use of specialized electrodes and electrolytes to drive the breakdown of PE polymers. Studies show that electro-reforming can effectively handle mixed plastic waste streams, including PE, and yield high-purity hydrogen with minimal by-products (Zhao et al., 2021). Additionally, the integration of renewable energy sources, such as wind or solar, to power the electro-reforming process can further enhance its sustainability. Recent advancements in electrode materials and cell designs have improved the scalability and efficiency of this method, positioning it as a viable technology for future hydrogen production from plastic waste (Mao et al., 2022).

2.6.2 Polypropylene

Polypropylene (PP) can be considered a viable source of hydrogen, particularly through innovative processes that enhance its production alongside hydrogen. Recent research highlights methods such as electrochemical dehydrogenation and catalytic hydrogenation, which not only yield propylene but also generate hydrogen efficiently. The major methodologies used for the production of hydrogen from polypropylene are discussed later.

The catalytic **pyrolysis-gasification** of PP for hydrogen production has been extensively studied. Key factors influencing hydrogen yield include water flow rate, gasification temperature, and catalyst ratio (Wu & Williams, 2008, 2009). Increasing gasification temperature significantly enhances hydrogen production, with yields rising from 13.4 wt% to 52.0 wt% as temperature increases from 600°C to 900°C (Wu & Williams, 2008). The addition of a $Ni/CeO_2/Al_2O_3$ catalyst dramatically improves hydrogen yield, with optimal results at a catalyst-to-polypropylene ratio of 1:1 (Wu & Williams, 2008). Studies using an activated carbon-based nickel catalyst achieved hydrogen yields of 134.91 mmol/gPP with water vapor introduction (Wang et al., 2022b). The process typically involves two steps: pyrolysis of plastics followed by catalytic steam reforming of pyrolysis gases (Czernik & French, 2006). Under optimized conditions, hydrogen yields of up to 80% of the theoretical potential have been reported (Czernik & French, 2006).

Catalytic degradation of PP for hydrogen production has shown promising results in recent studies. Pyrolysis followed by catalytic steam reforming can convert polypropylene into hydrogen with up to 80% efficiency (Czernik & French, 2006). Nickel-based catalysts have demonstrated high hydrogen yields, especially when combined with water vapor, achieving up to 134.91 mmol H_2/g polypropylene (Wang et al., 2022b). Fe-Ni catalysts prepared by sol–gel methods have also shown excellent performance, producing 25.14 mmol H_2/g plastic along with valuable carbon nanomaterials (Yao & Wang, 2020). Factors influencing hydrogen production include catalyst composition, preparation method, gasification temperature, and catalyst-to-plastic ratio. Increasing the gasification temperature from 600°C to 900°C significantly enhances hydrogen yield, while higher catalyst ratios improve hydrogen production and reduce coke deposition (Wu & Williams, 2008). These findings suggest that optimizing catalyst properties and reaction conditions can greatly enhance hydrogen generation from polypropylene degradation.

Several studies have explored innovative methods for generating hydrogen from PP through **electrochemical and thermal processes**. Solar thermo-coupled electrochemical depolymerization has shown promise, achieving high conversion rates and increased hydrogen production compared to conventional pyrolysis (Jiang et al., 2020). Pyrolysis combined with catalytic steam reforming has demonstrated the potential to produce hydrogen at 80% of the theoretical yield from PP (Czernik & French, 2006). Electrochemical approaches utilizing transition metal electrocatalysts have been investigated for selective functionalization and deconstruction of polyolefins, offering improved process control and reduced use of hazardous reagents (Haque et al., 2023). Additionally, pyrolysis-catalytic reforming over activated carbon-based nickel catalysts has achieved high hydrogen yields, particularly when introducing water vapor, reaching up to 134.91 mmol/gPP (Wang et al., 2022). These methods present promising alternatives for converting plastic waste into valuable hydrogen fuel, addressing both energy and environmental challenges.

2.6.3 Polystyrene

The production of hydrogen from polystyrene (PS) can be achieved through various innovative methods, primarily focusing on thermal processing and photo-reforming techniques. These approaches not only facilitate hydrogen production but also address environmental concerns associated with plastic waste.

Pyrolysis of PS waste has emerged as a promising technique for energy recovery and waste management (Maafa, 2021). This process generates various products, including hydrogen, especially at high temperatures (Trojer, 1981). The pyrolysis of polystyrene can yield styrene oil, which can be further converted to biodegradable polyhydroxyalkanoate (PHA) through bacterial fermentation (Ward et al., 2006). Recent research has explored the catalytic pyrolysis of polystyrene under a CO_2 atmosphere, using steel slag as a catalyst. This approach not only reduces the formation of polycyclic aromatic hydrocarbons but also significantly increases hydrogen and carbon monoxide production, allowing for control of the H_2/CO ratio (Lee et al., 2020). Pyrolysis-catalytic steam reforming is an effective method for converting PS into hydrogen, where heating PS with a catalyst results in significant hydrogen yield. A study demonstrated a maximum yield of 125 mmol/g of PS at 900°C using a nickel/aluminum oxide catalyst; the efficiency of the catalyst plays a crucial role in determining hydrogen output, with nickel/aluminum oxide showing superior catalytic activity compared to other alternatives (Aminu et al., 2022). These studies demonstrate the potential of pyrolysis as a sustainable method for converting polystyrene waste into valuable products, addressing both environmental concerns and energy recovery needs (Aminu et al., 2022).

Photocatalytic degradation of PS has been studied for hydrogen production through photo-reforming by developing nanocomposite photocatalysts, such as Au/TiO_2, Pt/TiO_2, and Fe/TiO_2, to enhance light utilization efficiency and catalytic activity (Thiloka Edirisooriya et al., 2023; He et al., 2022). These catalysts effectively degrade PS while simultaneously producing hydrogen gas.

For instance, FeB/TiO_2 achieved a 92.3% reduction in PS particle size and generated 103.5 μmol of H_2 in 12 hours (He et al., 2023). The photo-reforming process also yields valuable organic compounds like benzaldehyde and benzoic acid. Hydroxyl radicals play a crucial role in the degradation mechanism, as confirmed by density functional theory calculations and radical quenching experiments (He et al., 2022). This approach offers a promising solution to mitigate plastic pollution in aquatic environments while simultaneously producing clean hydrogen fuel (Thiloka Edirisooriya et al., 2023; He et al., 2023).

2.7 CHALLENGES IN HYDROGEN PRODUCTION FROM PLASTIC WASTE

There is immense potential of converting plastic waste into hydrogen, providing a dual solution to environmental concerns and energy demands. With global plastic waste reaching alarming levels, the ability to transform this waste into a valuable energy resource such as hydrogen could revolutionize sustainability efforts. Thermo-catalytic processes, including pyrolysis and gasification, have emerged as promising methods for hydrogen production from plastics (Al-Fatesh et al., 2023). These processes are influenced by several critical factors, including catalyst properties, operational temperatures, and steam-to-plastic ratios, making optimization a key focus for researchers (Al-Fatesh et al., 2023; Sharma & Batra, 2019).

Among thermo-catalytic approaches, pyrolysis and gasification stand out for their efficiency in breaking down plastic polymers into hydrogen and other by-products. Plasma gasification, in particular, has shown significant promise, achieving energetic efficiency values ranging from 0.28 to 0.43 (Midilli et al., 2021). These methods rely heavily on catalyst efficiency, which remains a critical challenge in scaling up to industrial levels. The development of durable, high-efficiency catalysts is essential to improve yield and reduce costs (Al-Fatesh et al., 2023).

Beyond thermo-catalytic routes, innovative methods such as photocatalytic and electrocatalytic processes are also gaining traction. These processes leverage light and electricity to break down plastic waste into hydrogen, offering a potentially cleaner alternative to traditional methods. However, significant technological hurdles must be overcome to make these methods viable for large-scale applications. Key challenges include improving catalyst longevity and ensuring energy efficiency during operation (Chen et al., 2024).

One of the most exciting aspects of plastic-to-hydrogen conversion is the production of valuable by-products. For instance, carbon nanotubes can be generated alongside hydrogen during these processes, adding economic value and creating opportunities for secondary applications (Sharma & Batra, 2019). The integration of by-product utilization into the process design can further enhance the economic feasibility of these technologies.

Photocatalytic methods, such as those involving poly(vinyl alcohol) (PVA), are highly sensitive to process conditions. The efficiency of these methods depends on factors like the type of plastic used and the irradiation conditions. Moreover, minimizing CO_2 emissions and resource wastage during these processes is a significant challenge (Miao et al., 2024). Gasification processes require precise control of variables such as temperature and pressure to optimize hydrogen yields. The integration of machine learning technologies into gasification systems holds potential for enhancing efficiency and reliability. However, the complexity of integrating these advanced technologies into existing systems presents a barrier to widespread adoption (Sofian et al., 2023; Lahafdoozian et al., 2024).

While gasification and photoreforming technologies can reduce plastic waste, they must be carefully managed to prevent secondary pollution. Uncontrolled emissions and improper disposal of by-products could negate the environmental benefits of these processes. Additionally, the economic viability of these methods is a concern, as the costs of advanced technologies and storage solutions can be prohibitive (Sofian et al., 2023; Thiloka Edirisooriya et al., 2023). Transitioning to a circular economy, where plastic waste is converted into valuable resources, is critical for sustainable development. However, scaling up these technologies for industrial applications requires substantial investment and infrastructure development.

The high capital costs and technical expertise needed for such transitions pose significant challenges (Ogundana et al., 2024). Pretreatment methods play a crucial role in determining the efficiency of hydrogen production processes. For example, the use of sacrificial agents in photo-reforming can enhance catalytic efficiency. However, careful selection and management of by-products are necessary to minimize environmental impact and maximize economic returns (Thiloka Edirisooriya et al., 2023).

2.8 CONCLUSIONS AND PERSPECTIVES

The sustainable production of hydrogen from plastic waste represents a dual opportunity: mitigating the environmental burden posed by plastic pollution and contributing to the global transition toward clean energy. Hydrogen, as a versatile and clean fuel, has gained prominence in decarbonizing industries and fostering energy resilience. Simultaneously, plastic waste, often regarded as a symbol of environmental degradation, offers an untapped feedstock for sustainable energy production. This synergy establishes a compelling pathway for advancing both environmental and energy goals. Current research has demonstrated the feasibility of converting plastic waste into hydrogen through thermochemical, biochemical, and catalytic processes. Technologies such as pyrolysis, gasification, and plasma-assisted methods have shown promise in breaking down polymers into hydrogen-rich syngas. Catalysts leveraging transition metals or nanomaterials have played pivotal roles in enhancing efficiencies. Despite these advances, challenges remain, particularly in optimizing reaction conditions and ensuring the economic scalability of these technologies.

One key limitation lies in the variability of plastic waste streams, which consist of mixed and often contaminated materials. These inconsistencies can negatively impact process efficiency, yield, and hydrogen quality. Emerging technologies such as photocatalysis and plasma-based methods hold significant potential, but their energy-intensive nature presents scalability barriers. Addressing these limitations requires advancements in catalyst design and improvements in waste management systems to ensure a consistent feedstock supply. Integrating hydrogen production into existing waste management systems offers a practical pathway for scaling up. Co-locating plastic-to-hydrogen facilities with recycling plants could improve economic and logistical feasibility. Furthermore, automation in sorting and preprocessing can ensure feedstock quality, enhancing the reliability of hydrogen generation processes.

From a policy perspective, governments must actively promote plastic-to-hydrogen technologies through subsidies, carbon-pricing mechanisms, and incentives for green hydrogen initiatives. International collaboration is also critical for establishing global standards and fostering knowledge exchange. Future prospects highlight the importance of interdisciplinary collaboration and emerging technologies. Advances in material science and nanotechnology promise more efficient and stable catalysts. Integrating renewable energy sources into hydrogen production systems can further reduce the carbon footprint, ensuring sustainability. Another key focus is the life cycle assessment (LCA) of plastic-to-hydrogen technologies. Holistic evaluations encompassing environmental, economic, and social dimensions are essential to determine sustainability. LCAs provide insights for optimizing resource use and identifying potential bottlenecks, informing both policy and technology development. Community engagement also plays a vital role. Public awareness about the benefits of converting plastic waste into clean energy can foster participation in segregation initiatives and support hydrogen infrastructure. Empowering communities to act as stakeholders in the hydrogen economy ensures inclusive and sustainable growth.

In conclusion, the sustainable production of hydrogen from plastic waste addresses two critical global challenges: plastic pollution and the energy transition. With continuous advancements in research, technology, and infrastructure, this approach holds immense potential for achieving a circular economy powered by clean hydrogen. Collaboration across sectors and disciplines is imperative to realize this vision and ensure a sustainable future.

REFERENCES

Adeniyi, A. G., Iwuozor, K. O., Emenike, E. C., Ajala, O. J., Ogunniyi, S., & Muritala, K. B. (2024). Thermochemical co-conversion of biomass-plastic waste to biochar: A review. *Green Chemical Engineering*, *5*(1), 31–49.

Al-Fatesh, A. S., Al-Garadi, N. Y., Osman, A. I., Al-Mubaddel, F. S., Ibrahim, A. A., Khan, W. U., Alanazi, Y. M., Alrashed, M. M., & Alothman, O. Y. (2023). From plastic waste pyrolysis to Fuel: Impact of process parameters and material selection on hydrogen production. *Fuel*.

Al-Qadri, A., Ahmed, U., Abdul Jameel, A. G., Zahid, U., Usman, M., & Ahmad, N. (2022). Simulation and modeling of hydrogen production from waste plastics: Technoeconomic analysis. *Polymers*, *14*(10), 2056.

Alshareef, R., Nahil, M., & Williams, P. (2023). Hydrogen production by three-stage processing of waste plastic. *Energy & Fuels*.

Amin, A. K., Wijaya, K., & Trisunaryanti, W. (2018). Catalytic performance in hydrocracking of LDPE waste into liquid fuels. *Oriental Journal of Chemistry*.

Aminu, I., Nahil, M. A., & Williams, P. T. (2022). High-yield hydrogen from thermal processing of waste plastics. *Proceedings of the Institution of Civil Engineers: Waste and Resource Management*, *175*(1), 3–13.

Banerjee, D., Kushwaha, N., Shetti, N., Aminabhavi, T., & Ahmad, E. (2022). Green hydrogen production via photo-reforming of bio-renewable resources. *Renewable & Sustainable Energy Reviews*.

Bhuyan, M. S. (2022). Effects of microplastics on fish and in human health. *Frontiers in Environmental Science*, *10*, 827289.

Borrelle, S., Ringma, J., Law, K. L., Monnahan, C. C., Lebreton, L., McGivern, A.,. . . Rochman, C. M. (2020). Predicted growth in plastic waste exceeds efforts to mitigate plastic pollution. *Science*.

Campanale, C., Massarelli, C., Savino, I., Locaputo, V., & Uricchio, V. F. (2020). A detailed review study on the potential effects of microplastics and additives of concern on human health. *International Journal of Environmental Research and Public Health*.

Chen, L., Li, H., Wang, H., & Liu, K. (2023a). Performance of red mud oxygen carriers in chemical-looping hydrogen production using different components of plastic waste pyrolytic gas. *Journal of Cleaner Production*.

Chen, Z., Wei, W., Chen, X., Liu, Y., Shen, Y., & Ni, B. (2024). Upcycling of plastic wastes for hydrogen production: Advances and perspectives. *Renewable and Sustainable Energy Reviews*.

Chunakiat, P., Panarmasar, N., & Kuchonthara, P. (2023). Hydrogen production from glycerol and plastics by sorption-enhanced steam reforming. *Industrial & Engineering Chemistry Research*, *62*(49), 21057–21066.

Czernik, S., & French, R. J. (2006). Production of hydrogen from plastics by pyrolysis and catalytic steam reform. *Energy & Fuels*, *20*, 754–758.

Du, L., Li, Z., Zhou, S., & Liu, Z. (2022). Photoreforming of plastic waste for sustainable hydrogen production: Advances in catalysts and mechanisms. *Journal of Energy Chemistry*, *64*, 325–335.

Eriksen, M., Lebreton, L. C. M., Carson, H. S., Thiel, M., Moore, C. J., Borerro, J. C., Galgani, F., Ryan, P. G., & Reisser, J. (2014). Plastic pollution in the world's oceans. *PLOS One*.

Fakhri, M., Shahryari, E., & Ahmadi, T. (2022). Investigate the use of recycled polyvinyl chloride (PVC) particles in improving the mechanical properties of stone mastic asphalt (SMA). *Construction and Building Materials*.

Ganza, P. E., & Lee, B. (2023). A novel method for industrial production of clean hydrogen (H_2) from mixed plastic waste. *International Journal of Hydrogen Energy*.

Gaur, V., Gupta, S., Sharma, P., Gupta, P., Varjani, S., Srivastava, J. K., Chang, J-S., & Bui, X.-T. (2022). Metabolic cascade for remediation of plastic waste: A case study on microplastic degradation. *Current Pollution Reports*, *8*, 30–50.

Gewert, B., MacLeod, M., & Breitholtz, M. (2021). Variability in toxicity of plastic leachates. *The Biological Bulletin*.

Haq, Md. Z. U., Sood, H., Kumar, R., Dixit, S., Srinivas, T., Ahuja, S., Lingam, B., & Kumar, K. (2023). Circular economy enabler: Enhancing high-performance bricks through geopolymerization of plastic waste. *E3S Web of Conferences*.

Haque, Z., Ortega Ramos, J., & Botte, G. G. (2023). (General student poster award winner—2nd place) electrochemical routes for polymer upcycling. *ECS Meeting Abstracts*.

He, J., Han, L., Ma, W., Chen, L., Ma, C., Xu, C., & Yang, Z. (2023). Efficient photodegradation of polystyrene microplastics integrated with hydrogen evolution: Uncovering degradation pathways. *iScience*, *26*.

He, J., Han, L., Ma, W., Chen, L., Ma, C., & Yang, Z. (2022). Efficient photocatalytic degradation of polystyrene microplastics integrated with hydrogen evolution over a Feb/Tio2 catalyst: Uncovering photodegradation pathways. *SSRN Electronic Journal*.

Hossain, S. T., & Mahmud, M. P. (2024). Optimizing process parameters and materials for the conversion of plastic waste into hydrogen. *Engineering Research Express*, *6*(4), 045319.

Hren, R., Vujanović, A., Fan, Y. V., Klemeš, J. J., Krajnc, D., & Čuček, L. (2023). Hydrogen production, storage and transport for renewable energy and chemicals: An environmental footprint assessment. *Renewable and Sustainable Energy Reviews*.

Hsu, P. H., Li, K., & Tsou, C. Y. (2023). The pollution premium. *The Journal of Finance*, *78*(3), 1343–1392.

Ibrahim, S. A., Ekinci, E., & Karaman, B. (2021). Gasification of HDPE for sustainable hydrogen production. *International Journal of Hydrogen Energy*.

Islam, M. R., Ruponti, S. A., Rakib, A., Nguyen, H. Q., & Mourshed, M. (2022). Current scenario and challenges of plastic pollution in Bangladesh. *Frontiers of Environmental Science & Engineering*, *17*.

Ismail, M. M., & Dincer, I. (2023a). Development and evaluation of an integrated waste to energy system based on polyethylene plastic wastes pyrolysis for production of hydrogen fuel and other useful commodities. *Fuel*.

Ismail, M. M., & Dincer, I. (2023b). Development of an integrated waste-to-energy system based on polyethylene pyrolysis for hydrogen fuel production. *Fuel*.

Jambeck, J. R., Geyer, R., Wilcox, C., Siegler, T. R., Perryman, M., Andrady, A., Narayan, R., & Law, K. L. (2015). Plastic waste inputs from land into the ocean. *Science*, *347*(6223), 768–771.

Jeong, S. J., Kim, J. H., Park, Y. J., & Lee, J. W. (2020). Gasification of low-density polyethylene for hydrogen production using supercritical water: Effect of operational parameters. *International Journal of Hydrogen Energy*, *45*(22), 12391–12400.

Jiang, T., Xule, Z., Gu, D., Yan, C., Jiang, H., Wu, H., Wang, B., & Wang, X. (2020). STEP polymer degradation: Solar thermo-coupled electrochemical depolymerization of plastics to generate useful fuel plus abundant hydrogen. *Solar Energy Materials and Solar Cells*, *204*, 110208.

Kaydouh, M., & El Hassan, N. (2022). Thermodynamic simulation of the co-gasification of biomass and plastic waste for hydrogen-rich syngas production. *Results in Engineering*.

Khan, A., Niazi, M. B., Ansar, R., Jahan, Z., Javaid, F., Ahmad, R., Anjum, H., Ibrahim, M., & Bokhari, A. (2023). Thermochemical conversion of agricultural waste to hydrogen, methane, and biofuels: A review. *Fuel*.

Kibria, M. G., Masuk, N. I., Safayet, R., Nguyen, H. Q., & Mourshed, M. (2023). Plastic waste: Challenges and opportunities to mitigate pollution and effective management. *International Journal of Environmental Research*.

Kim, D.-K., Lee, S. H., Kwon, T. H., Kwon, I., Han, D., Park, H., Han, S., Lee, D., & Yu, S. (2023). Study on high-temperature and high-voltage insulation characteristics of polypropylene blend with highly packed elastomeric domains for power cable applications. *Polymer Testing*.

Lahafdoozian, M., Hossein, K., Sharif, H., Sharif, Z., & Jalil, A. A. (2024). Hydrogen production from plastic waste: A comprehensive simulation and machine learning study. *International Journal of Hydrogen Energy*.

Lampitt, R. S., Fletcher, S., Cole, M., Kloker, A., Krause, S., O'Hara, F., Ryde, P., Saha, M., Voronkova, A., & Whyle, A. (2023). Stakeholder alliances are essential to reduce the scourge of plastic pollution. *Nature Communications*.

Lan, K., & Yao, Y. (2022). Feasibility of gasifying mixed plastic waste for hydrogen production and carbon capture and storage. *Communications Earth & Environment*, *3*(1), 1–12.

Lee, T., Jung, S., Park, Y., Kim, T., Wang, H., Moon, D. H., & Kwon, E. E. (2020). Catalytic pyrolysis of polystyrene over steel slag under CO_2 environment. *Journal of Hazardous Materials*, *395*, 122576.

Leslie, H. A., van Velzen, M. J. M., Brandsma, S. H., Vethaak, A. D., Garcia-Vallejo, J. J., & Lamoree, M. H. (2022). Discovery and quantification of plastic particle pollution in human blood. *Environment International*.

Li, J., Yang, D., Yao, X., Zhou, H., Xu, K., & Geng, L. (2023a). Slow pyrolysis experimental investigation of biomass tar formation and hydrogen production by tar reforming. *International Journal of Hydrogen Energy*.

Li, Y., Tao, L., Wang, Q., Wang, F., Li, G., & Song, M. (2023b). Potential health impact of microplastics. *Environment & Health*.

Maafa, I. M. (2021). Pyrolysis of polystyrene waste: A review. *Polymers, 13*.

MacLeod, M., Arp, H. P. H., Tekman, M. B., & Jahnke, A. (2021). The global threat from plastic pollution. *Science*.

Mahmood, Q., Zeng, Y., Wang, X., Sun, Y., & Sun, W.-H. (2017). Advancing polyethylene properties by incorporating NO2 moiety in 1,2-bis(arylimino) acenaphthylnickel precatalysts: Synthesis, characterization, and ethylene polymerization. *Dalton Transactions*, *46*(21), 6934–6947.

Mao, H., Shi, W., Zhang, P., & Wang, T. (2022). Electrochemical reforming of polyethylene plastic waste for hydrogen production: A critical review. *Chemical Engineering Journal*, *442*, 136127.

Miao, P., Jiaqing, Z., Pu, W., Run, S., & Tierui, Z. (2024). Enhanced photocatalytic hydrogen production from poly(vinyl alcohol) plastic-dissolved wastewater. *ACS Materials Letters*, *6*(2), 590–597.

Midilli, A., Kucuk, H., Hacıosmanoğlu, M., Akbulut, U., & Dincer, I. (2021). A review on converting plastic wastes into clean hydrogen via gasification for better sustainability. *International Journal of Energy Research*, *46*, 4001–4032.

Mishra, R., Ong, H. C., & Lin, C.-W. (2023). Co-processing of biomass and plastic waste for hydrogen production. *Energy Conversion and Management*.

Napier, I., & Thompson, R. C. (2020). Plastic debris in the marine environment. *Global Challenges*.

Ncube, L., et al. (2021). An overview of plastic waste generation and management in food packaging industries. *Recycling*.

Ogundana, A. K., Afolabi, O. O., Ilevbare, M., & Falae, P. O. (2024). *Green hydrogen generation from plastic waste: A review* (pp. 1–11). IEEE,

Pandey, P., Dhiman, M., Kansal, A., & Subudhi, S. P. (2023). Plastic waste management for sustainable environment: Techniques and approaches. *Waste Disposal & Sustainable Energy*.

Patlolla, S. R., Katsu, K., Sharafian, A., Wei, K., Herrera, O. E., & Mérida, W. (2023). A review of methane pyrolysis technologies for hydrogen production. *Renewable and Sustainable Energy Reviews*, *181*, 113323.

Peng, J., Wang, Q., Xu, L., & Liu, C. (2023). Plasma-assisted gasification of polyethylene for clean hydrogen production: An experimental study. *Energy Conversion and Management*, *277*, 116602.

Pinheiro, H., MacDonald, C., Santos, R. G., Ali, R., Bobat, A., Cresswell, B. J., . . Rocha, L. A. (2023). Plastic pollution on the world's coral reefs. *Nature*.

Plastic pollution by country 2024. (n.d.). https://worldpopulationreview.com/country-rankings/plastic-pollution-by-country

Prata, J. C., da Costa, J. P., Lopes, I., Duarte, A. C., & Rocha-Santos, T. (2019). Environmental exposure to microplastics. *Science of the Total Environment*.

Ren, T., Yu, Z., Yu, H., Deng, K., Wang, Z., Li, X., Wang, H., Wang, L., & Xu, Y. (2023). Sustainable ammonia electrosynthesis from nitrate wastewater coupled to electrocatalytic upcycling of polyethylene terephthalate plastic waste. *ACS Nano*.

Royer, S. J., Greco, F., Kogler, M., & Deheyn, D. D. (2023). Not so biodegradable: Polylactic acid and cellulose/plastic blend textiles lack fast biodegradation in marine waters. *PLOS One*.

Sahani, K., Joshi, B. R., Khatri, K., Magar, A. T., Chapagain, S., & Karmacharya, N. (2022). Mechanical properties of plastic sand brick containing plastic waste. *Advances in Civil Engineering*.

Shah, S. S., Nasser, G. A., Basha, S. I., Buliyaminu, I. A., Rahman, S. M., & Aziz, M. A. (2024). Unlocking the potential of solid carbon: Synergistic production with hydrogen from oil and gas resources for innovative applications and a sustainable future. *Advanced Composites and Hybrid Materials*, *7*(6), 211.

Sharma, S. S., & Batra, V. S. (2019). Production of hydrogen and carbon nanotubes via catalytic thermo-chemical conversion of plastic waste: Review. *Journal of Chemical Technology & Biotechnology*.

Singh, S., Kumari, P., & Srivastava, S. (2018). Catalytic pyrolysis of polyethylene waste for hydrogen production: An overview. *Waste Management*, *78*, 236–245.

Sofian, A. D. A. B. A., Hooi, R., Lim, K., Wayne, C., Kuan, S., Khoo, I., Shi, T., Zengling, M., & Show, P. K. (2023). Hydrogen production and pollution mitigation: Enhanced gasification of plastic waste and biomass with machine learning & storage for a sustainable future. *Environmental Pollution*, 123024–123024.

Sousa, F. D. D. de. (2021). Management of plastic waste: A bibliometric mapping and analysis. *Waste Management & Research*, *39*, 664–678.

Thiloka Edirisooriya, E. M. N., Senanayake, P. S., Xu, P., & Wang, H. (2023). Hydrogen production and value-added chemical recovery from the photo-reforming process using waste plastics. *Journal of Environmental Chemical Engineering*.

Trojer, L. (1981). Hydrogen formation during pyrolysis of polystyrene. *Journal of Analytical and Applied Pyrolysis*, *2*, 353–359.

Turner, A. (2020). Foamed polystyrene in the marine environment: Sources, additives, transport, behavior, and impacts. *Environmental Science & Technology*.

Walker, T., & Fequet, L. (2023). Current trends of unsustainable plastic production. *Trends in Analytical Chemistry*.

Wang, J., Li, X., Zhang, T., Chen, Y., Wang, T., & Zhao, Y. (2022a). Electro-reforming polyethylene terephthalate plastic to co-produce valued chemicals and green hydrogen. *Journal of Physical Chemistry Letters*.

Wang, J., Zhang, C., & Wang, Y. (2025). Current advances in the photoconversion of plastics: The catalysts and reaction pathways. *ChemSusChem*, *18*(7), e202401700.

Wang, S., Zhang, Y., Shan, R., Gu, J., Huhe, T., Ling, X., Yuan, H., & Chen, Y. (2022b). High-yield H2 production from polypropylene through pyrolysis-catalytic reforming over activated carbon based nickel catalyst. *Journal of Cleaner Production*.

Ward, P. G., Goff, M., Donner, M., Kaminsky, W., & O'Connor, K. E. (2006). A two step chemo-biotechnological conversion of polystyrene to a biodegradable thermoplastic. *Environmental Science & Technology*, *40*(7), 2433–2437.

Wei, R., Breite, D., Song, C., Gräsing, D., Ploss, T. N., Hille, P., Schwerdtfeger, R., Matysik, J., Schulze, A., & Zimmermann, W. (2019). Biocatalytic degradation efficiency of postconsumer polyethylene terephthalate packaging determined by their polymer microstructures. *Advanced Science*, *6*, 1900491.

Wu, C., & Williams, P. T. (2008). Effects of gasification temperature and catalyst ratio on hydrogen production from catalytic steam pyrolysis-gasification of polypropylene. *Energy & Fuels*, *22*, 4125–4132.

Wu, C., & Williams, P. T. (2009). Hydrogen production from the pyrolysis-gasification of polypropylene: Influence of steam flow rate, carrier gas flow rate and gasification temperature. *Energy & Fuels*, *23*, 5055–5061.

Wu, S.-L., Kuo, J.-H., & Wey, M. (2020). Design of catalysts for hydrogen production from plastic waste gasification. *Catalysis Science & Technology*.

Xu, Y., Zhang, Y., & Wang, R. (2023). Advances in polyethylene pyrolysis for hydrogen production: Catalysts, processes, and perspectives. *Fuel*, *333*, 126067.

Yan, Z., Xu, L., & Zhu, H. (2023). Advances in electrocatalytic semi-hydrogenation of acetylene in aqueous electrolyte: Progress, challenges, and opportunities. *Renewables*, *1*, 278–293.

Yang, R.-X., Jan, K., Chen, C.-T., Chen, W.-T., & Wu, K. (2022). Thermochemical conversion of plastic waste into fuels, chemicals, and value-added materials: A critical review and outlooks. *ChemSusChem*, *15*(6), e202200160.

Yao, D., & Wang, C. (2020). Pyrolysis and in-line catalytic decomposition of polypropylene to carbon nanomaterials and hydrogen over Fe- and Ni-based catalysts. *Applied Energy*, *265*, 114819.

Zhang, S., Li, H., Wang, L., Liu, J., Liang, G., Davey K., Ran, J., & Qiao, S.-Z. (2023a). Boosted photoreforming of plastic waste via defect-rich NiPS3 nanosheets. *Journal of the American Chemical Society*.

Zhang, Y., Wu, P., Xu, R., Wang, X., Lei, L., Schartup, A. T., . . Zeng, E. Y. (2023b). Plastic waste discharge to the global ocean constrained by seawater observations. *Nature Communications*, *14*.

Zhao, W., Chen, H., & Liu, J. (2021). Recent advancements in electrochemical reforming of plastic waste for hydrogen production. *Energy Reports*, *7*, 3876–3890.

Section II

Sustainable Technologies for Plastic Waste to Value-Added Products, Chemicals, and Material

3 Production of Aviation Fuel from Plastic Waste

Aneesh Raj[1], *Rohidas Bhoi*[2,3], *and Praveen Kumar Ghodke*[1*]
[1]Department of Chemical Engineering, National Institute of Technology Calicut, Kozhikode, India
[2]Department of Chemical Engineering, Malaviya National Institute of Technology Jaipur, India
[3]Centre for Rural Development, Malaviya National Institute of Technology Jaipur, India

3.1 INTRODUCTION

The need for plastics has increased due to their diverse applications in the building, healthcare, manufacturing, agriculture, automotive, and packaging sectors. Plastic use is projected to account for 20% of world crude oil production by 2050, which is an increase from the current level by 5%. The worldwide plastic generation report indicates that plastic production reached 430 million metric tons in 2024, reflecting approximately a 20% increase over the preceding five years. This value is anticipated to exceed 500 million metric tons by 2050 if the existing rate of plastic production continues. The significant manufacture of plastic has increased plastic waste generation, roughly 12% in 2016 (Al-Shihabi et al., 2023) and 12.2% in 2018 (Baca et al., 2023). Consequently, the substantial volume of plastic waste has significantly strained disposal management systems and surrounding ecosystems. This has resulted in the quest for treatment solutions that are not only effective but also cost-efficient and environmentally friendly (Nayanathara Thathsarani Pilapitiya & Ratnayake, 2024; Shahabuddin et al., 2020).

In addition to issues related to plastic waste disposal, rapid population growth in the past few decades has prompted critical debates on energy security as a global concern. McKinsey Global Institute forecasts that the worldwide energy demand will increase by triple by 2050 (Valizadeh et al., 2024a). Conversely, the overdependence on fossil fuels to fulfil the energy demands has led to the exhaustion of these resources. In addition to escalating prices and inequitable resource distribution, the overconsumption of fossil fuels has resulted in significant environmental impacts, such as excessive greenhouse gas emissions, deforestation, and acidification of oceans (Alawa et al., 2024). Consequently, various studies have been conducted to identify suitable alternatives to fossil fuels including nuclear energy, wind energy, and biomass sources. Nevertheless, these alternative energy sources may not completely resolve the issue of increasing energy consumption. The conversion of plastic waste into value-added fuels by thermochemical technologies, such as pyrolysis, has progressed, tackling the rising energy demand and the issues of plastic waste disposal. Significantly, fuels generated from plastic waste possess properties identical to those of fossil fuels, confirming the suitability of fuel derived from plastic waste as a substitute for fossil fuels. The liquid products from plastic waste (PW) pyrolysis can be categorized by carbon number, denoting gasoline ($C_5 - C_{12}$), jet fuel ($C_9 - C_{16}$), and diesel fuel ($C_{12} - C_{24}$) (Culsum et al., 2024; Valizadeh et al., 2024b).

In practical situations, plastic waste, which is found in various waste streams, is composed of a mixture of many different kinds of plastics, including HDPE, LDPE, PP, PS, PVC, and PET. The utilization of PVC and PET among other plastic wastes in pyrolysis to produce high-quality liquid products are limited. Reports indicate that HCl (~50 wt%), along with minor quantities of toluene,

[*] Corresponding author's email: praveenkg@nitc.ac.in

DOI: 10.1201/9781003581000-5

benzene, and other hydrocarbons are the primary products of PVC degradation during pyrolysis. The generation of hazardous HCl and other chlorinated intermediates and hydrocarbons via PVC pyrolysis requires supplementary systems and poses significant environmental challenges (Hu et al., 2024). Furthermore, prior research has investigated the pyrolysis of PET, revealing that the yield of liquid product is lower (<40 wt%) as compared to the gaseous product (>60 wt%). For instance, it was claimed that a minimal yield of liquid oil (~20 wt%) was achieved from the pyrolysis of PVC. Simultaneously, the oil derived from PET pyrolysis predominantly consists of benzoic acid and terephthalic acid, and both of the aromatic acids possess corrosive characteristics that might damage the reactor and significantly elevate the total acid number (TAN) of the liquid output. From the oxygen-containing polymers, significant amounts of oxygenates were also produced from PET, resulting in the formation of organic acids. As a result, PVC and PET generate only a small amount of liquid oil while producing a significant amount of harmful by-products, including benzoic acid and HCl, which makes them unsuitable for fuel applications (Saha et al., 2024; Wu et al., 2024). Consequently, from the plastic waste generated, polyethylene (HDPE and LDPE), polypropylene (PP) and polystyrene (PS) are favored for oil generation through pyrolysis. Studies demonstrated that the pyrolysis of polyethylene and polypropylene, due to their long-chain branching structure, produces liquid fuels that are primarily composed of paraffinic and olefinic compounds, with limited amount of aromatic chemicals. Among the different types of plastic waste, polystyrene is the most preferred material with a higher liquid yield due to the presence of styrene monomers with phenyl groups. The pyrolysis liquid product from polystyrene predominantly consists of aromatic compounds, when compared to the liquid products from other plastics. The pyrolysis of polyethylene generally yields olefins, paraffins, cycloalkanes, and aromatics, out of which olefins and paraffins are predominant. Compared to the pyrolysis of polyethylene, polypropylene yields iso-olefines and iso-paraffins, which are preferred for fuel production (Nguyen et al., 2024; Wang et al., 2023).

Compared to other plastic waste types, polystyrene (PS) exhibits the highest potential for generating substantial liquid oil yields, accompanied by a significant quantity of aromatic chemicals, primarily styrene, and benzene, toluene, and ethylbenzene (BTE). The review articles on PW pyrolysis have consolidated the scientific findings from all forms of PW pyrolysis and have offered an in-depth analysis of the processes. This chapter examines the current state of plastic waste management approaches, providing a detailed overview of the thermo-catalytic PW pyrolysis process. The chapter also summarizes the advancements and challenges over the past decade in PW pyrolysis, with a focus on LDPE, HDPE, PP, and PS materials. This background information provides a comprehensive explanation of the impact of experimental parameters and catalysts on the distribution of plastic waste pyrolysate, along with relevant reaction pathways and mechanisms. The techno-economic feasibility of commercializing the pyrolysis method for plastic waste is also assessed, along with the potential for fuel production through this process (Paucar-Sánchez et al., 2023; Royuela et al., 2024; Saha et al., 2024; Stančin et al., 2023).

This chapter discusses the pyrolysis behavior of polyethylene (LDPE, HDPE), PP, and PS, including the impact of various process variables and the efficacy of various catalysts, to provide a theoretical foundation along with sustainable perspectives for the execution of commercial PW pyrolysis processes.

3.2 DIVERSE METHODOLOGIES FOR THE SYNTHESIS OF AVIATION FUEL

The recent years have witnessed the emergence of several promising technologies for converting biomass, waste materials, and other renewable resources into high-quality aviation fuel. These methods are summarized in Figure 3.1, and each of these technologies offers unique advantages and challenges in terms of requirements and complexity.

Catalytic hydro-thermolysis (CH), a patented technology invented by Applied Research Associates, can produce "renewable, aromatic, and drop-in" fuels from plant or algal oils. CH is a hydrothermal process that can convert triglycerides into a blend of straight chain, branched chain,

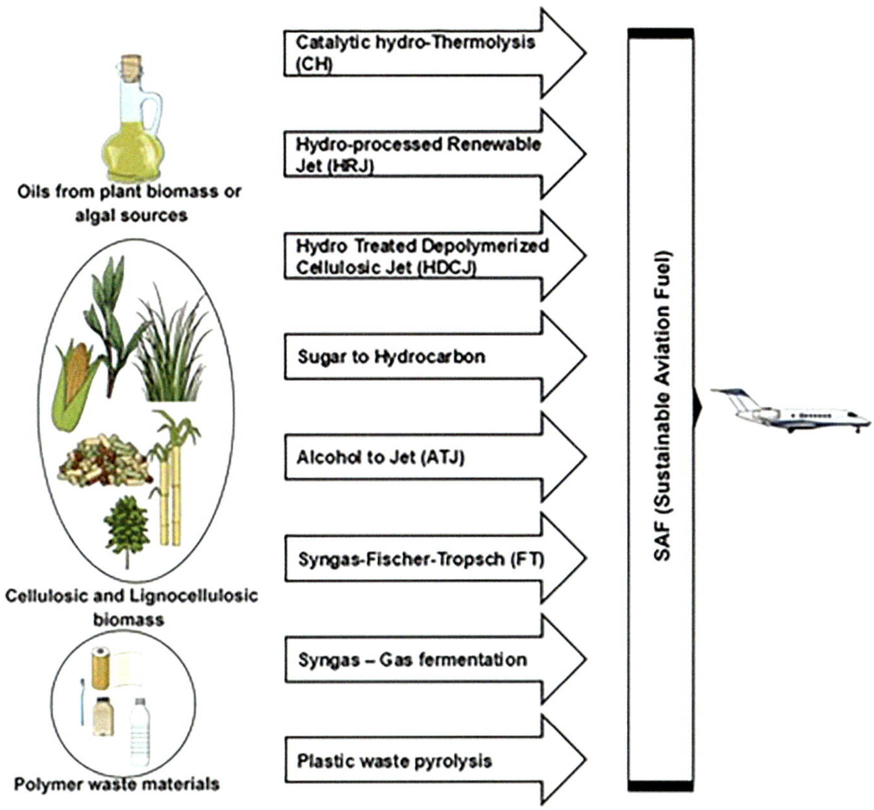

FIGURE 3.1 Methodologies for the synthesis of aviation fuel.

and aromatic hydrocarbons (Goh et al., 2022). The mechanisms involved in the CH process include hydrolysis, cracking, decarboxylation, isomerization, and cyclization.

The fuel generated from the catalytic hydro-thermolysis process complies with ASTM and MIL (military) specifications for jet fuel and exhibits superior stability, cold flow characteristics, and combustion performance. Hydro-Treated Depolymerized Cellulosic Jet (HDCJ) is another process, developed by Kior, for the production of hydrocarbons in the range of jet fuel, diesel, and gasoline from cellulosic biomass. The bio-oils obtained from pyrolysis are subjected to hydrotreatment to yield jet fuel properties. The integrated biorefinery process utilizes rapid thermal processing (RTP) pyrolysis and catalytic hydro-conversion technologies from biomass feedstocks to produce jet-fuel-range hydrocarbons. The feedstock includes switchgrass, cane bagasse, maize stover, guinea grass, forest waste, and algal biomass. The bio-oil from the pyrolysis and hydro conversion process contains 42.4–44.2 wt% jet fuel-range hydrocarbons and is separated through batch vacuum distillation. For the production of military (MIL) grade jet fuel, hydro-processed renewable jet (HRJ) technology has been developed. This advanced pyrolysis technology is highly developed and already commercialized. HRJ fuel shares the same qualities as conventional petroleum-based fuel, yet exhibits a higher cetane number, lower aromatic content, reduced sulfur content, and potentially decreased greenhouse gas emissions (Hussain et al., 2022; Wang & Tao, 2016).

Over the past six decades, several technologies have been developed and commercialized for the upgradation of pyrolytic bio-oil. Hydro-isomerization, hydrogenation, deoxygenation, and hydrocracking are among the few commercialized technologies. Another approach for producing transportation fuels is a metabolic pathway, where carbohydrates are biologically and catalytically converted into hydrocarbons suitable for vehicle fuel. In the natural sugar-to-hydrocarbon process, enzymatic hydrolysis is used to extract hemicellulose sugars from the biomass cell wall (Lin et al., 2024). The removal of particles via enzymatic hydrolysis usually proceeds by the concentration of liquid sugars and is generally accomplished by decreasing the water content by 50%. The concentrated sugar is then fermented in an aerobic batch or continuous reactor to produce hydrocarbon intermediates. Jet fuel from the resulting product is further recovered in downstream separation and purification units. Aqueous phase reforming (APR) is another process used for the conversion of lignocellulosic sugar obtained from biomass into hydrocarbon molecules and hydrogen. Biomass sugars separated through the pretreatment and enzymatic hydrolysis process necessitate purification and concentration prior to the catalytic conversion of sugar into hydrocarbons. The resultant product from the APR process is then fractionated to separate jet fuel-range hydrocarbons from other fuel fractions, including gasoline and diesel. Closed-loop processes are required to improve the economics of the conversion systems to recover the unreacted components through recycling. The economics of the process also depends upon the extraction of value-added co-products such as para-xylene for the production of polyester. The gas fermentation process can also be coupled with the conversion process, which enables the conversion of syngas into liquid biofuels (Goh et al., 2022; Hussain et al., 2022; Justicia et al., 2024; Wang & Tao, 2016).

In the alcohol-to-jet (ATJ) process, when lignocellulosic biomass is employed as feedstock, the biomass is initially undergoing gasification to generate syngas. Acetogenic bacteria are utilized to ferment the cooled syngas to produce ethanol or butanol. In the commercial alcohol-to-jet process, ATJ jet fuel is produced by converting alcohols such as butanol and ethanol. The process is commonly known as alcohol oligomerization. Butanol oligomerized using Amberlyst-70 catalyst yielded 71% of $C_8 - C_{16}$ alkenes for jet fuel synthesis, attaining a 99% conversion of butanol. For producing drop-in or easily adaptable fuels within the jet fuel sector, conversion of ethanol to jet fuel is the most viable solution. The requirement for high-purity ethanol in the conversion to aviation fuel products remains ambiguous. The processes to convert the mixed alcohol to jet fuel-range hydrocarbons and ATJ include dehydration, hydrogenation, oligomerization, and distillation processes. These phases have undergone commercial testing, hence reducing the risk for scale up (Huerta-Rosas et al., 2024). Gas fermentation offers numerous potential advantages. It can produce a wider range of products than traditional biochemical and thermochemical methods, with an overall energy efficiency of 57%. The approach necessitates reduced temperatures and pressures, together with more economical enzymes. The Fischer-Tropsch (FT) process can be utilized to convert the syngas into hydrocarbon fuels. FT fuels are generally devoid of sulfur and possess lower aromatic content compared to conventional fossil fuels, leading to decreased emissions during jet engine operation. Recent studies suggest that FT technology, for transforming biomass into transportation fuels, may provide a viable, sustainable substitute for traditional fossil fuels, including jet fuel, diesel, kerosene, and gasoline (Goh et al., 2022; Khalifa et al., 2024).

Coal, natural gas, and biomass are the most common sources for aviation fuel, with biomass being the only renewable option. Various routes for the production of aviation fuel are available based on renewable resource, each with specific advantages and disadvantages. Triglyceride-containing materials necessitate cost-effective processing, but the feedstock is expensive. Conversely, inexpensive raw materials like lignocellulosic biomass necessitate multiple processing steps, hence increasing costs (Sharma et al., 2021).

When selecting the most effective method of production, it is essential to consider the availability of raw materials. This is because the pricing of these resources and the expenses of transportation will have an impact on the supply chain, which is very important for the long-term viability of the bio-jet fuel manufacturing process. Plastic, a paramount accomplishment of the 20th century, may now be transformed into jet fuel, petroleum, and other fuels through the pyrolysis process. The process is conducted at elevated temperatures in the presence of a suitable catalyst in the absence of oxygen. The principal products of the pyrolysis process are oil, gas, and solid residue. Pyrolysis is an efficient technique for processing diverse plastic feedstocks, even with those that are somewhat

TABLE 3.1
Fuel Properties for Sustainable Aviation Fuel in Comparison with Jet A-1

Fuel Property	Jet A-1 (ASTM D1655)	Bio-Jet Fuel (ASTM D7566)
Density at 15°C (g/cm^3)	0.75–0.84	0.75–0.84
Kinematic viscosity (mm^2/s)	1.3–1.5 at 40°C	8 Max. at −20°C
Lower heating value (MJ/kg)	42–43	Min. 42.8
Flash point (°C)	Min. 38	Min. 38
Boiling point (°C)	Max. 300	150–300
Freezing point (°C)	−47	−47

polluted. Furthermore, they can be altered to provide elevated quantities of specific chemicals for designated applications (Mousavi & Dehaghani, 2024; Saha et al., 2024; Valizadeh et al., 2024b). This chapter concentrates on the conversion of waste polymers into aviation fuel via thermochemical processes.

3.3 SUSTAINABLE AVIATION FUEL SPECIFICATIONS

Sustainable aviation fuel must adhere to stringent specifications to guarantee safety, performance, and compatibility with current aircraft and infrastructure. The primary standard is ATM D7566, the internationally recognized standard for synthetic aviation fuels. It delineates sanctioned production methods and mixing restrictions with standard Jet A-1 fuels. The key sustainable aviation fuel properties in comparison with commercial Jet A-1 fuel are provided in Table 3.1 (Aghamiri & Lahijani, 2024).

3.4 THERMOCHEMICAL CONVERSION METHODS FOR TRANSFORMING POLYMERS INTO JET FUEL

The thermochemical conversion techniques for converting polymers to jet fuel-range hydrocarbons encompass high-temperature processes that break down plastic waste into hydrocarbons in the range of aviation fuel. The major techniques involved in the thermochemical conversion methods are pyrolysis, hydro-liquefaction, microwave-assisted thermal degradation, methanolysis, hydrogenation, co-hydrogenation, thermal cracking, and aqueous phase hydrodeoxygenation. The aforementioned techniques are elaborated in the subsequent sections.

3.4.1 Thermal Pyrolysis of Plastic Waste

Thermal pyrolysis of plastic waste is a sophisticated process used to decompose waste plastic materials under controlled conditions. The process operates at elevated temperatures (300–900°C) in an inert environment, typically in a nitrogen environment. The thermal decomposition of the polymer waste materials under these conditions yields monomers, and the resultant product stream contains oil, gas, and char, as depicted in Figure 3.2.

The liquid product (crude oil) obtained on the thermal pyrolysis of plastic waste contains $C_5 - C_{18}$ hydrocarbons, which is in the range of gasoline, diesel, and jet fuels. The pyrolysis process also emits gaseous hydrocarbons in the range of $C_1 - C_4$. The crude oil derived from plastics is economically viable and has good market value due to its applications in transportation fuel as well as in manufacturing. In 2018, the pyrolysis oil from plastic waste had a market worth of USD0.81

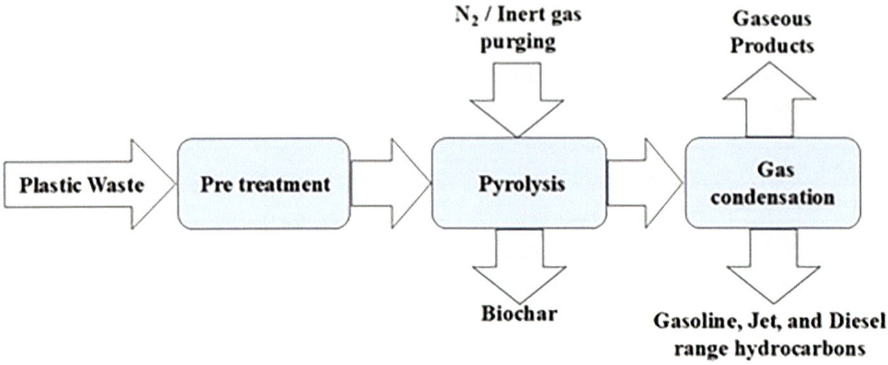

FIGURE 3.2 Flow diagram for the thermal pyrolysis process of plastic waste.

TABLE 3.2
Pyrolysis of Plastic Waste to Yield Different Liquid Fuels Equivalent to Near Aviation Fuel

Type of Plastic	Source	Conditions	Products/Yield (%)	Reference
PP	Packaging industry Manufacturing industries Automotive sector Special devices such as living hinges, textiles	T = 500°C–550°C	83.16% yield of pyrolysis oil	Saha et al., 2024
PS	Food service containers Packaging and insulation Garbage and grocery bags Wire, cable, bottles, and toys	T = 300°C–500°C	83%, $C_5 - C_{12}$ liquid oils	Hussain et al., 2022
LDPE	Containers Dispensing bottles Wash bottles Tubing Plastic bags	T = 500°C	80.4% yield of liquid oils	Huo et al., 2020
HDPE	Pipes Food and beverage containers	T = 575°C	57.6% liquid oils	Zhou et al., 2021
PET	Food and beverage packaging	T = 400°C–450°C	$C_{10} - C_{14}$ kerosene	Saha et al., 2024
Mixed plastics		T = 500°C	87.0% yield of $C_5 - C_{16}$ hydrocarbons	Akin et al., 2023

billion. This signifies the feasibility of employing PW-derived liquid fuel as an alternative energy source. Besides transportation fuel, pyrolysis of PW produces value-added compounds used for synthesizing new polymers and chemicals. The gaseous by-product of pyrolysis can also be utilized as an excellent energy source for heating applications or for producing other value-added chemicals such as ethylene or propylene. The diverse usage of the PW pyrolysis process will enhance the self-sufficiency and efficiency of the pyrolysis facility. The solid residue obtained from pyrolysis can also be used as a fuel source. The solid char can also be used in wastewater treatment and in soil conditioning (Hussain et al., 2022). Pyrolysis of plastic waste to yield different liquid fuels equivalent to near-aviation fuel is given in Table 3.2.

The plastic wastes mentioned here, with volatile matter content as high as 87 wt% and relatively lower ash content of less than 2 wt% demonstrate their significant potential for the production of high-quality liquid fuels via pyrolysis. The characteristics of the pyrolytic liquid from plastic waste, including high octane number and limited carbon number range, render the product appropriate for direct application as fuel. The research octane number (RON) of the produced oil can be further enhanced by augmenting the proportions of the components, especially the naphthenes, isoparaffins, and BTEX components benzene, toluene, ethylbenzene, and xylene. One of the drawbacks of the pyrolytic oil is the presence of waxes, olefinic, and naphthenic compounds. The presence of the undesired components is detrimental, as they may create substantial engine problems when utilized as fuel feedstocks (Valizadeh et al., 2024a).

The operating parameters of PW pyrolysis influence the distribution of PW pyrolyze products. The key operational elements in PW pyrolysis are operating temperature, heating rate, and residence time. The other major parameters are the pyrolysis medium and the type of plastic feedstock. All of these parameters influence the reaction kinetics, yield of the products obtained, and carbon number distribution of the products. The efficiency of the process relies on the optimization of operational parameters, and optimal conditions may result in enhanced energy recovery from PW (Kartik et al., 2022).

3.4.2 Catalytic Pyrolysis of Plastic Waste

The pyrolysis of plastic waste and the subsequent thermal breakdown of long-chain molecules is considerably endothermic and requires substantial energy input. The thermal cracking process yields wide range of products. The generation of wax is regarded as a considerable drawback in the process due to the risk of equipment damage and obstruction in the pipeline system. The difficulties mentioned here can be addressed by the catalytic pyrolysis process, which reduces the activation energy, minimizes the energy requirements, and optimizes the product distribution. Figure 3.3 illustrates the process of catalytic pyrolysis.

The application of catalysts in the pyrolysis process accelerates polymer decomposition, resulting in increased gas production. Catalysts also increase the concentration of lighter hydrocarbons and elevate the octane rating of the resulting liquid product. For PW pyrolysis, based on the size of the substituents, the effectiveness of catalysts in reducing the required temperature and required energy for the pyrolysis reaction follows this order: polyethylene > polypropylene > polystyrene. The reduction in energy consumption and better product distribution demonstrate that the incorporation of a catalyst is advantageous in the conversion of plastic waste into liquid fuel by quick pyrolysis. After incorporating the catalyst, the oil produced from PW shows improved quality, even with lower energy input, reaction temperature, and residence time. This helps reduce the increasing energy consumption of the pyrolysis process when compared to other treatment methods. Diverse

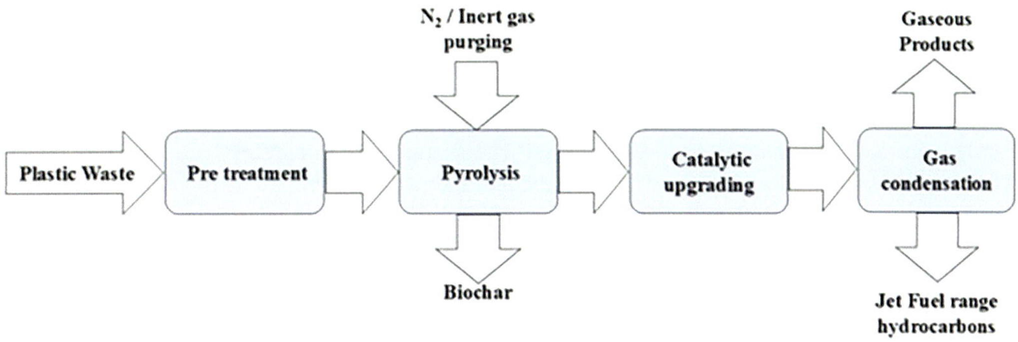

FIGURE 3.3 Flow diagram of catalytic pyrolysis process of plastic waste.

TABLE 3.3
Catalytic Pyrolysis of Plastic Waste to Produce Aviation Fuel-Range Hydrocarbons

Type of Plastic	Conditions	Catalyst and Products	References
HDPE	T = 350–390°C, Fixed bed reactor	• $CuCo_3$ catalyst: 14.7 wt% gases and 85 wt% C_9-C_{25} liquid oil • $BaTiO_3$/Ni-Al_2O_3 catalyst: C_6-C_{16}/C_7-C_{27} liquid oil	Saha et al., 2024
PS	T = 450–500°C, Tubular reactor, Batch-type	• Al-Al_2O_3: 7.31 wt% gases and 92.69 wt% of liquid fuels • Natural zeolite: 60.8 wt% ethylbenzene • Cu, FCC: 3–6 wt% gases and 90–97 wt% liquid fuels	Hussain et al., 2022
LDPE	T = 250–500°C, Spouted bed reactor, Fixed bed reactors	• Pt-promoted SZ: 55% of olefins • HZSM-5/Zeolites: 25 wt% (C_5-C_{11} of paraffins range:15 wt%, aromatics: 10 wt%) and 75 wt% (C_2-C_4) olefins. • Activated carbon: 23.7 wt% of gas (H_2 47% and 53% C_1-C_4) and 72 wt% C_8-C_{16} (including alkanes and aromatics)	B. Valizadeh et al., 2024a
PP	T=360–450°C, Fixed bed reactor, Stirred semi-batch reactor	• SiO_2/Al_2O_3: 92 wt% of C_3-C_{25} hydrocarbons • USY zeolite: 82.0 wt% of C_5-C_{30} and 16% of gaseous products • FCC: 85 wt% of liquid fuel and 15 wt% gaseous products	Goh et al., 2022
Polycarbonate	T = 500°C, Fixed bed reactor	• ZSM-5: gases:19 wt% and liquid fuels: 57 wt% • γ-Al_2O_3: gases:22 wt% and liquid fuels: 38 wt%	Vijayakumar & Pandurangan, 2017
Mix (PP/PS/PET)	T = 500°C, Fixed bed reactor	• Commercial equilibrium catalyst: 84.6 % of combustible gaseous products	Saha et al., 2024

catalysts, including FCC, spent FCC, HZSM-5, ZSM-5, Cu-Al_2O_3, Co−Mo/Z, Zeolite-β, Red Mud, $Al(OH)_3$, $Ca(OH)_2$, and Fe_2O_3, have been utilized to enhance the catalytic pyrolysis yield of plastics (Ali et al., 2022; Mousavi & Dehaghani, 2024). As shown in Table 3.3, the catalytic pyrolysis of plastic waste can yield liquid products in the range of aviation fuel.

The fixed bed catalytic pyrolysis of PP, HDPE, LDPE, HIPS, and GPPS utilizing a Fe/Al_2O_3 catalyst at temperatures ranging from 500°C to 800°C yielded hydrocarbons within the jet fuel range (C_8-C_{16}) and hydrogen. The pyrolysis of high-impact polystyrene demonstrated a hydrogen yield of 74 vol%, whereas general-purpose polyethylene exhibited a hydrogen yield of 70 vol%. Conversely, PP, HDPE, and LDPE exhibited yields of jet fuel-range hydrocarbons with 80.6%, 81.4%, and 80.2%, respectively. This is 20% higher than the yields obtained from HIPS and GPPS, which can be attributed to their lower carbon content (~81 wt%) in contrast to HIPS and GPPS (carbon content ~91 wt%). The catalytic pyrolysis of shopping bags and packaging trays using a biochar catalyst derived from nano-cellulose produced jet fuel-range hydrocarbons (C_8-C_{16}) at 600°C. The resultant product was consisted of aliphatic and monocyclic aromatic compounds 48.57% and 42.05%, respectively (Chen et al., 2024).

3.4.3 Microwave-Assisted Catalytic Pyrolysis of Plastic Waste

A two-step methodology is used to transform LDPE into jet fuel-quality hydrocarbons utilizing microwave conditions (Luo et al., 2024). Figure 3.4 illustrates a schematic representation of microwave-assisted pyrolysis.

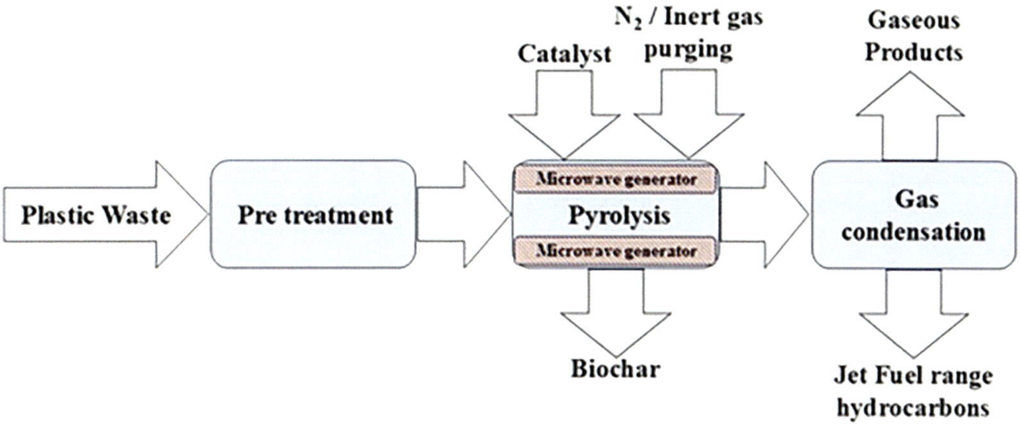

FIGURE 3.4 Flow diagram of microwave-assisted pyrolysis process of plastic waste.

The initial step involves microwave-assisted catalytic degradation of LDPE using ZSM-5, yielding up to 67% organic liquid products, predominantly heavier hydrocarbons. The subsequent hydrotreatment in the presence of Raney Nickel Catalyst at 250°C produced a product mixture comprising of aliphatic hydrocarbons, cycloalkanes, and aromatics in the ranges of 31–32%, 53–54%and 14–15%, respectively. The oil generated from PW had properties akin to JP-5 fuel and was compatible for blending with regular jet fuel in airplanes.

Studies indicate that microwave/hydrotreatment at 200°C yields 66.18 wt% plastic-derived oil from LDPE (Ali et al., 2022). The synthesized Raney nickel catalyst was utilized to hydrogenate oil produced from polymers. The hydrocarbon production was compared to the commercially available Raney Ni 4200. The selectivity of aliphatic and cycloalkanes was 8.14% and 76.08%, respectively, for the in-house-produced Raney nickel (Ni), surpassing that of commercial Raney Ni 4200, which exhibited selectivity of 7.14% for aliphatic and 56.28% for cycloalkanes. The cost-efficient homemade catalyst showed an enhanced performance as compared to the commercial catalyst in the synthesis of jet fuels from LDPE. The microwave-assisted pyrolysis of LDPE with ZSM-5 catalyst, followed by hydrogenation using a Raney nickel catalyst, also produces aliphatic alkenes and aromatic hydrocarbons, with carbon range between C_8 and C_{13}, suitable for aviation fuel (Khalifa et al., 2024). Table 3.4 summarizes the different operating conditions and product yields of microwave-assisted pyrolysis in the production of aviation fuel-range hydrocarbons.

3.4.4 Hydro-Liquefaction of Plastic Waste

Hydro-liquefaction is an advanced thermochemical process that can be used to convert plastic waste into valuable liquid hydrocarbons. This process uses hydrogen and catalysts to produce liquid fuel under high-pressure and moderate-to-high temperature conditions. This method enhances the quality of the resulting oil by reducing oxygen, sulfur, and nitrogen impurities, making it more suitable for fuel applications. Hydro-liquefaction can be utilized to produce jet fuel-grade hydrocarbons from plastic waste including high-density polyethylene, polycarbonate, and polyethylene. The hydro-liquefaction of HDPE, conducted without solvent over Ni/Al_2O_3 at 400°C, yielded hydrocarbons within the jet fuel range ($C_7 - C_{16}$). The catalyst (Ni/Al_2O_3) exhibited favorable selectivity, more than 72%, for $C_7 - C_{16}$ jet fuel hydrocarbons, which can be attributed to the lower acidity of the alumina support, resulting in a reduced cracking of larger molecular chain hydrocarbons. In a subsequent work, a similar group conducted the hydro-liquefaction of HDPE using Ni/HZSM-5 to enhance aromatics yield in the resulting combination. Jet fuel-range hydrocarbons up to 32% were obtained from

TABLE 3.4
Microwave-Assisted Pyrolysis to Produce Aviation Fuel-Range Hydrocarbons from Plastic Waste

Plastic Type	Conditions	Catalyst/Absorber	Product/Yields	References
PP	• Microwave power: 300–600 • Temperature: 600°C	Graphite	63% of liquid oils ($C_8 \sim C_{12}$ hydrocarbons, includes cycloalkanes, and aromatic) and 32% of gases including H_2	(Dai et al., 2024; Emmanouilidou et al., 2023; Hussain et al., 2022; Shahabuddin et al., 2020; Zhou et al., 2021)
PS	• Microwave power: 1200–6000 • Temperature: 360–600°C	Carbon/SiC	41–86% of liquid oils ($C_8 \sim C_{10}$ hydrocarbons, includes aliphatic styrene, and aromatic) and 9–30% of gases	
PE	• Microwave power: 3000 • Temp: 350–550°C	Tire, carbon with Fe powder	31% of liquid oils (linear alkanes) and 24% of gases	
LDPE	• Microwave power: 3000 • Temperature: 350–550°C	MgO/SiC	37% liquid fuels containing $C_5 - C_{12}$, aromatic and alkenes	
PP and PS	• Microwave power: 900 • Temperature: 900°C	Activated Carbon	85% of liquid fuels mainly aromatic and styrene	
PP and PET	• Microwave power: 800 • Temperature: 550°C	HZSM-5/SiC	25–36% of liquid fuels containing $C_6 - C_{12}$ hydrocarbon range	

HDPE on adding 15 wt% Ni/HZSM-5 at an operating temperature of 400°C. Figure 3.5 illustrates the process of hydro-liquefaction into hydrocarbons suitable for aviation fuel.

The direct approach of employing catalytic hydrotreatment effectively transforms PW into transportation fuels. The hydro-liquefaction of PW produced as high as 85% yields and the hydrocarbons in jet/diesel range around 47% by utilizing a $Pt/WO_3/ZrO_2$ catalyst in conjunction with HY zeolite at 225°C. Additionally, the PC waste was transformed into jet fuel-range hydrocarbons by a single-step APHDO method. The aqueous phase hydrolysis of PW conducted at 35 bar pressure and 200°C using 0.1 g Rh/C + 0.1 g H-USY catalyst demonstrated a product yield of more than 94%. The products, predominantly containing propane-2 and 2-diyldicyclohexane, exhibited properties comparable to the jet fuel. The density (0.925 g/mL) and net heat of combustion (39.04 MJ/L) values compared between the synthesized compound and established aviation fuels suggest its potential as a sustainable alternative fuel source (Aghamiri & Lahijani, 2024; Ali et al., 2022; Emmanouilidou et al., 2023).

3.4.5 Alcoholysis Succeeded by Hydrogenation of Plastic Waste

Research has been performed on the alcoholysis and hydrodeoxygenation of polycarbonate waste (PCW) to convert it into hydrocarbons appropriate for jet fuel. Methanolysis was conducted at 180°C to generate the intermediate product bisphenol from polycarbonate waste. Hydrogenation of bisphenol at 30 bar pressure and 140°C in the presence of Pt/C on zeolite catalyst generated polycycloalkanes in the $C_{13} - C_{15}$ carbon range, appropriate for jet fuel. The higher power density of the resultant product permits its utilization as it is or in combination with traditional jet fuel. A similar methodology of methanolysis and hydrodeoxygenation was employed to transform PET into $C_7 - C_8$ range hydrocarbons.

The waste PET was converted into dimethyl terephthalate via methanolysis at 200°C, and the dimethyl terephthalate formed is further hydrogenated to dimethyl cyclohexane-1,4-dicarboxylate. Hydrogenation was carried out at 140°C and 50 bar pressure and in the presence of a catalyst. Out

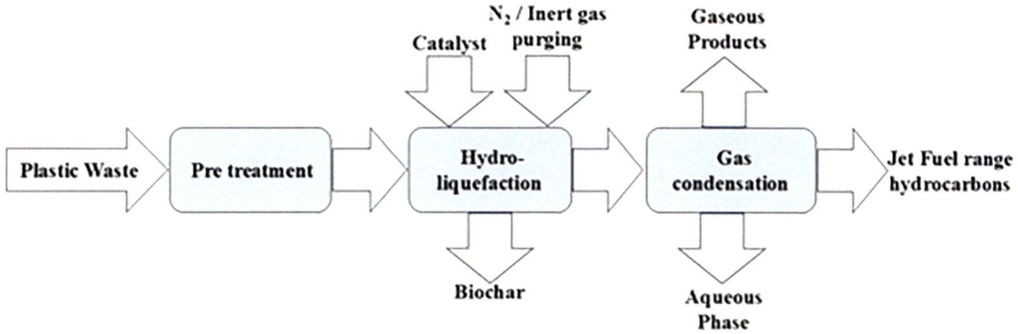

FIGURE 3.5 Flow diagram of hydro-liquefaction of polymers into the jet fuel process.

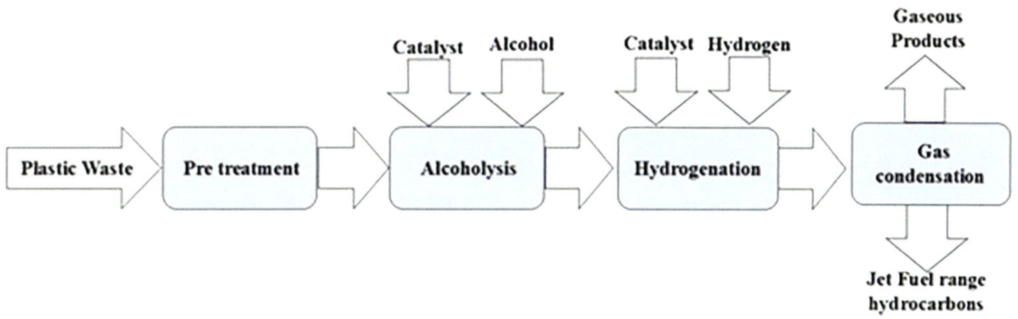

FIGURE 3.6 Flow diagram of alcoholysis succeeded by hydrogenation of polymers into the jet fuel process.

of the several catalysts experimented including Pt, Ru, and Pd, Pt supported on carbon catalyst was found to be the suitable catalyst for the production of dimethyl cyclohexane-1,4-dicarboxylate. The product yield was more than 97 wt%, however, the dimethyl cyclohexane-1,4-dicarboxylate yield was up to 70 wt%. The hydrodeoxygenation of the dimethyl cyclohexane-1,4-dicarboxylate in the presence of ruthenium-copper catalyst on silica support at 40 bar pressure and 400°C resulted in the production of aromatics and cycloalkanes. The yield of cycloalkanes thus produced in the range of 60–65% made this another potential pathway for converting PET plastic waste to jet fuel-range hydrocarbons (Ali et al., 2022; Hussain et al., 2022; Liu et al., 2023). A flow diagram for alcoholysis succeeded by hydrogenation process is represented in Figure 3.6.

3.4.6 Thermal Cracking of Plastic Waste with and without Hydrogenation

Thermal cracking followed by co-hydrogenation is an effective technique for producing jet fuel from a polyethylene and polypropylene blend. Research demonstrates that jet fuel-range hydrocarbons with yields of up to 11% were generated from a blend of polyethylene and polypropylene waste under pyrolysis conditions of 40 bar and 300°C, utilizing a nickel–molybdenum catalyst supported on alumina/phosphorus. In a separate study, polypropylene waste with varying molecular weights of 12,000, 250,000, and 340,000 was pyrolyzed into hydrocarbons. On pyrolysis, the polypropylene with lower molecular weight 12,000 produced more of diesel range hydrocarbons, whereas the higher molecular weight PP 250,000 and 340,000 produced more of jet fuel-range hydrocarbons. Figure 3.7 indicates the flow diagram of thermal cracking with hydrogenation to produce jet fuel-range hydrocarbons.

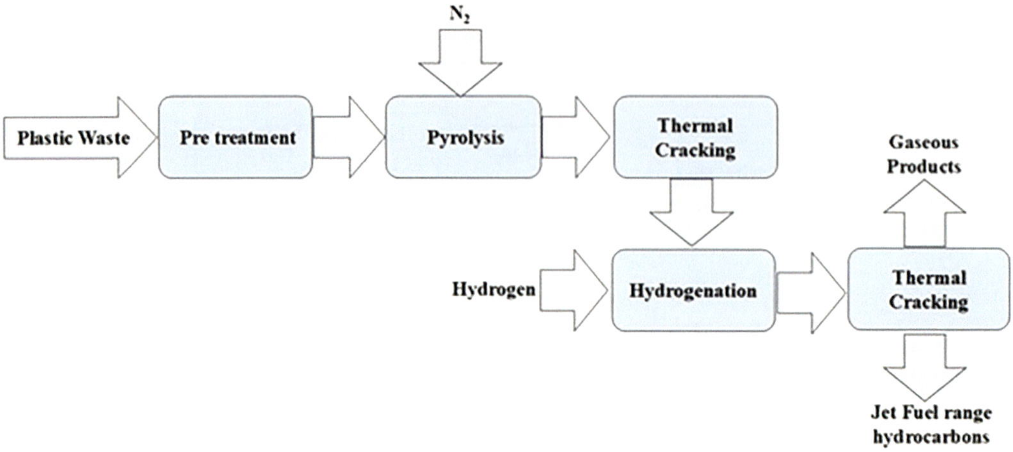

FIGURE 3.7 Flow diagram of thermal cracking succeeded by hydrogenation to produce jet fuel-range hydrocarbons from plastic waste.

Studies indicated that the thermal cracking of LDPE and HDPE feedstock and subsequent hydrogenation produced entirely jet fuel-range hydrocarbons. The optimal working conditions for producing jet-range products from LDPE and HDPE feedstock in a single step are reported at a temperature of 400°C, a hydrogen pressure of 10 bar, and a duration of 2 hours. Another study devised a two-step methodology in which separate reactors are used for cracking and hydrogenation with different operating conditions. Cracking was conducted at a temperature of 375°C at 1 bar pressure, and hydrogenation at a pressure above 34 bar at a temperature of 250°C. Although a 100% output of jet fuel-range hydrocarbons was attained using this methodology, the overall process was excessively costly. A separate investigation performed at a high temperature of 850°C demonstrates that the whole process is energy-intensive and expensive. Thus, the aforementioned single-step process aligns most accurately with the optimal route for jet fuel generation from HDPE and LDPE. The co-pyrolysis of plastic waste and biomass to generate specific transportation fuels is an alternative approach. Aromatic hydrocarbons are mostly generated throughout these operations, forming the essential constituents for transportation fuel. Consequently, the present data indicate that only LDPE as the feedstock may deliver the highest amount of jet fuel hydrocarbons (Aghamiri & Lahijani, 2024; Lee et al., 2024; Park & Lee, 2021). LDPE consists primarily of long, unbranched, and small, branched hydrocarbon chains with a high proportion of $C_{10} - C_{30}$ alkanes and alkenes, which fall within the jet fuel range ($C_8 - C_{16}$).

Numerous methods are available for the conversion of PW to aviation fuel; however, a thorough comprehension of the catalytic conversion mechanism is lacking. Analytical methods such as atmospheric pressure photoionization mass spectrometry (APPI-MS) and mechanistic modeling are necessary to elucidate the accurate mechanism of PW degradation into hydrocarbons suitable for transportation fuel (Ali et al., 2022). The feasibility of converting PW to aviation fuel necessitates further research for commercial-scale applications, addressing intricate reaction kinetics and transport phenomena. Emphasis must also be placed on the utilization of non-noble metal catalysts, which are economically advantageous. Additionally, optimizing operating parameters is crucial for the commercial viability of the PW-to-aviation fuel conversion process. The economic evaluation of PW-to-jet fuel conversion requires further investigation; nonetheless, initial data suggest the economic feasibility of the pyrolysis process, considering the capital and operational costs, compared to alternative thermochemical and biochemical conversion processes. Moreover, in the microwave processing of plastics, the risk of uncontrolled heating and scale-up challenges cannot be disregarded; similar to hydrothermal processes, the increased pressure and temperature might be harmful. Plastic pyrolysis transpires at elevated temperatures, posing a possible thermal hazard. The interactions among volatiles generated by plastic pyrolysis can occasionally become unregulated, leading to excessive pressure build-up inside the system.

3.5 ADVANTAGES OF PYROLYSIS OVER ALTERNATIVE PLASTIC WASTE TREATMENT METHODS

Approximately 98% of precursors for synthetic plastic production originate from fossil origin, which is non-renewable, and hence the careful disposal of PW is essential to avert environmental damage. Recycling PW is one of the most preferred methods for managing environmental damage; however, there are a number of factors that contribute to the poor recycling rate of PW around the world. These factors include the complex properties of plastic polymers, limited infrastructure, transport, logistics, and economic hurdles. As per the available data in 2022, only 9% of the produced waste was recycled, 19% was incinerated, and a substantial 72% was either landfilled or released into the natural environment (Lee et al., 2025). The challenges on landfilling of the plastic waste include the leaching of heavy metals, microplastics, and pathogens into the soil and water bodies, posing a significant risk to the ecosystem and human health. Incineration of the plastic waste generates significant health hazards than land filling due to the release of hazardous pollutants, including acid gasses, NOx emissions, greenhouse gas emissions, dioxins, and particulate matter. Therefore, there is an urgent necessity to enhance PW management measures and transition toward more sustainable practices (Liu et al., 2024; Valizadeh et al., 2024b).

The sole method to avert PW from reaching hazardous disposal methods is the adoption of advanced methods for processing with strict emission controls. Recycling has emerged as an effective strategy for managing municipal solid waste, particularly post-consumer plastic waste. The efficacy of PW recycling can be enhanced by implementing suitable sorting and recycling methodologies. Out of the several recycling methods including mechanical, thermochemical, and biological methods, mechanical recycling is environmentally beneficial; however, the varied compositions of materials in the plastic waste renders the sorting and recycling process difficult or rather impractical. The microplastics and larger plastic polymers are resistant to microbial degradation, hence limiting the applicability of biological recycling techniques for processing PW. Conversely, thermochemical conversion technologies, especially pyrolysis, among other technologies, have been extensively researched and suggested as sustainable alternatives to the hazardous methods of PW management. Thermochemical processes can be operated under modifiable circumstances to mitigate environmental impact and to diminish the necessity for lengthy procedures.

Pyrolysis has been emphasized as an efficient and eco-friendly process for generating value-added transportation fuels and value-added chemicals from PW. In contrast to other thermochemical processes, which require elevated temperatures (~600–1000°C), leading to substantial reactor fabrication and heating costs, pyrolysis technology represents a more economical and favorable approach for the valorization of PW into transportation fuels. From an environmental standpoint, pyrolysis may reduce the generation of dioxins and greenhouse gas emissions (Mishra et al., 2024; Saha et al., 2024).

3.6 CONVERSION OF PLASTIC WASTE INTO JET FUEL

Pyrolysis of PPE plastic waste can yield over 75 wt% of plastic-derived oil, signifying a notable progression in waste-to-energy transformation and the reduction of PPE plastic waste volume. Pyrolysis of face masks into oil produced from plastic is a feasible recycling option. Empirical evidence indicates that the pyrolysis of face masks at temperatures ranging from 350°C to 600°C can provide hydrogen (H_2), lighter olefins ($C_1 - C_2$), and long-chain hydrocarbons ($C_8 - C_{46}$). The yield and content of pyrolysis-derived oil are profoundly affected by operational parameters, with yields reaching up to 81% for the pyrolysis of filter masks. The oil produced from plastic consists of branched hydrocarbons and holds considerable potential for liquid fuel production (Ali et al., 2022; Gurram et al., 2024).

Pyrolysis followed by hydrothermal upgrading is a favored method for processing any other plastic waste, including polypropylene, polyethylene, and polycarbonate materials, as alternatives for jet fuel production. The domain is unexplored, requiring extensive research to guarantee that the final product requirements correspond to the operational and statutory requirements. Utilizing these technologies, PW can be converted into plastic-derived oil rather than being disposed of, thereby mitigating environmental damage (Luo et al., 2024; Wang & Tao, 2016).

In addition to medical waste, plastics disposed of in landfills, aquatic ecosystems, and oceans can be processed by hydrothermal techniques and pyrolysis instead of incineration, thereby mitigating environmental contamination and generating useful by-products. Nonetheless, the field of plastic-to-jet fuel conversion is still in its infancy, requiring substantial research attention to develop a simple, economical, environmentally sustainable, and scalable solution that can gain broad acceptance (Auersvald et al., 2024). Other constraints include the necessity for the development of economically feasible catalysts and ideal process conditions suitable for diverse plastic waste feedstocks. Research predominantly suggests that the operational parameters should be maintained between 400–600°C and a hydrogen pressure above 34 bar; these parameters require optimization to enhance economic viability and assure process scalability (Emmanouilidou et al., 2023; Liu et al., 2024). To convert the plastic waste to jet fuel, the process requires further investigation due to the presence of many competing reactions during the process, resulting in the formation of gaseous products and waxy compounds in addition to jet fuel-range hydrocarbons. To render the conversion of PW to jet fuel-range hydrocarbons feasible, molecular-level simulations are necessary to identify the governing and non-governing processes involved in the conversion, which presents a scope for future research.

3.7 CONCLUSIONS AND PERSPECTIVES

Plastic waste poses a significant threat to the environment, contributing to microplastic pollution, ocean litter, and disruptions to the carbon cycle. Converting this plastic waste into sustainable fuel is a promising option to reduce plastic waste and also reduce the carbon footprint of the aviation industry. Over the years, several technologies have been developed for producing sustainable aviation fuel from oil, plant biomass, algal biomass, lignocellulosic biomass, and polymer waste materials. When selecting the most effective production method, it is essential to consider the availability of raw materials. The abundant availability of plastic waste and the necessity to process it to avoid environmental damage make sustainable aviation fuel from plastics a promising option compared to other technologies. Thermal pyrolysis, catalytic pyrolysis, microwave-assisted pyrolysis, hydro-liquefaction, alcoholysis, followed by hydrogenation, and thermal cracking of plastic waste are some of the thermal chemical conversion technologies that can be used to convert plastic waste into sustainable aviation fuels. Among the various technologies available, catalytic pyrolysis has been highlighted as an efficient method for producing a sustainable range of hydrocarbons suitable for aviation fuel. Based on the current available knowledge, the conversion of plastic waste to jet fuel-range hydrocarbons requires further in-depth study for commercialization to be successful.

Aviation fuels derived from plastics are a promising option for the aviation sector to achieve sustainability, where the solution is only a drop in the fuel alternative. To commercialize the available technologies, several challenges must be overcome. Some of the key issues that need to be addressed include the price gap between commercial jet fuel and sustainable aviation fuel. Studies on efficient production processes, including process intensification, are crucial for the cost-effective, alternative, and sustainable production of fuels. Optimizing catalyst performance and improving the efficiency of catalysts through efficient catalyst design are other areas in which further improvements are needed. Other potential problems that need to be addressed include sustainability concerns and the presence of pollutants in the waste streams. To address all these challenges, a comprehensive understanding of the combustion properties of sustainable fuels is required. Further research is needed to develop the fuel combustion models with improved predictability. Comprehensive life-cycle assessment, sustainability matrices, and certification processes will also be required in the future for the commercialization of sustainable aviation fuel technologies.

REFERENCES

Aghamiri, A. R., & Lahijani, P. (2024). Catalytic conversion of biomass and plastic waste to alternative aviation fuels: A review. *Biomass and Bioenergy*, 183(August 2023), 107120. https://doi.org/10.1016/j.biombioe.2024.107120

Akin, O., Varghese, R. J., Eschenbacher, A., Oenema, J., Abbas-Abadi, M. S., Stefanidis, G. D., & Van Geem, K. M. (2023). Chemical recycling of plastic waste to monomers: Effect of catalyst contact time, acidity and pore size on olefin recovery in ex-situ catalytic pyrolysis of polyolefin waste. *Journal of Analytical and Applied Pyrolysis*, 172(June). https://doi.org/10.1016/j.jaap.2023.106036

Al-Shihabi, S., Aydin, R., & Al Nahlawi, M. (2023). Forecasting the effects of municipal solid plastic waste generation and streaming using system dynamics: A case study in Dubai. *Science of the Total Environment*, 897(June). https://doi.org/10.1016/j.scitotenv.2023.165204

Alawa, B., & Chakma, S. (2024). Feasibility and techno-economic analysis of hydrocarbon liquid fuels production via catalytic pyrolysis of waste plastic materials. *Carbon Capture Science & Technology*, 100337. https://doi.org/10.1016/j.ccst.2024.100337

Ali, S. S., Zamir, S., Shakeelur Raheman, A. R., Ansari, K. B., Qyyum, M. A., Al Mesfer, M. K., Danish, M., Khan, M. S., & Al-Muhtaseb, A. H. (2022). Current progress in thermochemical conversion of plastics into jet-fuel hydrocarbons and recommendations for COVID-19 waste management. *Process Safety and Environmental Protection*, 166(January), 535–557. https://doi.org/10.1016/j.psep.2022.08.050

Auersvald, M., Šiman, M., Lyko Vachková, E., Kroufek, J., Straka, P., Barzallo, G., & Vozka, P. (2024). A comparative study of aromatic content in pyrolysis oils from waste plastics and tires: Assessing common refinery methods. *Fuel*, 369(January). https://doi.org/10.1016/j.fuel.2024.131714

Baca, D., Monroy, R., Castillo, M., Elkhazraji, A., Farooq, A., & Ahmad, R. (2023). Dioxins and plastic waste: A scientometric analysis and systematic literature review of the detection methods. *Environmental Advances*, 13(October). https://doi.org/10.1016/j.envadv.2023.100439

Chen, X., Cheng, L., Gu, J., Yuan, H., & Chen, Y. (2024). Chemical recycling of plastic wastes via homogeneous catalysis: A review. *Chemical Engineering Journal*, 479(December 2023), 147853. https://doi.org/10.1016/j.cej.2023.147853

Culsum, N. T. U., Kismanto, A., Zuldian, P., Supriatna, N. K., Yarsono, S., Nainggolan, L. M. T., Raksodewanto, A. A., Fariza, O., Ermada, F. J., Saribu, D. G. D., Khalil, M., & Kadja, G. T. M. (2024). An overview of catalytic pyrolysis of plastic waste over base catalysts. *Journal of Analytical and Applied Pyrolysis*, 183(October), 106828. https://doi.org/10.1016/j.jaap.2024.106828

Dai, L., Lata, S., Cobb, K., Zou, R., Lei, H., Chen, P., & Ruan, R. (2024). Recent advances in polyolefinic plastic pyrolysis to produce fuels and chemicals. *Journal of Analytical and Applied Pyrolysis*, 180(May), 106551. https://doi.org/10.1016/j.jaap.2024.106551

Emmanouilidou, E., Mitkidou, S., Agapiou, A., & Kokkinos, N. C. (2023). Solid waste biomass as a potential feedstock for producing sustainable aviation fuel: A systematic review. *Renewable Energy*, 206(October 2022), 897–907. https://doi.org/10.1016/j.renene.2023.02.113

Goh, B. H. H., Chong, C. T., Ong, H. C., Seljak, T., Katrašnik, T., Józsa, V., Ng, J. H., Tian, B., Karmarkar, S., & Ashokkumar, V. (2022). Recent advancements in catalytic conversion pathways for synthetic jet fuel produced from bioresources. *Energy Conversion and Management*, 251(November 2021). https://doi.org/10.1016/j.enconman.2021.114974

Gurram, H. K. V., Pamu, S. H., & Singh, S. A. (2024). Parametric studies on fractionation column design for the separation of plastic waste pyrolysis oil into valuable fuels. *Journal of Environmental Chemical Engineering*, 12(2), 112390. https://doi.org/10.1016/j.jece.2024.112390

Hu, E., Zhang, Y., Liu, Z., Yu, J., Li, M., Xiong, Q., & Zeng, Y. (2024). Insight into dechlorination of pyrolysis oil during fast co-pyrolysis of high-alkali coal and polyvinyl chloride (PVC). *Chemical Engineering Journal*, 494(June), 153016. https://doi.org/10.1016/j.cej.2024.153016

Huerta-Rosas, B., Sánchez-Ramírez, E., Contreras-Zarazua, G., Quíroz-Ramírez, J. J., Luevano-Rivas, O. A., & Segovia-Hernández, J. G. (2024). Design and optimization of an intensified sustainable plant to produce biojet fuel using the ATJ process. *Chemical Engineering and Processing—Process Intensification*, 200(March). https://doi.org/10.1016/j.cep.2024.109774

Huo, E., Lei, H., Liu, C., Zhang, Y., Xin, L., Zhao, Y., Qian, M., Zhang, Q., Lin, X., Wang, C., Mateo, W., Villota, E. M., & Ruan, R. (2020). Jet fuel and hydrogen produced from waste plastics catalytic pyrolysis with activated carbon and MgO. *Science of the Total Environment*, 727, 138411. https://doi.org/10.1016/j.scitotenv.2020.138411

Hussain, I., Ganiyu, S. A., Alasiri, H., & Alhooshani, K. (2022). A state-of-the-art review on waste plastics-derived aviation fuel: Unveiling the heterogeneous catalytic systems and techno-economy feasibility of catalytic pyrolysis. *Energy Conversion and Management*, 274(October), 116433. https://doi.org/10.1016/j.enconman.2022.116433

Justicia, J., Heras, F., Moreno, I., Baeza, J. A., Calvo, L., Fermoso, J., & Gilarranz, M. A. (2024). Understanding the relationship between catalytic pyrolysis conditions and hydrogen production by aqueous phase reforming of the water-soluble fractions of bio-oils. *Energy Conversion and Management*, 320(June). https://doi.org/10.1016/j.enconman.2024.118999

Kartik, S., Balsora, H. K., Sharma, M., Saptoro, A., Jain, R. K., Joshi, J. B., & Sharma, A. (2022). Valorization of plastic wastes for production of fuels and value-added chemicals through pyrolysis—A review. *Thermal Science and Engineering Progress*, 32(March), 101316. https://doi.org/10.1016/j.tsep.2022.101316

Khalifa, R., Alherbawi, M., Bicer, Y., & Al-Ansari, T. (2024). Fueling circularity: A thorough review of circular practices in the aviation sector with sustainable fuel solutions. *Resources, Conservation and Recycling Advances*, 23(June), 200223. https://doi.org/10.1016/j.rcradv.2024.200223

Lee, J. E., Lee, D., Lee, J., & Park, Y. K. (2025). Current methods for plastic waste recycling: Challenges and opportunities. *Chemosphere*, 370(November 2024). https://doi.org/10.1016/j.chemosphere.2024.143978

Lee, T., Jung, S., Lee, S., Tsang, Y. F., Lee, K. H., & Kwon, E. E. (2024). Production of aviation fuel via thermal cracking of plastic waste. *Energy Conversion and Management*, 315(July), 118827. https://doi.org/10.1016/j.enconman.2024.118827

Lin, W., Ning, D., Xu, D., Wu, S., Zou, Q., Mondal, A. K., & Huang, F. (2024). Study on the effect of combined and free lignin to the cellulose enzymatic hydrolysis. *Industrial Crops and Products*, 222(P1), 119516. https://doi.org/10.1016/j.indcrop.2024.119516

Liu, J., Wei, J., Feng, X., Song, M., Shi, S., Liu, S., & Liu, G. (2023). Ni/HZSM-5 catalysts for hydrodeoxygenation of polycarbonate plastic wastes into cycloalkanes for sustainable aviation fuels. *Applied Catalysis B: Environmental*, 338(June), 123050. https://doi.org/10.1016/j.apcatb.2023.123050

Liu, Z., Xie, M., Zhou, T., Yang, J., Yang, Y., Liu, T., Dai, S., Huang, Q., Cen, Q., Xiao, P., Li, B., & Du, H. (2024). A review on liquid fuel produced from microwave-assisted pyrolysis of plastic waste. *Process Safety and Environmental Protection*, 187(February), 833–844. https://doi.org/10.1016/j.psep.2024.05.035

Luo, T., Zhou, W., Wang, Y., Jiang, H., Wu, J., Hu, J., Wang, M., Wang, W., Wang, Q., Hu, Y., & Fan, L. (2024). Integrating microwave pyrolysis and hydrotreating for converting low-density polyethylene into jet fuel. *Renewable Energy*, 236(September), 121432. https://doi.org/10.1016/j.renene.2024.121432

Mishra, R., Shu, C. M., Gollakota, A. R. K., & Pan, S. Y. (2024). Unveiling the potential of pyrolysis-gasification for hydrogen-rich syngas production from biomass and plastic waste. *Energy Conversion and Management*, 321(August), 118997. https://doi.org/10.1016/j.enconman.2024.118997

Mousavi, S. A. H. S., & Dehaghani, A. H. S. (2024). Catalytic pyrolysis of plastic waste to gasoline, jet fuel and diesel with nano MOF derived-loaded Y zeolite: Evaluation of temperature, zeolite crystallization and catalyst loading effects. *Energy Conversion and Management*, 299(November 2023), 117825. https://doi.org/10.1016/j.enconman.2023.117825

Nayanathara Thathsarani Pilapitiya, P. G. C., & Ratnayake, A. S. (2024). The world of plastic waste: A review. *Cleaner Materials*, 11(August 2023), 100220. https://doi.org/10.1016/j.clema.2024.100220

Nguyen, Q. Van, Choi, Y. S., Jeong, Y. W., Han, S. Y., & Choi, S. K. (2024). Catalytic co-pyrolysis of coffee-grounds and waste polystyrene foam by calcium oxide in bubbling fluidized bed reactor. *Renewable Energy*, 224, 120124. https://doi.org/10.1016/j.renene.2024.120124

Park, C., & Lee, J. (2021). Pyrolysis of polypropylene for production of fuel-range products: Effect of molecular weight of polypropylene. *International Journal of Energy Research*, 45(9), 13088–13097. https://doi.org/10.1002/er.6635

Paucar-Sánchez, M. F., Martín-Lara, M. A., Calero, M., Blázquez, G., Solís, R. R., & Muñoz-Batista, M. J. (2023). Towards fuels production by a catalytic pyrolysis of a real mixture of post-consumer plastic waste. *Fuel*, 352(March). https://doi.org/10.1016/j.fuel.2023.129145

Royuela, D., Veses, A., Martínez, J. D., Callén, M. S., López, J. M., García, T., & Murillo, R. (2024). Thermochemical recycling of polystyrene waste by pyrolysis using a pilot-scale auger reactor: Process demonstration in a relevant environment. *Resources, Conservation and Recycling*, 211(May). https://doi.org/10.1016/j.resconrec.2024.107869

Saha, B., Vedachalam, S., Dalai, A. K., Saxena, S., Dally, B., & Roberts, W. L. (2024). Review on production of liquid fuel from plastic wastes through thermal and catalytic degradation. *Journal of the Energy Institute*, 114(October 2023), 101661. https://doi.org/10.1016/j.joei.2024.101661

Shahabuddin, M., Alam, M. T., Krishna, B. B., Bhaskar, T., & Perkins, G. (2020). A review on the production of renewable aviation fuels from the gasification of biomass and residual wastes. *Bioresource Technology*, 312(March), 123596. https://doi.org/10.1016/j.biortech.2020.123596

Sharma, A. K., Kumar Ghodke, P., Manna, S., & Chen, W. H. (2021). Emerging technologies for sustainable production of biohydrogen production from microalgae: A state-of-the-art review of upstream and downstream processes. *Bioresource Technology*, 342(August), 126057. https://doi.org/10.1016/j.biortech.2021.126057

Stančin, H., Strezov, V., & Mikulčić, H. (2023). Life cycle assessment of alternative fuel production by co-pyrolysis of waste biomass and plastics. *Journal of Cleaner Production*, 414(May). https://doi.org/10.1016/j.jclepro.2023.137676

Valizadeh, B., Valizadeh, S., Kim, H., Choi, Y. J., Seo, M. W., Yoo, K. S., Lin, K. Y. A., Hussain, M., & Park, Y. K. (2024a). Production of light olefins and monocyclic aromatic hydrocarbons from the pyrolysis of waste plastic straws over high-silica zeolite-based catalysts. *Environmental Research*, 245(November 2023), 118076. https://doi.org/10.1016/j.envres.2023.118076

Valizadeh, S., Valizadeh, B., Seo, M. W., Choi, Y. J., Lee, J., Chen, W. H., Lin, K. Y. A., & Park, Y. K. (2024b). Recent advances in liquid fuel production from plastic waste via pyrolysis: Emphasis on polyolefins and polystyrene. *Environmental Research*, 246(November 2023), 118154. https://doi.org/10.1016/j.envres.2024.118154

Vijayakumar, G., & Pandurangan, A. (2017). Up-gradation of A-tetralone to jet-fuel range hydrocarbons by vapour phase hydrodeoxygenation over Pd–Ni/SBA-16 catalysts. *Energy*, 140, 1158–1172. https://doi.org/10.1016/j.energy.2017.09.038

Wang, J., Ma, Y., Li, S., & Yue, C. (2023). Study of HDPE plastic pyrolysis characteristics using high pressure autoclave. *Journal of the Energy Institute*, 108(March), 101244. https://doi.org/10.1016/j.joei.2023.101244

Wang, W. C., & Tao, L. (2016). Bio-jet fuel conversion technologies. *Renewable and Sustainable Energy Reviews*, 53, 801–822. https://doi.org/10.1016/j.rser.2015.09.016

Wu, S., Li, J., Sun, Q., & Li, Z. (2024). Experimental and kinetic study of PET pyrolysis under fast and slow heating rates using a visualized Macro TGA. *Thermochimica Acta*, 739(May), 179812. https://doi.org/10.1016/j.tca.2024.179812

Zhou, N., Dai, L., Lyu, Y., Li, H., Deng, W., Guo, F., Chen, P., Lei, H., & Ruan, R. (2021). Catalytic pyrolysis of plastic wastes in a continuous microwave assisted pyrolysis system for fuel production. *Chemical Engineering Journal*, 418(December 2020), 129412. https://doi.org/10.1016/j.cej.2021.129412

4 Valorization of Plastic Waste to Materials

Kartik S.[1], *Sivasankar Kakku*[2], *Abhishek Sharma*[2,3], *and Manisha Sharma*[2*]

[1]Department of Environmental Science & Technology, UPL University of Sustainable Technology, Bharuch, India

[2]Department of Biotechnology and Chemical Engineering, Manipal University Jaipur, India

[3]School of Engineering, RMIT University, Melbourne, Australia

4.1 INTRODUCTION

With production of 400 million tons in 2022, plastics are recognized as the most important innovations of the 21st century (Plastics Europe, 2023). Pipes, fitting segments, the fast-moving consumer goods sector, and the packaging industries rely heavily on plastics due to their low density and corrosion-resistant properties. There are more than 30 types of primary plastics and their derivatives are present with high-density polyethylene (HDPE), low-density polyethylene (LDPE), polyvinyl chloride (PVC), polyethylene terephthalate (PET), polypropylene (PP), polystyrene (PS), acrylonitrile butadiene styrene (ABS), polyamide (PA), polycarbonates (PC), polybutylene terephthalate (PBT), and mixed layer plastics (MLP) as the major contributors (Plastics Europe, 2023; Silvestre, 2013). Plastics are synthesized through the polymerization of petroleum derivatives and can be molded repeatedly. When mixed with composite materials, plastics offer more strength, durability, and inertness. Figure 4.1 illustrates the production of plastics over the world in various categories from 2018 to 2022.

Rapid industrialization and contemporary lifestyles substantially increased plastic usage, leading to high fossil fuel utilization and the accumulation of plastic trash in the environment. Approximately 4% of fossil fuel resources are utilized in plastic manufacturing with liquefied natural gas (LNG) and lower gas fractions from the petroleum industry as the principal feedstocks (Lebreton and Andrady, 2019). The demand for plastics is projected to attain 1800 million tons by 2050, with a utilization rate double that of the baseline year 2040 (Lusher et al., 2017). Plastic materials offer an array of products across the world, with China commanding a market share of 32%. European Union and North America, which account for 31% of global production, rank second in the category (Figure 4.2).

Several industries are heavily dependent on plastics. These include food and preservation, textiles, packaging, electrical and electronics, and transportation. The plastic packaging industry has made it easy to transport items from one place to another. In India, packaging items constitute more than 80% of plastic waste, including polyolefins and polystyrene (Marico Innovation Foundation, 2023). The packaging industry accounts for 35.8% of plastic production and 46% of plastic waste (Tsakona, 2020).

In low-income countries, plastics constitute approximately 6% of the waste stream, which eventually increases with economic advancements (Kaza et al., 2018). For instance, in underdeveloped

* Corresponding author's email: manisha.sharma@jaipur.manipal.edu; msharma.mnit@gmail.com

 DOI: 10.1201/9781003581000-6

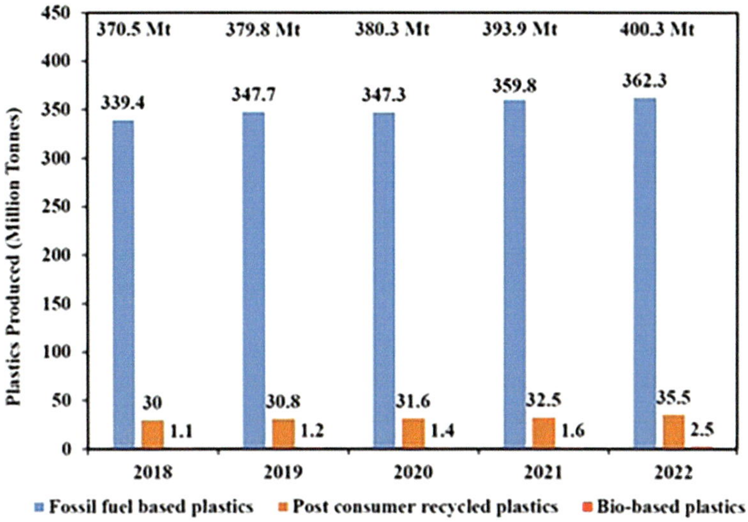

FIGURE 4.1 World production of plastics: 2018–2022. (From Plastics Europe, 2023.)

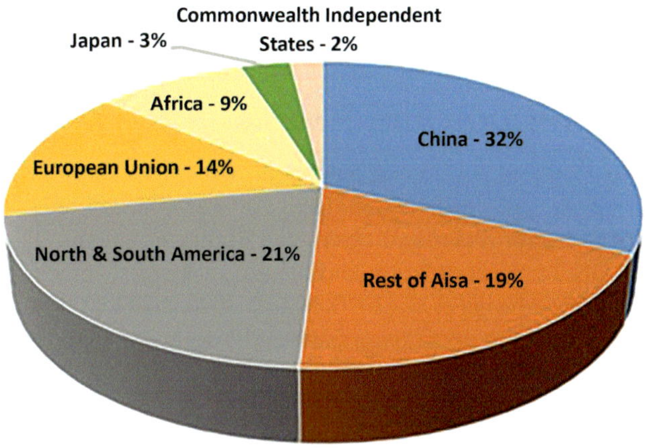

FIGURE 4.2 Share of different nations producing plastics. (From Plastics Europe, 2023.)

countries, about 50–60 % of plastics are discarded as waste. Only 30% of the plastic garbage generated annually is recycled, while the remainder is dumped in landfills or waterbodies without any treatment. It, therefore, poses health risks and degrades the quality of the surrounding air, water, and soil (Nichols and Smith, 2019). Thermosetting plastics manufactured through condensation polymerization of monomers cannot be reprocessed. On the other hand, thermoplastics, which include polyethylene (PE), PP, and PS, are recyclable. They contain alkyd, epoxy, ester, melamine and phenolic formaldehyde, silicon, urea formaldehyde, and polyurethane. It has been estimated that around 33 million tons of plastic are thrown into the aquatic ecosystem, which highlights the inability of existing waste management and recycling systems (Portilla, 2024). Natural degradation of plastics occurs over centuries, resulting in microplastics and nanoplastics, which adversely affect marine ecosystems (Mohanan et al., 2020). The lifespan of the key contributors in the plastic industry are presented in Table 4.1. Consequently, it is expected that by 2050, oceans will contain more plastic than fishe (World Economic Forum, 2016).

TABLE 4.1
Lifetime of Commonly Used Plastic Fractions

Polymer	Life Span (years)
LDPE	10–600
HDPE	>600
PS	50–80
PP	10–600

Source: Mohanan et al., 2020.

To reduce or eliminate the adverse impacts of plastic waste pollution, a circular economy can be actualized by accomplishing Sustainable Development Goals SDG1, SDG8, SDG9, and SDG12. This chapter examines numerous sustainable technologies that transform plastic waste into value-added products, serving as an essential element of the circular economy. A comprehensive assessment of pyrolysis as a waste valorization technique and the substantial influence of parameters on the process is provided. The methodologies for creating an effective process scheme that incorporates kinetic, particle, and reactor scale models are outlined, emphasizing commercially recognized pyrolysis processes along with their technology readiness level (TRL). Furthermore, factors to be considered in the economic and environmental aspects of the plastic waste valorization process are presented.

4.2 ADVANCED METHODOLOGIES FOR PLASTIC WASTE CONVERSION

Most of the underdeveloped and developing nations utilize open dumping, landfilling, and incineration as the primary waste processing techniques. The uncontrolled buildup of plastic debris in the environment leads to health issues and obstructed drainage systems. The scarcity of land for garbage dumping and GHG emissions from landfills are substantial impediments to the extensive adoption of this technology (Abiriga et al., 2020; Verma et al., 2016).

The challenge of managing plastic garbage can be efficiently addressed by the following:

1. Reducing the volume of waste.
2. Inappropriate evaluation of waste generation and treatment methods.
3. Inappropriate recycling initiatives to alleviate the consequences of landfilling
4. Effective extraction of power and thermal energy from waste streams.

For this purpose, the principles of circular economy can be implemented through policy to minimize the use of specific resources while adopting recycling procedures for others. The processing of waste to recover energy, either as heat or as electricity, entails substantial investments and incurs significant running costs. These contain food, organics, paper, and non-biodegradable substances as the primary constituents. Nevertheless, a substantial quantity of plastic waste is present, which takes years to decompose. The bulk of plastic waste is recyclable; however, municipal authorities lack the requisite infrastructure, hindering the widespread implementation of this energy-intensive process.

Diverse recycling techniques are implemented for the processing of waste to optimize the recovery of valuable materials with enhanced efficacy. *Primary recycling* transforms waste material into a product with similar characteristics at par with its original. The conversion of plastics into materials possessing distinct qualities from the original plastic products is referred to as *secondary recycling*. *Tertiary recycling* refers to the conversion of waste to useful chemicals and fuels, whether from a segregated waste stream or from municipal solid trash. Furthermore, *quaternary recycling* entails the combustion of trash or abandoned polymers to reclaim their energy content.

4.2.1 Pretreatment Techniques for Plastic Waste

For efficiently carrying out plastic waste recycling, it is crucial to remove contaminants (organic residues, additives, stabilizers, colorants, labels, and inks) along with the feedstock. Pretreating the feedstock may overcome the challenges encountered during various downstream processing operations. At the initial stages, plastics undergo a crude shredder to reduce their particle size. The shredded plastics are then cleaned in a rotary drum washer to separate them from pebbles, metals, and glassware. A friction washer is utilized to cleanse plastic surfaces of organic residues and adhesives by vigorous agitation. Subsequently, the washed plastic waste is conveyed through a screw conveyor system, while dissolved contaminants, particulates, and process water are expelled through sieves. Milling further reduces particle size, facilitating subsequent separation by density. Polymers with a density < 1 kg/L are segregated from denser materials, including PET, PS, and PVC, through a float and sink method. Thereafter, flakes are dried in a mechanical dryer, and melt filtration is performed to eliminate wood, paper, and rubber materials. Some pretreatment facilities utilize hot water combined with NaOH or detergent for water washing to decrease surface tension, thus reducing contamination from dirt, labels, and adhesive residues.

Deodorization of waste streams is an additional pretreatment method that must be implemented with compost bins and containers containing insecticides and pigments. Volatile organic compounds containing esters and alkanes constitute a major group with odor and are found in plastics. Detergent washing, in some cases, has proved ineffective in bringing down the odor below a specific threshold value. Alternatively, solvents such as ethyl acetate and polyethylene glycol (PEG) are employed to eliminate odor. Enhanced removal effectiveness for VOCs is observed following treatment with PEG at 100°C. Reduced volatility, toxicity, and enhanced miscibility with organic molecules facilitate superior extraction of VOCs via PEG (Cabanes and Fullana, 2021). Unfortunately, solvent-based cleansing incurs capital expenditures for facility setup and operational costs associated with chemicals, utilities, and energy, hence reducing the project's financial appeal. Consequently, solvent-based cleaning should be applied to a stream with a relatively higher extent of contamination to enable the removal of a broader range of impurities. Other than solvent extraction, hot air stripping, steam stripping, degassing system, and probiotic bacterial treatment are used for eliminating odors.

4.2.2 Treatment Techniques for Plastic Waste

Landfilling and incineration are traditional options for plastic waste disposal, although they are directly linked to health concerns and environmental damage (Abiriga et al., 2020; Acomb et al., 2014). Substantial obstacles to the use of this technology include the shortage of land space for waste disposal and the GHG emissions generated by landfills (Verma et al., 2016). Furthermore, methane generated in landfills is highly flammable at concentrations ranging from 5% to 15% of the total air volume (World Economic Forum, 2016; Candelaresi and Spazzafumo, 2021). The various techniques used for waste valorization are shown in Figure 4.3 (Agarwal, 2014).

Two prevalent treatment procedures that harness intrinsic energy content include incineration and recycling, albeit to varying degrees. This technique is one of the most advanced for waste disposal, achieving a 90% reduction in volume and a 75% reduction in weight for landfilling (Rand et al., 1999). Substantial capital and operational expenditures are associated with incinerators that release harmful mercury, arsenic, and dioxins.

4.2.2.1 Mechanical Recycling

This method involves recycling waste plastics in an industry to manufacture new products with the same quality as the original. Additives are also added to enhance the durability, applicability, and desired level of modification. Secondary molding processes, including rotational molding and thermoforming, are also utilized. Calendaring involves exerting pressure and heat on molten polymer in

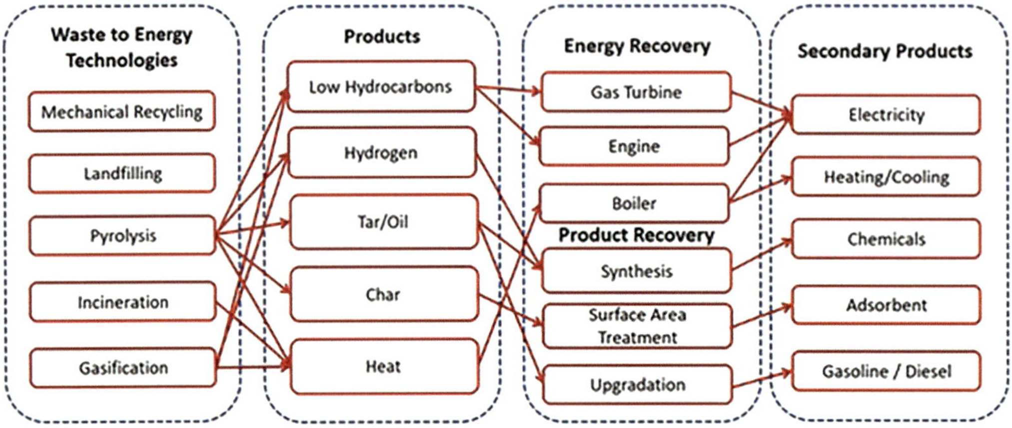

FIGURE 4.3 Pathways for waste valorization.

an extruder to produce high-quality plastic films, sheets, and large quantities of plastic. It is utilized primarily for polyvinyl acetate (PVA). In casting, a polymer is melted and molded into products of a specified shape. In mechanical recycling, minimizing energy consumption is difficult, and adequate quality assessments downstream are necessary to ensure product quality.

4.2.2.2 Chemical Recycling

Constraints and financial limitations of mechanical recycling highlight the potential of chemical recycling, a tertiary treatment technique (Schyns and Shaver, 2021). In this process, the waste is washed, shredded into small-sized particles, agglomerated, and then subjected to extrusion and granulation. The plastic is depolymerized into monomers, which may then be used to produce new polymers. Finally, the end products are purified to remove impurities. The process may involve a catalyst to break down polymers either chemically or thermally, producing hydrocarbons, which are used as a source of fuel or chemicals. The treatment of pre-consumer waste can be efficiently executed after selection and segregation, which must adhere to the IS codes.

Enzymatic depolymerization decomposes high molecular weight (MW) polymers into low MW polymers at relatively low temperatures. The procedure results in reduced energy usage. During hydrogenation, polymeric chains cleave to provide highly saturated liquids that are immediately utilized for energy-related applications without further processing. Furthermore, the utilization of hydrogen aids in the extraction of heteroatoms from polymeric waste materials. However, the technology incurs a financial drawback owing to the elevated expenses associated with hydrogen and high-pressure operations. Syngas, in conjunction with other volatile organic compounds (VOCs) such as polycyclic aromatic hydrocarbons (PAHs), is primarily produced, which can either be utilized for energy generation or be converted into ethanol and methanol (Ahmed et al., 2020).

4.2.2.3 Gasification

Gasification is crucial for minimizing and recycling plastic waste, as it recovers the energy content from polymers and diminishes emissions at elevated temperatures (500–1300°C) in the absence of oxygen. Primarily, syngas, along with VOCs including PAHs, is produced, which are employed to generate energy or be processed into ethanol and methanol (Acomb et al., 2014; Navarro et al., 2012). Gasification converts hydrocarbons into gases with beneficial thermal characteristics. This process, unlike combustion, does not yield a product stream with a consistent calorific value. Syngas, comprising different proportions of CO and H_2, is generated through the partial oxidation of feedstocks such as coal, oil, or natural gas. Steam, air, or pure oxygen may all be present in the oxidant stream. The relative ease of handling gas compared to solid fuel is crucial to the technology's success.

In gasification, saturated steam is typically used as a moderator. Since steam may condense in lines during expansion, temperatures of 300–400°C and pressures >30 bar are preferable. Elevated temperatures and pressures (10 bar–100 bar) are advisable to operate a gasifier satisfactorily. In the process, energy consumption is reduced significantly by 70% during compression, and small equipment is required at increased pressure. Similarly, the temperature at 1300°C is optimum to produce syngas with reduced methane concentration. Beyond this point, the overall efficiency of the process will decrease, and oxygen consumption will increase. Additionally, at optimum conditions, an enhanced yield of syngas is obtained. In gasification, solid carbonaceous contents (coal, coke, and char) undergo a series of reactions as mentioned:

Combustion Reactions

$$C + \frac{1}{2}O_2 \rightarrow CO \qquad \Delta H = -111\,MJ/mol \tag{4.1}$$

$$CO + \frac{1}{2}O_2 \rightarrow CO \qquad \Delta H = -283\,MJ/mol \tag{4.2}$$

$$H_2 + \frac{1}{2}O_2 \rightarrow H_2O \qquad \Delta H = -242\,MJ/mol \tag{4.3}$$

Boudouard Reaction

$$C + CO_2 \rightleftharpoons CO \qquad \Delta H = 172\ MJ/mol \tag{4.4}$$

Water Gas Shift Reaction

$$C + H_2O \rightleftharpoons CO + H_2 \qquad \Delta H = 131\ MJ/mol \tag{4.5}$$

Methanation Reaction

$$C + 2H_2 \rightleftharpoons CH_4 \qquad \Delta H = -75\ MJ/mol \tag{4.6}$$

The mechanism frequently relies on an equilibrium between the water-gas shift reaction and the partial oxidation of carbon. Standard heat of formation and the ultimate analysis of fuel must be considered in order to assess the energy generation.

Carbon conversion and cold gas efficiency (CGE) are two criteria for determining energy efficiency in gasification. For CGE computations, a consistent basis of heating values must be accurately defined.

$$\text{Cold gas efficiency (CGE)}(\%) = \frac{\text{Heating value in product gas}}{\text{Heating value in feedstock}} \times 100 \tag{4.7}$$

Similarly, carbon conversion is defined as

$$\text{Carbon conversion}(\%) = \left[1 - \frac{\text{Carbon in gasified residue}}{\text{Carbon in feedstock}}\right] \times 100 \tag{4.8}$$

Although these measurements indicate the efficacy of gasification, a process yielding greater methane will exhibit a superior CGE value. Moreover, biomass feedstocks will provide greater quantities of syngas, which can be effectively employed for the extraction of high-value chemicals. Given that gasification is energy intensive, it is essential to ensure that the products can compensate for the whole production costs. Pyrolysis, as an alternate method, employs an inert atmosphere and a moderate temperature of 500°C to transform plastics into useful compounds.

4.2.2.4 Pyrolysis

Pyrolysis decomposes long polymer chains into smaller fragments at 400°C to 500°C in an inert atmosphere (Rand et al., 1999; Schyns and Shaver, 2021; Whitty et al., 2008). The endothermic nature of the process necessitates operating temperatures of around 700–900°C to facilitate the formation of the appropriate product stream (Gao, 2010; Murata et al., 2004). Pyrolysis with thermal energy is termed as thermal cracking, while pyrolysis in the presence of a catalyst is referred to as catalytic cracking. Pyrolytic oil is derived from the condensation of vapors produced during pyrolysis and can serve as fuel oil in numerous applications. The non-condensables that exit the chamber can serve as a heat source for the pyrolysis process. The produced char can serve as an adsorbent due to its high carbon content and intrinsic porous structure. Char briquettes are employed as an energy source (Jamradloedluk and Lertsatitthanakorn, 2014). The relative yield of products can be improved by improving the parameters of pyrolysis, particularly pyrolysis temperature and heating rate. For example, slow pyrolysis (heating rate 0.1–1°C/sec and temperature approximately 400°C) produces char as the primary product. In conventional pyrolysis, liquid oil is the major product at moderate heating rates (1–1000°C/sec) and at a temperature range of 400–600°C. Similarly, to obtain gases, fast pyrolysis with a high heating rate is preferable (Jahirul et al., 2012; Bridgwater, 2018). Besides, temperature and heating rate, residence time determines the product yield. A prolonged residence time at elevated temperatures promotes secondary reactions, which enhance the selectivity of lighter gas fractions, whereas a short residence time at intermediate temperatures yields liquid-based products.

Pyrolysis is advantageous for generating high-energy-density liquid fuels, as well as reducing production costs and facilitating transportation to pertinent application locations. The capability to modify product distribution by suitably adjusting process parameters is an additional benefit.

4.2.2.5 Biochemical Conversion Systems

Biochemical conversion processes plastic waste under very moderate circumstances and is regarded as an eco-friendly method (Lear et al., 2021; Muneer et al., 2020). In the process, microorganisms and enzymes are involved to interact and degrade the polymer (Amobonye et al., 2021; Siracusa, 2019). This method determines both high product stream selectivity and low temperatures. However, the process occupies a large space and requires time and robust control, which makes it expensive.

Bioplastics are known to reduce CO_2 emissions as compared to fossil fuel utilization. Bioplastics are biodegradable and derived from biomass, organic matter, and any biological waste stream. Biological decomposition of the waste leads to compost, water, and carbon dioxide formation as the primary products. Furthermore, products such as ethanol, lactic acid, n-butanol, acetone, and polyhydroxyalkanoates are produced, exhibiting properties akin to those of petroleum-derived feedstocks. Regulatory bodies must ensure the quality of final products, revise regional waste management strategies, and promote the market adoption of bio-based products, despite the sustainability of this product category. It may additionally reduce emissions from incineration and landfills about ordinary plastics.

4.2.2.6 Hydrocracking

The hydrocracking of polymers is conducted in hydrogen-rich environments at increased pressures of 70 atm and temperatures between 375°C and 400°C. Nickel-based catalysts are prevalent for enhancing the hydrogen-to-carbon ratio and reducing aromatic content. Furthermore, heteroatoms in the feedstock are effectively managed to prevent the formation of hazardous compounds. The technique yields high-quality naphtha from a blend of polymers. The process entails significant operational expenses related to hydrogen, along with huge capital requirements.

4.3 SUSTAINABLE TECHNOLOGIES FOR PLASTIC WASTE MANAGEMENT

The continual rise in plastic trash generation, along with the necessity for sustainable waste management strategies, has stimulated study in this domain. Various treatment options have been explored to alleviate the detrimental impacts of landfills, including a range of methods from basic recycling to energy production approaches. Consequently, the thermochemical conversion of plastic waste was looked upon as sustainable and feasible method for waste reduction and the generation of fuel components. It has the capacity to reduce the dependency on conventional fuel resources. The feedstock quality of the plastic significantly influences the efficacy of the process, as it is a measure of substantial investment and processing expenses. The succeeding sections emphasize the pyrolysis of waste plastics, detailing numerous types and process parameters that influence the procedure.

4.3.1 Plastic Waste to Fuel Technologies

4.3.1.1 Pyrolysis

Pyrolysis, also known as thermolysis, involves the thermal and chemical decomposition of waste in an oxygen-limited environment. This technology is crucial for transforming low-quality feedstocks into value-added products. Endothermic pyrolytic reactions necessitate a continual energy input to dissociate bonds, a phenomenon referred to as thermal cracking, where polymer chain complexes disintegrate into smaller fragments, producing char, oil, and gas components. The calorific value of the oil and gas streams is equivalent to that of conventional fuels in the market. The process is employed in refineries, including delayed coking and visbreaking, to crack residual hydrocarbons into higher-value products. Pyrolysis has the additional benefit of tailoring product distribution by adjusting process parameters appropriately. It emits reduced levels of harmful waste into the environment. Additionally, because of the absence of oxygen, the approach will be devoid of dioxins and furans. Figure 4.4 illustrates a schematic block flow diagram depicting the various steps of pyrolysis.

The cracking procedures are classified as mentioned in the following sections.

4.3.1.1.1 Thermal Cracking

The essential stages of thermal cracking encompass initiation, the generation of unstable secondary molecules, and termination through radical recombination or disproportionation. Due to structural

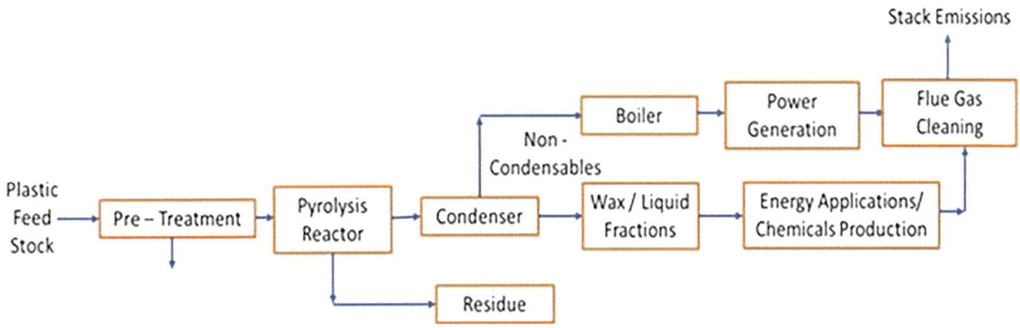

FIGURE 4.4 Block diagram representing the pyrolysis process.

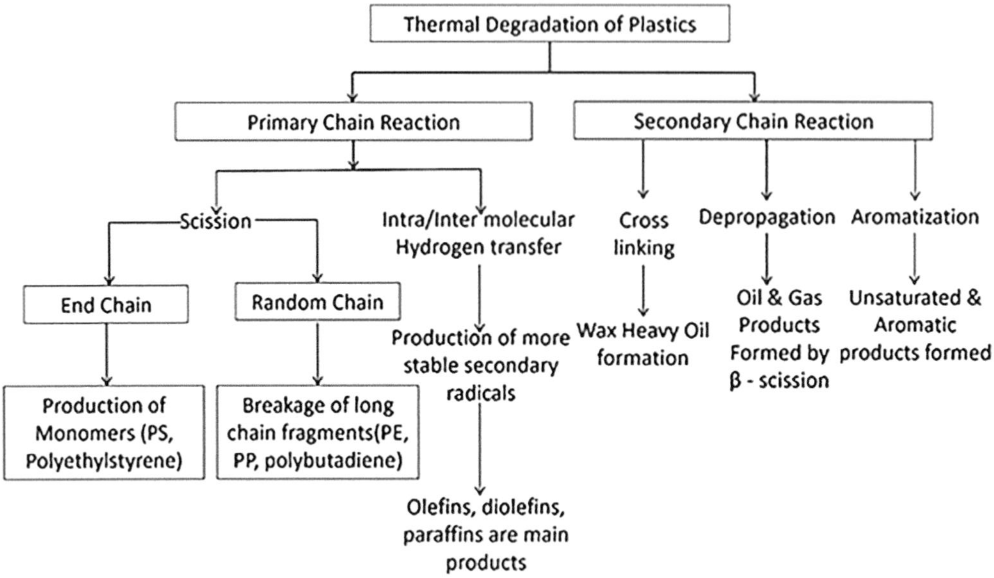

FIGURE 4.5 Reactions involved in the thermal cracking process.

deformation and electron cloud reconfiguration, bonds within the plastic polymer chain are disrupted during initiation reactions. The predominant mechanism involves the cleavage of C—C bonds, necessitating energy for the dissociation of C—H bonds. The resulting free radicals engage in polymerization, and alpha and beta scission and are characterized by strong reactivity and transience. Although alpha scission is not thermodynamically preferable, it is a process that extracts a hydrogen atom from a neutral HC molecule, producing methane and a secondary or tertiary free radical. The series of reactions' products have a minimal amount of olefins and are high in C1 and C2 fractions. Thermal cracking reactions and their various products are summarized in Figure 4.5.

Temperature is the paramount operational parameter in thermal cracking, governing the degradation rate and ultimately affecting the stability of the end product. For instance, liquids with a greater preference for coke are formed at temperatures <400°C, whereas simple gas fractions are produced at temperatures >600°C. In addition to temperature, residence time is an important parameter to regulate product yield and quality. A prolonged residence time facilitates the generation of stable products, whereas a reduced residence time encourages the production of primary products, particularly monomers.

4.3.1.1.2 Catalytic Cracking

Catalytic cracking is an alternate method for obtaining high-value products via pyrolysis (Whitty et al., 2008). The presence of basic and acidic sites on the surface of catalyst enhances selectivity, affects the reaction mechanism, and accelerates the reaction rate. In terms of operational qualities, they provide the following advantages:

1. Pyrolysis processes can be conducted at reduced temperatures to minimize energy consumption.
2. Reducing the kinetics of secondary reaction by prioritizing primary cracking reactions.
3. A limited product distribution can be obtained by promoting the selectivity of reaction products toward VAP.
4. With the use of catalysts, polyolefin feedstock can be converted into branching and aromatic hydrocarbon molecules that positively impact octane number.

5. Undesired products can be limited.
6. Catalyst regeneration and reuse affect the viability of the process.

Carbocations, positively charged atoms of carbon or carbenium, are formed during catalytic cracking. The addition of protons to a paraffin molecule or the removal of hydrogen and two electrons results in the formation of a carbenium ion through the addition or removal of electrons at Brønsted and Lewis sites. The stability of a carbocation is contingent upon the nature of the alkyl group associated with the positive charge; the inductive action confers the greatest stability to tertiary ions. Three primary processes—cleavage of the C—C bond, isomerization, and hydrogen transfer—stabilize the carbenium ions (Gao, 2010). The characteristics and intensity of acidic sites on the catalyst's surface directly affect the selection of the reaction. The cracking reaction occurs by beta scission of carbon–carbon bonds, resulting in the formation of new carbenium ions and an olefin. Small carbon chain olefins will be generated by the freshly synthesized carbenium ion, undergoing the same sequence of events. The subsequent stage is isomerization, when carbocations often undergo fragmentation to generate branched molecules and rearrange to produce tertiary ions. It produces elevated oxygenated feedstock in the C3 and C4 fractions, as well as gasoline with a high octane rating. Methyl tertiary butyl ether (MTBE) and tertiary amyl methyl ether (TAME), utilized with gasoline to reduce emissions, are derived from these components (Murata et al., 2004). The presence of iso-paraffins elevates the cloud point of fuel oil. Hydrogen transfer reactions transpire when olefins adsorbed on proximate active sites undergo reaction. At the active site, one olefin molecule changes into a paraffin, and the other molecule into a cycloid-olefin, which reorganizes to produce aromatics. When carbocations undergo a cyclization reaction brought on by the abstraction of hydride ions from olefins, olefinic carbonium ions are generated. When the double bond of the carbonium ion breaks, aromatic molecules are produced.

4.3.2 Effect of Process Parameters

4.3.2.1 Reactor Configuration

During the pyrolysis processing stage, the physical characteristics of polymers that are to be treated provide a significant challenge. Given the limited heat conductivity of solid plastics and plastic melts, maintaining an adequate temperature distribution is essential for continuous operation. Furthermore, because melted plastics are sticky, they must be handled carefully before being fed into the reactor. An appropriate reactor architecture is vital for achieving the desired product distribution, with transport effects significantly influencing this distribution. A high heat flux is essential for the processing of waste plastics, as heat permeates the medium within the reactor and facilitates the melting of the polymers. Typically, this procedure starts at a temperature between 120°C and 160°C. As a result, having an appropriate treatment method is essential to achieving the intended product distribution with a particular percentage of goods.

Batch reactors are extensively utilized with an external heating source. At the start of experimentation, an inert gas is circulated through the reactor to purge oxygen. Vapors are continuously emitted from the reactor. Batch or semi-batch reactor systems offer significant advantages in terms of product handling, but they frequently encounter challenges such as variations in product quality, extended residence times, and difficulties in removing char.

Fluidized bed reactors can be utilized to investigate the effects of temperature and residence time distribution on product quality. In this reactor, the gas velocity is increased beyond a specific limit when the behavior of a packed bed of solids approaches that of a fluidized bed. Heat transfer properties are improved by inducing mixing that is created inside the bed by the motion of fluidized particles. This process controls the time required for decomposition. Additionally, considering the low thermal conductivity of plastics and their melts, heat transfer for this reactor system is highly beneficial. Plastic waste pyrolysis can be conducted using the bubbling fluidization technique, which is superior to circulating fluidized bed reactors, as they avoid feedstock entrainment in the fluidizing medium. Non-uniform particles with diverse shapes and sizes require different velocities for fluidization. However, practical difficulties associated with handling non-uniform particle sizes and the scale-up of this technology pose challenges for its commercialization on an industrial scale.

These limitations can be effectively addressed by employing a *conical spouted bed reactor* (CSBR). This reactor is particularly suitable for managing sticky solids and regulating vapor residence time, enabling the control of the product stream. The crystalline structure of sand utilized as a heat transfer medium enhances heat transfer to polymer melts surrounding the particles, thereby mitigating de-fluidization issues. Furthermore, the efficiency of this reactor in handling molten plastics makes it potentially effective for the treatment of plastic wastes on a larger scale, if a pre-heater to melt plastics is installed prior to the CSBR.

For industrial pyrolysis of plastic waste, a *rotary screw kiln reactor* is a preferred choice. This reactor can manage non-uniform particles with diverse heat capabilities. The technology provides benefits for thermal energy input to the feedstock material. The heat input can be controlled by modifying the residence time by changes through screw speed, which influences the product distribution. Angled screw kiln reactors can facilitate the control of mixing intensity, thus enhancing product distribution.

4.3.2.2 Residence Time

Residence time refers to the duration it takes for a solid material to undergo complete degradation after traversing the entire length of a reactor. Alternatively, it can be expressed in terms of vapor residence time. Variations in residence time can be achieved by adjusting the feed flow rate to the reactor. A longer residence time of solid material within the reactor generally favors secondary reactions, leading to the formation of char and stable products. For instance, when examining the decomposition temperature of plastics at 500°C, it is important to note that plastics are composed of volatile compounds that are effluxed from the sample. Consequently, a longer residence time for the vapor phase results in the formation of fewer volatile components. This typically encompasses non-condensable petroleum gases and aromatics, resulting in lower oil yields and higher char yields.

A change in reactor geometry also significantly affects the yield of gas and oil in a fixed bed reactor (FBR) system. At temperatures above 685°C, there was negligible variation in the effect of oil yield. However, increasing the residence time at temperatures below 685°C resulted in higher oil yield, where reactor hydrodynamics play a key role in determining product composition. For reactor operation at temperatures below 450°C, residence time is an important factor to be considered (Mastral et al., 2003).

4.3.2.3 Heating Rate

In general, product yields for plastic pyrolysis at 600°C result in an increase in oil yield associated with a decrease in gas and char products. This is primarily due to faster heating rates, accelerating the thermal volatile release from the feedstock. When primary pyrolysis products are produced, further reduction in the residence time of oil results in suppressing secondary degradation reactions. This results in an increase in the yield of oil products at 600°C, where a further increase in temperature results in an enhancement of gas yield. An increase in temperature results in rapid cracking of polymer chains, which could be the primary reason for the increase in gas yield at the expense of a decrease in oil yield.

Pyrolysis reactions conducted at a slow heating rate, combined with a higher residence time, result in oil with a lower density. Higher residence time promotes the conversion of straight-chain hydrocarbons into branched-chain hydrocarbons with the same number of carbon atoms as their homologues. The heating rate can also be considered a parameter that results in the classification of pyrolysis reactions into three categories: slow, fast, and flash pyrolysis. Slow pyrolysis process was traditionally used in the production of charcoal, where carbonization is employed to obtain this valuable product. Adopting faster heating rates allows β-scission reactions to proceed, followed by intermolecular hydrogen transfer reactions that result in the production of shorter-chain

TABLE 4.2
Types of Pyrolysis Processes and Major Products

Technology	Residence Time (min)	Heating Rate (°C/sec)	Temperature (°C)	Major Products
Slow Pyrolysis	Hours to days	0.1–1	400	Char
Conventional Pyrolysis	5–60	1–1000	400–600	Char, liquid, and gas
Fast Pyrolysis	<0.1	>1000	400	Liquid and gas

Sources: Jahirul et al., 2012; Bridgwater, 2018.

hydrocarbons (Singh et al., 2019). Table 4.2 provides a summary of the pyrolysis process classification based on temperature, heating rate, and residence duration.

Rapid pyrolysis at elevated temperatures results in a decrease in the yield of pyrolysis liquid. This can be further substantiated by the early onset of an exothermic phase, where the maximum weight loss is achieved. The exothermic phase is preceded by an endothermic phase, during which the sample is exposed to shorter time scales, with degradation occurring primarily in the liquid phase. Since the extent of gas-phase reactions is negligible, the volatile compounds produced are also minimal.

A comprehensive understanding of the reactions and their sequence are essentially required for developing an efficient process scheme for plastic waste treatment. Kinetics of pyrolysis and the associated transport mechanisms significantly influence their capacity to extract fuel and valuable components. Considering that polymer degradation transpires via multiple pathways, the incorporation of additional physical processes like melting and evaporation complicates the modeling of degradation at both the kinetic and particle scales. Thus, polymer degradation is essential for the decomposition of reactants (plastics), the kinetics of product synthesis, and waste management.

4.4 PILOT SCALE STUDIES AND CASE EXAMPLES

Plastics feedstock recycling solutions have been presented to meet the circular economy and sustainability ideas that are crucial to meet global carbon-neutral targets. Numerous life cycle assessment studies indicate that treatment technologies like pyrolysis and gasification of mixed plastic trash can alleviate the adverse environmental impacts of GHG emissions and climate change. The pyrolysis of single resin plastic waste, including PS, PP, and PE, can markedly diminish its adverse environmental impacts. The yield and composition of the pyrolysis process may fluctuate depending on the input plastic waste, in addition to being significantly influenced by the intended process scheme and the environmental impacts of the feedstock recycling process.

4.4.1 Formulation of an Efficient Process Framework

The formulation of an efficient process scheme necessitates a profound understanding of mechanisms and sequences of waste treatment reactions, facilitating the successful design of technology.

Kinetic scale studies are conducted via TGA, wherein the mass loss of samples is assessed as a function of time and temperature. Rate equation for such a decomposition is given by

$$\frac{d\alpha}{dt} = \beta \frac{d\alpha}{dT} = k(T) f(\alpha) \tag{4.9}$$

where, α is given by

$$\alpha = \frac{m_0 - m_t}{m_0 - m_f} \tag{4.10}$$

Here, m_0, m_t, and m_f represent initial mass, mass at any time t, and final mass of the sample, respectively. $f(\alpha)$ specifies the differential form of kinetic decomposition model. The rate constant parameter in Equation (4.9) may be denoted by

$$k(T) = A\exp\left(-\frac{E_a}{RT}\right) \tag{4.11}$$

Combining Equations (4.9) and (4.11) results in the expression of reaction rate written as

$$\frac{d\alpha}{dt} = \beta\frac{d\alpha}{dT} = A\exp\left(-\frac{E_a}{RT}\right)f(\alpha) \tag{4.12}$$

Rearranging Equation (4.10) we get

$$g(\alpha) = \int_0^{\alpha}\frac{d\alpha}{f(\alpha)} = \frac{A}{\beta}\int_0^{T}\exp\left(-\frac{E_a}{RT}\right)dT \tag{4.13}$$

Here $g(\alpha)$ is the integral form of decomposition functions. Kinetic models in solid state decomposition reactions are presented in Table 4.3.

In solid-state decomposition kinetics, mechanistic interpretations necessitate the identification of the reaction model and is crucial for reactor designing and operation. The decomposition model estimates the time required for the complete degradation. Furthermore, a model as a function of temperature and heating rate can be employed to determine the operational temperature at a specific heating rate where continuous reactor operation can be ensured with optimal heat input. Excessive

TABLE 4.3
Reaction Models Used in Solid-State Kinetics

Serial Number	Reaction Model with Code	$f(\alpha)$	$g(\alpha)$
1	Power law (P4)	$4\alpha^{(3/4)}$	$\alpha^{(3/4)}$
2	Power law (P3)	$3\alpha^{(2/3)}$	$\alpha^{(2/3)}$
3	Power law (P2)	$2\alpha^{(1/2)}$	$\alpha^{(1/2)}$
4	Avarami Erofeyev (A2)	$2(1-\alpha)[-\ln(1-\alpha)]^{(1/2)}$	$[-\ln(1-\alpha)]^{(1/2)}$
5	Avarami Erofeyev (A3)	$3(1-\alpha)[-\ln(1-\alpha)]^{(2/3)}$	$[-\ln(1-\alpha)]^{(1/3)}$
6	Avarami Erofeyev (A4)	$2(1-\alpha)[-\ln(1-\alpha)]^{(3/4)}$	$[-\ln(1-\alpha)]^{(1/4)}$
7	Prout–Tompkins (B1)	$(\alpha)(1-\alpha)$	$\ln[a/(1-\alpha)]$
8	Contracting cylinder (R2)	$2[(1-\alpha)]^{(1/2)}$	$[1-(1-\alpha)^{(1/2)}]$
9	Contracting sphere (R3)	$3[(1-\alpha)]^{(2/3)}$	$[1-(1-\alpha)^{(1/3)}]$
10	1-D Diffusion (D1)	$1/2\alpha$	α^2
11	2-D Diffusion (D2)	$[-\ln(1-\alpha)]^{-1}$	$[(1-\alpha)\ln(1-\alpha)]+\alpha$
12	3-D Diffusion (D3)	$[3(1-\alpha)^{2/3}]/[2(1-(1-\alpha)^{1/3})]$	$[1-(1-\alpha)^{1/3}]^2$
13	Ginstling-Brounshtein (D4)	$3/[2((1-\alpha)^{-1/3}-1)]$	$1-(2a/3)-(1-\alpha)^{2/3}$
14	Zero order (F0/R1)	1	α
15	First order (F1)	$(1-\alpha)$	$-\ln(1-\alpha)$
16	Second order (F2)	$(1-\alpha)^2$	$(1-\alpha)^{-1}-1$
17	Third order (F3)	$(1-\alpha)^3$	$0.5[(1-\alpha)^{-2}-1]$

Source: Vyazovkin et al., 2011.

temperature levels result in energy losses, while operating at temperatures below the optimum level leads to incomplete conversion of the substrate. Therefore, it is essential to develop methods capable of accurately identifying the decomposition model. Thereafter, particle interactions, wall effects, and hydrodynamics at the reactor scale are analyzed to estimate the duration for full decomposition. This is important with respect to continuous reactor operation with optimal heat input, resulting in complete conversion of the feedstock. Figure 4.6 summarizes the approach that encompasses these three operational levels.

In the scaling up of pyrolysis, it is essential to account for heat and mass transmission resistances alongside the kinetics of feed material decomposition. The time from the kinetic front is assessed by effectively analyzing TGA data and applying an appropriate conversion function. To confirm economic and operational viability, the quantity of feedstock in commercial reactors is sometimes greater than that utilized in TGA. In a scale-up process, the transport resistances are added up, resulting in a longer time for complete conversion of particles. The low heat transfer properties of plastics and medium contribute to these resistances. (Mass transfer resistances can be neglected due to the higher volatile content in plastics.) This is based on the total amount of time required to overcome the particle's conductive resistance, which is where unsteady-state heat transfer takes place, as well as the moving medium's convective resistance in relation to the particle. The temperature profile achieved in the particle and reactor determines the degree of pyrolysis. To estimate the degradation behavior of the particle, under these conditions, dynamic state material and energy balance equations coupled with kinetics and transport properties must be solved. In a non-ideal reactor, residence time must be determined based on kinetic and heat transfer constraints, taking into account the energy input necessary for the complete conversion of plastic feed.

Plastics will advantageously utilize process streams to generate liquid fuel for energy purposes due to their relatively elevated volatile content. Intermediate temperature pyrolysis effectively resolves the challenges associated with chlorine in the feedstock (Sharuddin et al., 2016). A washing with a base solvent is suggested to remove chlorine. Lime is utilized to purify the gas stream containing HCL, facilitating the production of hydrochloric acid through the condensation of $CaCl_2$ or HCL vapors (López et al., 2011). Under ideal conditions of temperature, pressure, and residence time, pyrolysis products can be precisely customized. Furthermore, the catalyst type substantially influences process behavior and product selectivity (Kartik et al., 2022). In contrast to thermal cracking, catalytic cracking is considered as a process that produces high-quality products with the added benefit of using less energy. Care must be taken to treat feedstock containing chloride and acid components, which may affect the catalyst activity. Inorganic substances obstruct the catalyst's active sites. While the catalyst reduces the processing temperature, pre-processing is crucial for purification to achieve the required end products (Ragaert et al., 2017). Hydrocracking involves adding hydrogen to produce high-quality products at 70 atm hydrogen pressure and temperature

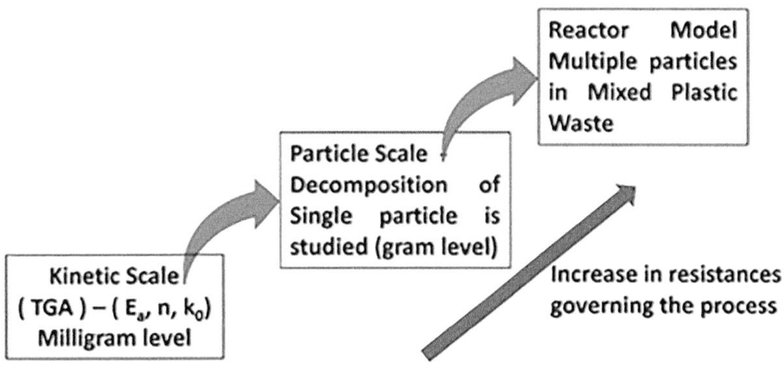

FIGURE 4.6 Progression levels of operations in the pyrolysis process.

350–500°C. Under ideal processing conditions, the polymer feedstock is liquefied and introduced through a catalyst bed. Nonetheless, the technology is prevalent in the petroleum business; yet, catalyst poisoning incurs significant operational expenditures for catalyst renewal. Moreover, the expense of hydrogen presents a substantial obstacle and elevates operational expenses at an industrial scale. Gasification, among other methods, functions at elevated temperatures to generate high-purity syngas and intermediate products. Nevertheless, precautions must be implemented to prevent tar formation resulting from the inherent high volatile organic compounds in plastics. An essential enhancement of the gas stream is necessary prior to its utilization for chemical value. Gasification exhibits considerable sensitivity to feedstock composition relative to other processes.

With respect to commercial implementation, a few plastic pyrolysis plants are reviewed in forthcoming sections.

4.4.2 Commercial Pyrolysis Process for Plastic Waste Conversion

The Texaco gasification process has a throughput capacity of 10 tons per day, treating mixed waste plastics at temperatures ranging from 1200°C to 1500°C in the presence of steam and oxygen. Syngas, the principal product, contains traces of CH_4, CO_2, water, and inert gases. By-products include sulfur, ammonium chloride, slag, and fine fly ash. Chlorine-containing waste and inorganic materials are managed without flue gas treatment. BP Chemicals introduces polymer cracking in a fluidized bed process. Mixed plastic trash is exposed to intermediate temperatures of 500°C, where the vapors are condensed to yield hydrocarbon oil, while the gas stream serves as a fluidizing medium. Chlorine in feedstock is treated by lime purification, and the $CaCl_2$ produced is transported for landfilling. Feedstock at BASF (Badische Anilin-und Sodafabrik) is subjected to pretreatment before feeding it into the pyrolysis process. In order to ensure a consistent rate of feedstock in the reactor, the feed plastic is first reduced in size before being separated from metals. A multistage procedure is designed to remove halogens selectively while preventing corrosion from PVC in the feedstock. The HCL production plant absorbs and processes the hydrogen chloride that is extracted from the process. Further, liquified plastics are heated at 400°C to yield oil (70%) and gas fractions (30%), which are separated in a distillation column. The process produces naphtha, which is then processed through a steam cracker to produce ethylene and propylene, which are raw materials used to make virgin polymers and support the circular economy. High-boiling oils are processed to produce syngas or may be converted into coke, which forms a good substitute as a reducing agent in blast furnaces.

Sekundärrohstoff Verwertungs Zentrum (SVZ) process utilizes wood, wastewater sludge, MSW, and paper fractions as feedstock. Additionally, shredded fractions of electronic equipment and auto accidents are added, with an average permitted chlorine level in feedstock of 2% to 6%. The plant was originally intended for gasification, but with the right modifications, it was also able to handle plastic waste. Annually, 460 kT of solid and liquid waste materials are treated in the plant. Lignite, waste oil, and mixed plastic waste are fed in a counter-current manner with oxygen and steam in a fluidized bed reactor for gasification, producing synthesis gas and liquid hydrocarbons. Gas fractions from fluidized bed reactors are filtered using the Rectisol method to eliminate sulfur compounds from the gas stream. Waste gases (10%) are burned after liquid hydrocarbons are gasified with gases; 30% of the syngas is utilized to generate power, while 70% is used to synthesize methanol. In comparison to alternative chemical recycling methods, this process is competitive owing to similar chlorine concentrations and particle sizes ranging from 20 to 80 mm.

Reentech Limited (Korea) transforms mixed plastic waste comprising polypropylene, polystyrene, and polyethylene into gasoline, kerosene, and diesel. Plastics undergo hydrogenation and melting processes at temperatures of 370°C. The melted polymer is then put through additional FCC testing to create diesel, kerosene, and gasoline. The technique uses alumina silicate–based solid acid catalysts. The unit is additionally outfitted with an air injection system, a catalytic regeneration system, and a moving bed catalytic cracker. The technology facilitates catalyst renewal through continuous catalytic cracking and recycling. Minimizing the heat duty of the system and functioning at reduced temperatures enhances process efficiency through continuous catalyst renewal. Following separation by distillation, non-condensable gases are directed through a water scrubber to remove any acidic compounds, including HCL. To concentrate lighter fractions, the stream is subsequently subjected to chilled petrol prior to being flared in a burner (Kwak, 2002). The schematic representation of the process is shown in Figure 4.7.

The Linde gasification method utilizes feedstocks with a relatively high concentration of chlorine. The pulverized plastic trash is subjected to treatment at temperatures ranging from 1400°C to 1600°C to generate synthesis gas and hydrochloric acid as primary by-products for chemical processing or energy applications.

4.4.3 Technology Readiness Level

TRLs evaluate the maturity of a technology across several stages of research, development, and implementation. The point-based evaluation system clearly defines three levels: (i) lab scale, (ii) pilot size, and (iii) commercial scale (Harmsen, 2019). TRLs for the proven process of converting plastic waste to fuel oil range from 6 to 9. The Agile process, with a processing capacity of 5 tons, requires considerable focus as it handles PET plastics. The Shell facility is functioning at a TRL of 9, yielding 70% oil, 20% gas, and 10% char. Co-pyrolysis is widely utilized for enhancing oil quality. Plastic trash, combined with a portion of biomass, serves as a feedstock. A conical spouted bed reactor has been shown to produce 10% yield of char fractions, 20% gas, and 65% bio-oil at pilot scale. A TRL of 4 has been assigned, and the work for progressing levels is currently being examined. Given that the majority of processes possess a TRL of 3, there is a demonstrable necessity for laboratory-scale testing. The TRL for the specified treatment methods, namely, catalytic cracking and conventional gasification, is 9. The hydrotreating process is classified as TRL 7. Policymakers, researchers, investors, and commercial entities must meticulously assess and make informed judgments on the establishment of plastic waste treatment plants.

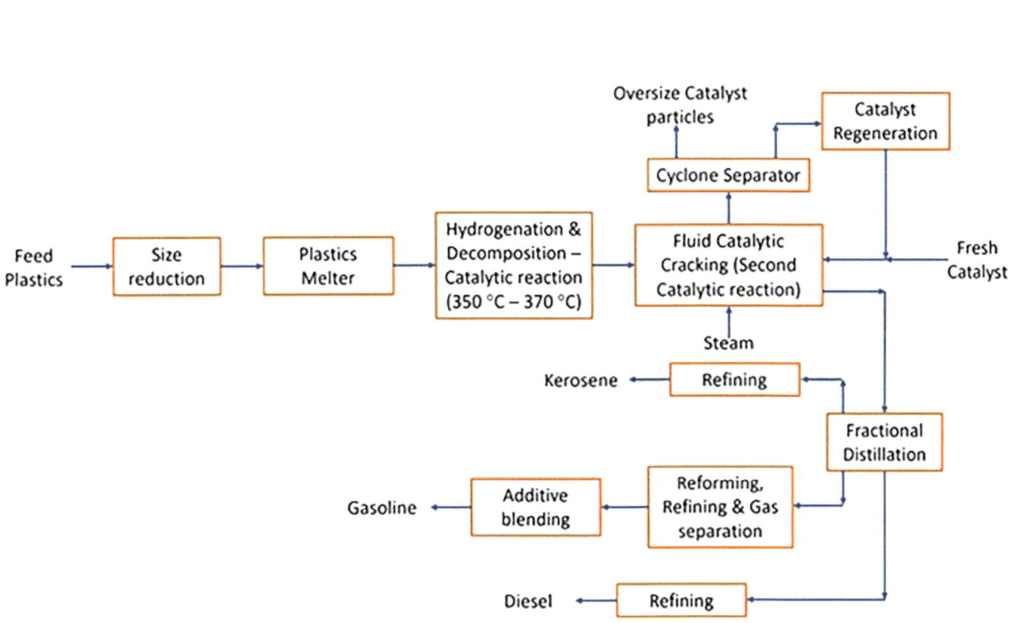

FIGURE 4.7 Reentech process to produce fuel oil from plastic waste.

4.4.4 Environmental and Economic Trade-Off for Plastic Waste Valorization

Selecting the most effective treatment method for the valorization of plastic waste always requires an understanding of the fundamentals of economic and environmental alternatives. Key indicators that can be chosen to enhance the attractiveness of a specific project primarily include gross profit estimated using a payback period as an economic indicator, and greenhouse gas emissions accounting for the environmental impact.

The payback period, a techno-economic feasibility report, evaluates its viability based on mass and energy balances and considering economic parameters. The parameters mainly include total capital investment, operational expenditures involving specific feed treatment costs, transportation to the site, and utility costs. Various alternatives can be analyzed in terms of the utilization of cost-effective fee streams available for processing from various material recovery facilities. The payback period decreases with the increase in the size of adaptable technology for the plastic waste treatment process. The environmental aspects of each treatment technology can be interpreted in terms of the greenhouse gas (GHG) emission factor and the avoided GHG emission factor. Furthermore, taking into account particular boundaries for each technology, a thorough life cycle assessment research can be carried out to determine the amount of ecological consequences of abandoned plastics or the influence of microplastics on human health. The optimization can be carried out through a multi-objective integer linear programming approach or any rational optimization algorithm (Crîstiu et al., 2024). The selection of technology or its combination can be made based on the gross profit of the proposed activity, and considering the market value of product streams generated. However, due consideration should be taken into account regarding greenhouse gas emissions and the positive environmental impacts of alternative processing streams involved in the overall activity.

4.5 CONCLUSIONS AND PERSPECTIVES

As plastic usage continues to surge, the challenge of managing plastic waste has become a critical concern. Future strategies for valorizing plastics will emphasize the development of sustainable and circular solutions to address environmental concerns while generating economic value. An essential component involves sophisticated recycling and upcycling methods, such as pyrolysis and catalytic processes, that transform plastic waste into value-added chemicals. Integrating a circular economy framework with sustainable waste management principles facilitates the development of systems for producing hydrocarbon products such as gasoline, diesel, and value-added chemicals. There is an immediate necessity to devise economical and effective pretreatment methods, as the quality of final products is directly dependent on the feedstock preparation.

The development of catalysts is another research area focused on reducing the energy demands of conversion and improving the product quality, thus making it economically viable and well aligned with the principles of green chemistry and the circular economy. Catalysts that retain effectiveness across multiple cycles offer economic and environmental benefits. Therefore, design and synthesis of novel catalysts that exhibit prolonged lifespans and improved efficiency is desirable.

Furthermore, the synergistic potential of plastics and biomass pyrolysis of mixed plastic waste or co-pyrolysis with locally produced biomass feedstocks may be explored to generate higher-value chemicals, including acids, aldehydes, and ketones. Oil produced from pyrolysis may require refinement to meet commercial gasoline standards. Therefore, developing a multistage condensation system can improve the separation of hydrocarbon vapors produced during pyrolysis. Hydrocarbons can be selectively condensed by controlling vapor residence time and selecting appropriate coolant fluids. Fractional condensation systems are especially advantageous in this direction (Papari and Hawboldt, 2018).

Ultimately, the valorization of plastics is expected to change the linear "produce-use-dispose" model into a circular system, where waste plastics are converted into valuable feedstocks for new products. The incorporation of innovative pretreatment, improved recycling, catalyst advancement, and sophisticated process design will advance the forthcoming generation of sustainable plastic valorization, augmenting worldwide resource efficiency and climate goals. Governments and industries must establish region-specific frameworks that tailor valorization processes to local plastic waste streams and market conditions. Collaborative initiatives across academic institutions, industries, and governments are essential for scaling these technologies.

REFERENCES

Abiriga, D., Vestgarden, L. S., Klempe, H., Groundwater contamination from a municipal landfill: Effect of age, landfill closure, and season on groundwater chemistry, *Sci. Total Environ.*, 737 (2020) 140307. https://doi.org/10.1016/j.scitotenv.2020.140307

Acomb, J. C., Wu, C., Williams, P. T., Control of steam input to the pyrolysis-gasification of waste plastics for improved production of hydrogen or carbon nanotubes, *Appl. Catal. B Environ.*, 147 (2014) 571–584. https://doi.org/10.1016/j.apcatb.2013.09.018

Ahmed, M. H. M., Batalha, N., Mahmudul, H. M. D., Perkins, G., Konarova, M., A review on advanced catalytic co-pyrolysis of biomass and hydrogen-rich feedstock: Insights into synergistic effect, catalyst development and reaction mechanism, *Bioresour. Technol.*, 313 (2020) 123457. https://doi.org/10.1016/j.biortech.2020.123457

Agarwal, M., An investigation on the pyrolysis of municipal solid waste (2014). Available at: https://core.ac.uk/download/pdf/32229084.pdf (Accessed September 23, 2024).

Aguado, J., Serrano, D. P., *Feedstock recycling of plastic wastes*, Royal Society of Chemistry (2007).

Amobonye, A., Bhagwat, P., Singh, S., Pillai, S., Plastic biodegradation: Frontline microbes and their enzymes, *Sci. Total Environ.*, 755 (2021) 143536. https://doi.org/10.1016/j.scitotenv.2020.143536

Bridgwater, B. T., Challenges and opportunities in fast pyrolysis of biomass: Part I, *Biofuels Bioprod. Bioref.*, 12(1) (2018) 118–130. https://doi.org/10.1002/bbb.1837

Cabanes, A., Fullana, A., New methods to remove volatile organic compounds from post-consumer plastic waste, *Sci. Total Environ.*, 758 (2021) 143661. https://doi.org/10.1016/j.scitotenv.2020.144066

Candelaresi, D., Spazzafumo, G., Introduction: The power-to-fuel concept, in: *Power to Fuel: How to Speed Up a Hydrogen Economy*, Academic Press (2021), pp. 1–15. https://doi.org/10.1016/B978-0-12-822813-5.00005-9

Crîstiu, D., d'Amore, F., Bezzo, F., Economic and environmental optimisation of mixed plastic waste supply chains in Northern Italy comparing incineration and pyrolysis technologies, *Comput. Chem. Eng.*, 180 (2024) 108503. https://doi.org/10.1016/j.compchemeng.2023.108503

Gao, F., Pyrolysis of waste plastics into fuels, Master's thesis, University of Canterbury, New Zealand (2010). Available at: https://www.ciel.org/wp-content/uploads/2017/09/Fueling-Plastics-Fossils-Plastics-Petrochemical-Feedstocks.pdf

Harmsen, J., *Industrial process scale-up: A practical innovation guide from idea to commercial implementation*, Elsevier, Amsterdam (2019). https://doi.org/10.1016/b978-0-444-64210-3.00001-9

Jahirul, M. I., Rasul, M. G., Chowdhury, A. A., Ashwath, N., Biofuels production through biomass pyrolysis—A technological review, *Energies*, 5(12) (2012) 4952–5001. https://doi.org/10.3390/en5124952

Jamradloedluk, J., Lertsatitthanakorn, C., Characterization and utilization of char derived from fast pyrolysis of plastic wastes, *Procedia Eng.*, 69 (2014) 1437–1442. https://doi.org/10.1016/j.proeng.2014.03.139

Kartik, S., et al., Valorization of plastic wastes for production of fuels and value-added chemicals through pyrolysis—A review, *Therm. Sci. Eng. Prog.*, 32 (2022) 101316. https://doi.org/10.1016/j.tsep.2022.101316

Kaza, S., Yao, L., Bhada-Tata, P., Van Woerden, F., *What a waste 2.0: A global snapshot of solid waste management to 2050*, World Bank Publications (2018).

Kwak, H.-J., System for continuously preparing gasoline, kerosene and diesel oil from waste plastics, US Patent 6,866,830 B2 (2002).

Lear, G., Kingsbury, J., Franchini, F., Gambarini, V., Maday, S. D., Weaver, L., Plastics and the microbiome: Impacts and solutions, *Microbiome Res. Rep.*, 9 (2021) 1–12. https://doi.org/10.1186/s40793-020-00371-w

Lebreton, L., Andrady, A., Future scenarios of global plastic waste generation and disposal, *Palgrave Commun.*, 5 (2019) 1–11. https://doi.org/10.1057/s41599-018-0212-7

López, A., De Marco, I., Caballero, B. M., Laresgoiti, M. F., Adrados, A., Dechlorination of fuels in pyrolysis of PVC containing plastic wastes, *Fuel Process. Technol.*, 92(2) (2011) 253–260. https://doi.org/10.1016/j.fuproc.2010.05.008

Lusher, A., Hollman, P., Mendoza-Hill, J., *Microplastics in fisheries and aquaculture: Status of knowledge on their occurrence and implications for aquatic organisms and food safety*, FAO, Rome (2017).

Marico Innovation Foundation, *Innovation in plastics: The potential and possibilities* (2023). Available at: https://www.praxisga.com/PraxisgaImages/ReportImg/innovation-in-plastics-the-possibilities-and-potential-Report-3.pdf (Accessed September 23, 2024).

Mastral, F. J., Esperanza, E., Berrueco, C., Juste, M., Ceamanos, J., Fluidized bed thermal degradation products of HDPE in an inert atmosphere and in air–nitrogen mixtures, *J. Anal. Appl. Pyrolysis*, 70(1) (2003) 1–17. https://doi.org/10.1016/S0165-2370(02)00068-2

Mohanan, N., Montazer, Z., Sharma, P. K., Levin, D. B., Microbial and enzymatic degradation of synthetic plastics, *Front. Microbiol.*, 11 (2020) 580709. https://doi.org/10.3389/fmicb.2020.580709

Muneer, F., Rasul, I., Azeem, F., Siddique, M. H., Zubair, M., Nadeem, H., Microbial polyhydroxyalkanoates (PHAs): Efficient replacement of synthetic polymers, *J. Polym. Environ.*, 28(9) (2020) 2301–2323.

Murata, K., Sato, K., Sakata, Y., Effect of pressure on thermal degradation of polyethylene, *J. Anal. Appl. Pyrolysis*, 71 (2004) 569–589. https://doi.org/10.1016/j.jaap.2003.08.010

Navarro, M. V., Murillo, R., Mastral, A. M., Gea, G., García-Bacaicoa, P., Callén, M. S., Application of a particle model to pyrolysis. Comparison of different feedstock: Plastic, tyre, coal and biomass, *Fuel Process. Technol.*, 103 (2012) 1–8. https://doi.org/10.1016/j.fuproc.2011.12.031

Nichols, W., Smith, N., Waste generation and recycling indices 2019: Overview and findings, *Verisk Maplecroft* (2019). Available at: https://www.circularonline.co.uk/wp-content/uploads/2019/07/Verisk_Maplecroft_Waste_Generation_Index_Overview_2019.pdf

Papari, S., Hawboldt, K., A review on condensing system for biomass pyrolysis process, *Fuel Process. Technol.*, 180 (2018) 1–13. https://doi.org/10.1016/j.fuproc.2018.08.001

Plastics Europe, *Plastics—the fast Facts 2023*, Plastics Europe, Brussels (2023). Available at: https://plasticseurope.org/knowledge-hub/plastics-the-fast-facts-2023/ (Accessed September 23, 2024).

Portilla, N., Plastic pollution in the ocean—facts and statistics, *RTS* (2024). Available at: https://www.rts.com/blog/plastic-pollution-in-the-ocean-facts-and-statistics/ (Accessed September 24, 2024).

Ragaert, K., Delva, L., Van Geem, K., Mechanical and chemical recycling of solid plastic waste, *Waste Manag.*, 69 (2017) 24–58. https://doi.org/10.1016/j.wasman.2017.07.044

Rand, T., Haukohl, J., Marxen, U., *Municipal solid waste incineration: A decision maker's guide*, World Bank, Washington, DC (1999).

Schyns, Z. O. G., Shaver, M. P., Mechanical recycling of packaging plastics: A review, *Macromol. Rapid Commun.*, 42(3) (2021) 2000415. https://doi.org/10.1002/marc.202000415

Sharuddin, S. D. A., Abnisa, F., Wan Daud, W. M. A., Aroua, M. K., A review on pyrolysis of plastic wastes, *Energy Convers. Manage.*, 115 (2016) 308–326. https://doi.org/10.1016/j.enconman.2016.02.037

Silvestre, S. C. C., *Ecosustainable polymer nanomaterials for food packaging*, Woodhead Publishing, Cambridge, UK (2013).

Singh, R. K., Ruj, B., Sadhukhan, A. K., Gupta, P., Impact of fast and slow pyrolysis on the degradation of mixed plastic waste: Product yield analysis and their characterization, *J. Energy Inst.*, 92(6) (2019) 1647–1657. https://doi.org/10.1016/j.joei.2019.01.009

Siracusa, V., Microbial degradation of synthetic biopolymers waste, *Polymers*, 11(6) (2019) 1066. https://doi.org/10.3390/polym11061066

Tsakona, M., Baseline report on plastic waste, *Basel Convention* (2020). Available at: http://www.basel.int/Implementation/Plasticwaste/PlasticWastePartnership/Meetings/PWPWG1/tabid/8305/ctl/Download/mid/23074/Default.aspx?id=7&ObjID=22886 (Accessed September 23, 2024).

Verma, R., Vinoda, K. S., Papireddy, M., Gowda, A. N. S., Toxic pollutants from plastic waste—A review, *Procedia Environ. Sci.*, 35 (2016) 701–708. https://doi.org/10.1016/j.proenv.2016.07.069

Vyazovkin, S., Burnham, A. K., Criado, J. M., Pérez-Maqueda, L. A., Popescu, C., Sbirrazzuoli, N., ICTAC kinetics committee recommendations for performing kinetic computations on thermal analysis data, *Thermochim. Acta*, 520(1–2) (2011) 1–19. https://doi.org/10.1016/j.tca.2011.03.034

Whitty, K. J., Zhang, H. R., Eddings, E. G., Emissions from syngas combustion. *Combust. Sci. Technol.*, 180(6) (2008) 1117–1136.

World Economic Forum, *The New Plastics Economy: Rethinking the future of plastics*, World Economic Forum, Geneva (2016).

5 Valorization of Plastic Waste to Chemicals

Avanish Kumar[1*], *Swati Dubey*[2], *and Munsaf Ali*[3]
[1,2]Department of Chemical Engineering, Marwadi University, Rajkot, India
[3]Department of Nanoscale Science and Engineering, University at Albany, New York, USA

5.1 INTRODUCTION

Plastics are getting popular day by day because of the various significant properties including, lighter mass, hydrophobicity, and thermal and electrical resistance (Schultz and Reinhardt 2023). Plastics are also applied as feedstock in textile industries, as packing materials, and furthermore as components of different machinery. Scientists are developing new types of plastics with unique properties. New plastics can be lighter, stronger, and more heat-resistant or even have built-in electrical conductivity. This opens the door for their use in fields like construction and electronics (Damayanti et al. 2022). Plastics play a surprisingly significant role in renewable energy. Although they traditionally used heavy steel, manufacturers are increasingly using fiberglass composites, which are lighter and stronger. This allows for larger blades that capture more wind and generate more electricity. Several key components in solar panels utilize plastics (Pol 2010). The encapsulant layer that protects the delicate silicon cells is often made from a special type of plastic.

A circular economy model for plastics is a very promising solution to the challenges we face with plastic use. Figure 5.1 depicts that implementing a circular economy for plastics can create jobs, stimulate economic growth, and reduce the costs associated with waste management. Traditionally, plastics follow a linear model—used once and thrown away. This creates massive plastic waste and pollution. A circular economy aims to keep plastics in a closed loop, minimizing waste. This can involve using recycled materials, creating products for disassembly and reuse, or even incorporating bioplastics that decompose naturally (Evode et al. 2021). This can be achieved through strategies like repair, refurbishment, remanufacturing, and recycling. At the heart of the circular economy model is the principle of "closing the loop"—an approach that seeks to keep materials in continuous use for as long as possible, extracting maximum value before recovering and regenerating them at the end of their service life. When applied to plastics, this principle represents a dramatic departure from the traditional, linear model. Rather than allowing plastics to become waste after a single use, the circular economy transforms them into resources that can be reused, repaired, remanufactured, or recycled within the system. This closed-loop strategy not only minimizes the generation of waste but also reduces the environmental, social, and economic costs associated with the current model of plastic use (Kibria et al. 2023).

In a circular plastics economy, the flow of materials follows a regenerative cycle aimed at maintaining the value of plastic throughout its entire life cycle. This begins with smart design and material innovation, where products and packaging are intentionally created to be durable, modular, and easy to disassemble. The choice of materials prioritizes recyclability and non-toxicity, with growing interest in bio-based and biodegradable alternatives to conventional fossil-fuel-based plastics (Sakao et al. 2024). Reuse and refill systems play a crucial role in extending the life of plastic

* Corresponding author's email: avanish.kumar@marwadieducation.edu.in

DOI: 10.1201/9781003581000-7

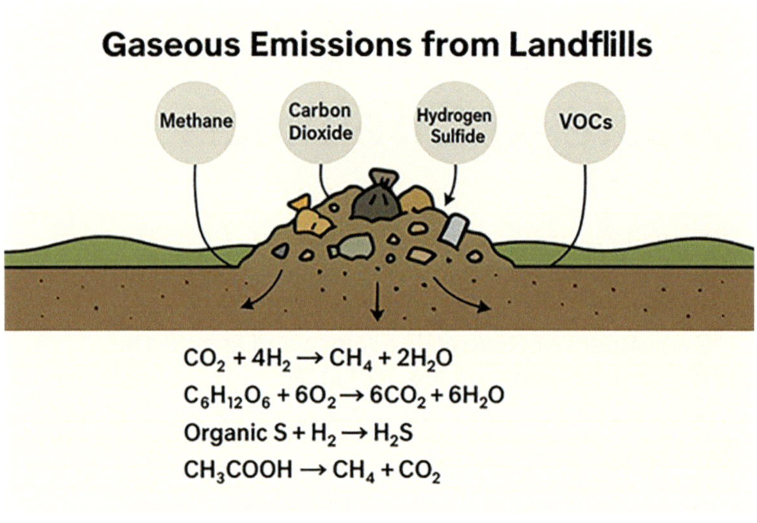

FIGURE 5.1 Gaseous emissions from landfills.

products through initiatives such as deposit-return schemes, refillable packaging, and reusable containers, particularly in retail and food services. These systems reduce the demand for new plastic production by encouraging multiple uses of the same material. Maintaining the integrity of plastics for future use depends heavily on efficient collection and sorting, which requires investment in advanced infrastructure, public education, and incentives for source separation. Proper sorting ensures the effectiveness of downstream recycling processes. Moving beyond traditional methods, advanced recycling technologies-including chemical recycling, enzymatic depolymerization, and pyrolysis enable plastics to be broken down into their molecular components and remanufactured into high-quality new products, theoretically allowing for infinite recycling (Saxena 2025). Finally, material recovery and reincorporation close the loop by reintegrating processed plastics into the production cycle, thereby reducing the need for virgin materials, lowering greenhouse gas emissions, and conserving energy and natural resources (Lange 2021).

Global plastic production is indeed massive, presently at 360 million tons annually and projected to double in the next 20 years. Unfortunately, plastic recycling rates are dismal. Only a small fraction of plastic waste actually gets recycled. The increasing accretion of plastic waste has led to global pollution. There are several challenges that demand the reduction and reprocessing of plastic waste (Kantarelis et al. 2009; Brenner 2010).

In order to lessen the feedstock and decrease plastic dumping in lands, the reutilization of waste is a potential choice. Several methods have been adopted for utilizing and managing plastic waste that is non-biodegradable in nature. These are categorized into traditional and advanced treatment methods. The traditional methods for plastic waste management include incineration and disposal in landfills. However, combustion of plastic waste requires a large amount of energy and releases harmful chemicals. Decomposition of plastic waste in landfills is a very slow process and demands a considerable amount of land for incremental waste dumping. Moreover, dumping of waste in land releases noxious chemicals and leads to the depreciation of the proximate soils. Therefore, conventional waste management approaches seemed to be unfavorable management option for plastic waste (Kibria et al. 2023).

Further, to tackle the shortcomings of the traditional management methods, various novel methods have been applied, such as pyrolysis, liquefaction, and construction purposes. Consequently, it becomes essential to look for sustainable valorization techniques to manage plastic waste. Plastics are mainly comprised of polymers, which makes them a potential raw material for converting them to chemicals. Currently, mechanical reutilization is extensively applied to valorize plastic waste. This technique constitutes the collection and reprocessing of plastic waste, followed by sorting, shredding, melting, and re-extrusion. This mechanical reuse of plastic waste is an environmentally favorable technique, but the properties of the final product depreciate due to chain scission, branching, and oxidation of the waste. Furthermore, this technique is largely dependent on the type of plastic waste. Mainly two categories of plastics—polyethylene terephthalate and polyethylene—are processed. This is because plastics that are more susceptible to elevated temperatures or become highly viscous upon melting are difficult to recycle with the mentioned approach.

Mechanical recycling, also referred to as mechanical reuse, is one of the most established and widely used methods for managing plastic waste. This process involves collecting, sorting, cleaning, shredding, melting, and reprocessing plastic materials into new products, often without significantly altering their chemical structure. It is considered an environmentally favorable technique as it helps divert plastic waste from landfills and incinerators, reduces the demand for virgin plastic, and lowers overall greenhouse gas emissions compared to producing plastic from fossil fuels (Kassab et al. 2023). However, mechanical recycling has several inherent limitations. One of the major challenges is the degradation of plastic properties during repeated processing. This degradation is primarily due to chain scission, branching, and oxidation that occur when plastics are exposed to heat, shear forces, and oxygen during reprocessing (Tang 2022). As a result, the mechanical strength, flexibility, color, and overall performance of the recycled material often decline with each cycle, making it less suitable for high-quality or demanding applications. This phenomenon, often referred to as downcycling, limits the range of products that can be made from mechanically recycled plastics. Moreover, the effectiveness of mechanical recycling is highly dependent on the type and purity of the plastic waste. It works best with thermoplastics—especially polyethylene terephthalate (PET) and polyethylene (PE)—which can be remelted and reshaped multiple times with relatively predictable results (Saxena 2025). However, many other plastics, such as polyvinyl chloride (PVC), polystyrene (PS), and certain multilayer or composite materials, pose significant challenges. These materials may degrade at lower temperatures, emit harmful gases, or exhibit high melt viscosity, making them unsuitable for conventional mechanical processes.

Contamination is another critical issue. Plastic waste streams often contain a mix of materials, labels, adhesives, dyes, food residues, and other additives. These contaminants must be carefully removed, as they can compromise the quality of the recycled output and damage equipment. As such, mechanical recycling requires clean, well-sorted, and homogeneous plastic inputs to function effectively, which adds to the operational cost and complexity (Sakao et al. 2024). Despite these limitations, mechanical recycling remains a vital component of the circular economy, particularly for widely used packaging plastics. Innovations such as improved sorting technologies, compatibilizers, and processing additives are helping to enhance the quality and range of mechanically recycled plastics. However, for more contaminated or complex plastic waste streams, chemical recycling and other advanced methods are gaining traction as complementary solutions (Biessey et al. 2023). On the contrary, reutilizing the plastic waste chemically is the new emerging technique academically and industrially. This technique emphasizes transforming plastics into comprising hydrocarbons.

In this chapter, recent developments in the field of valorization of plastic waste to produce chemicals and fuels are presented and discussed.

5.2 CONCERNS REGARDING PLASTIC WASTE

Plastic waste is considered one of the most severe environmental concern of the current times. It has been projected that a large amounts of plastic waste is produced, and the majority culminates in the sea. Moreover, large amounts of plastic waste are dumped in landfills, which leads to contaminating sea and land (Evode et al. 2021). The plastic waste problem has been recognized as a major environmental issue that needs to be addressed. During the production of plastics, fossil fuels are used, which emit greenhouse gases; hence, there are harmful consequences when plastic waste is disposed. Moreover, single-use plastics prove to be expensive on the basis of ecological influence and for atmospheric variation. These wastes accumulate in various water bodies and cause suffocation to aquatic species and different health issues. Various studies have been reported related to plastic waste being ingested by the aquatic species, which ultimately goes into the stomach. This leads to disruption in the life cycle.

The growing accumulation of plastic waste has emerged as one of the most pressing environmental challenges of the 21st century. Plastics, valued for their durability and versatility, persist in the environment for hundreds of years, posing long-term threats to ecosystems, wildlife, and human health. A significant portion of plastic waste ends up in landfills, waterways, and oceans, where it breaks down into microplastics that are ingested by marine organisms and, eventually, enter the food chain. Inadequate waste management infrastructure, especially in developing regions, exacerbates this problem, leading to open dumping and uncontrolled burning, which release toxic pollutants into the air, soil, and water. Additionally, the vast majority of plastic is derived from fossil fuels, linking plastic production and disposal to climate change through greenhouse gas emissions. These environmental and health concerns are compounded by the economic burden of managing plastic waste and the loss of valuable material resources, highlighting the urgent need for sustainable solutions, such as recycling, reuse, and the transition to a circular economy (Sheriff et al. 2025).

Beyond the environmental and health implications, plastic waste also raises significant socioeconomic concerns. In many low-income and densely populated regions, informal waste pickers often bear the brunt of plastic pollution, working under hazardous conditions with little protection or financial security. The rising tide of plastic pollution also threatens critical sectors, such as fisheries, tourism, and agriculture, as polluted beaches, clogged waterways, and degraded soil affect livelihoods and economic productivity (Kibria et al. 2023). Moreover, the proliferation of single-use plastics, especially in packaging, reflects a deeply embedded culture of convenience and disposability that fuels overconsumption and undermines sustainability efforts (Mihai et al. 2022). The global nature of the problem further complicates solutions, as plastic waste is frequently exported across borders, sometimes ending up in countries lacking the infrastructure to handle it responsibly. These interconnected concerns underscore the complexity of the plastic waste crisis—it is not merely an issue of disposal but a systemic challenge that demands coordinated action from governments, industries, and consumers alike. Addressing it effectively requires both technological innovation and behavioral transformation, supported by robust policy frameworks and public awareness initiatives.

5.3 PLASTIC WASTE MANAGEMENT APPROACHES

There has been a rapid increase in the production of plastic waste due to various factors, including industrialization, urbanization, and a growing population. There are various measures that have been developed by different regulating authorities to tackle the problem of plastic waste. These include various approaches that aim to manage and utilize plastic litter in a more scientific way, for instance, reusing, biological degradation, and combustion of the plastic waste (Figure 5.2). These strategies have been incorporated to achieve a sanitary atmosphere and sustainable dumping of plastic waste.

5.3.1 Complete Combustion of Plastic Waste

One of the current adopted approaches to manage plastic waste is complete combustion of the waste. Amongst the several combustion products, one of them is fuel in the form of liquid, solid, and gas applied in various transportation. This plastic waste management approach helps in energy

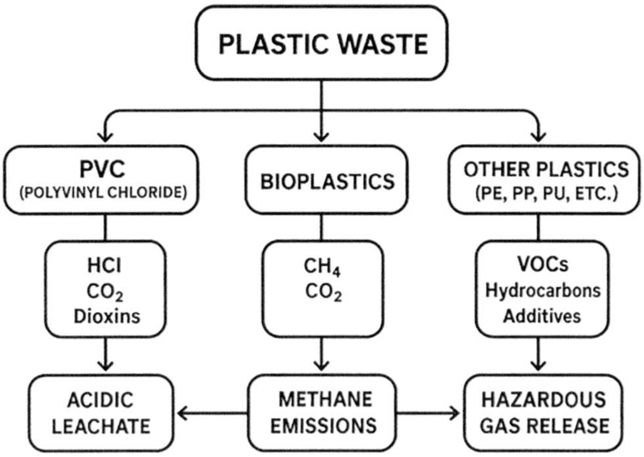

FIGURE 5.2 Schematic comparing gas pathways from different plastic types for further elaboration on gas capture technologies.

production, which is much required in the current urbanization. Subsequent heat recovery with combustion has been applied in various places, which proves to be a good strategy for plastic waste management. On the contrary, one of the disadvantages of this approach is the production of various impulsive chemicals and ash. These products harm the environment significantly. Largely, entire plastic waste is not considered to be an appropriate choice for incineration. Some of the plastic wastes are resilient to the heat and explosion, and therefore combustion approach requires proper attention while segregating the non-combustible waste. In addition, setting up the combustion experiment is costly.

Combustion, or incineration, is one of the most traditional methods for the disposal of plastic waste, particularly in regions where land availability is limited or where waste-to-energy (WTE) infrastructure is prioritized. This thermochemical process involves the oxidative degradation of plastics at high temperatures to convert waste into heat, carbon dioxide, water vapor, and residual ash. While combustion can significantly reduce the volume of plastic waste by up to 90%, its application raises concerns due to potential emissions of toxic compounds and the loss of material value that could be preserved through recycling (Vuppaladadiyam et al. 2024).

5.3.1.1 Process Details and Chemical Reactions

The combustion of plastics typically occurs in a **two-stage process**:

- *Primary combustion (pyrolysis and partial oxidation)*:
  - Plastic polymers are heated in the presence of limited oxygen. During this stage, thermal cracking leads to the breakdown of long-chain hydrocarbons into smaller volatile compounds, including carbon monoxide (CO), hydrocarbons, and carbonaceous char. This is followed by partial oxidation.
- *Secondary combustion (complete oxidation)*:
  - The volatile compounds and gases produced in the first stage are subjected to excess oxygen and high temperatures (typically 850–1100°C), allowing for complete oxidation into CO_2 and H_2O.

5.3.1.1.1 Representative Reactions

For PE:

$$(C_2H_4)_n + O_2 \rightarrow CO_2 + H_2O + \text{Heat}$$

For PVC, an additional concern is the emission of hydrogen chloride:

$$(C_2H_3Cl)_n + O_2 \rightarrow CO_2 + H_2O + HCl + \text{Other pollutants}$$

Incomplete combustion may result in harmful by-products such as dioxins, furans, and polycyclic aromatic hydrocarbons, particularly from chlorine-containing plastics.

5.3.1.2 Combustion Reactor Technologies

Different reactor configurations are used depending on scale, efficiency, and environmental regulations:

a. **Mass Burn Incinerators**
   - Direct combustion of unsorted municipal solid waste, including plastics.
   - Large-scale, simple design.
   - Lower efficiency in energy recovery and higher pollutant emissions if not properly controlled.

b. **Fluidized Bed Reactors**
   - Utilize hot, fluidized sand, or inert particles to ensure uniform heat distribution.
   - Operate at 800–1000°C with excellent mixing and combustion control.
   - Suitable for pretreated or homogenized plastic waste.
   - High thermal efficiency and better emission control.

c. **Rotary Kilns**
   - Cylindrical, rotating combustion chambers that allow for extended residence time.
   - Useful for co-processing of plastics with hazardous or medical waste.
   - Often used in cement kilns for energy recovery.

d. **Modular Incinerators (Small-Scale)**
   - Used for localized plastic waste disposal.
   - Lower capital investment but often lack advanced flue gas treatment.

5.3.1.3 Variations and Findings from Comparative Studies

Several studies have examined the performance, emissions, and energy efficiency of different combustion methods for plastics. Arena and Mastellone (2006) contrasted fluidized bed and grate furnaces, demonstrating that fluidized beds offer improved thermal uniformity, reduced NOx emissions, and increased energy efficiency for plastic-rich waste streams. Verma et al. (2016) highlighted the increased emissions of dioxins and furans from PVC-rich combustion unless flue gas scrubbing and de-chlorination techniques are utilized. Bujak (2015) did pilot rotary kiln combustion experiments on polypropylene and polyethylene and noted that pretreatment (i.e., drying, shredding) greatly enhances combustion completeness as well as reduces pollutant loading. Lin et al. (2020) emphasized the performance of advanced emissions control technologies like electrostatic precipitators, fabric filters, and scrubbers in minimizing particulate matter as well as acid gases from incinerators in plastics.

5.3.1.4 Environmental and Operational Considerations

Although combustion provides the advantage of energy recovery, it is also associated with serious environmental and public health issues, particularly when emissions are not properly controlled. Toxic gas, heavy metal, and fine particulate matter emissions require the application of advanced air pollution control equipment, which increases capital and operating costs substantially. Also, combustion is traditionally criticized as material—destroying it destroys the possibility of plastic recycling and recovery, which are more consistent with a circular economy. Hence, most sustainability models currently prioritize energy recovery only as an extreme measure, only after all attempts at reuse and recycling are depleted.

Different biomass wastes have been efficaciously applied as raw material in the co-pyrolysis process, including agricultural wastes, wooden waste, and municipal solid wastes (Abnisa and Wan Daud 2014). Plastic is also among the raw materials used in the process; various blends of raw materials have been summarized in Table 5.1, which have been investigated to produce pyrolytic oil. The primary advantage of plastic materials is the extremely low cost but the main drawback is that they are produced from non-renewable sources. Therefore, plastic waste can be applied as a raw material in the co-pyrolysis technique. The liquid product generated from pyrolyzed plastics has a high heating value of approximately 40 MJ/kg. Hence, the plastic could contribute significantly to giving pyrolytic oil a higher calorific value.

Biomass has a great potential to produce pyrolytic oil due to its ability to be converted into various fuel forms, including solids, liquids, and gases. It offers flexibility in production, with an abundance of diverse forms such as wood residues, agricultural residues, dedicated energy crops, and municipal solid waste. Pyrolysis is the most utilized method for biofuel production, resulting in a high liquid fuel yield of approximately 75% under moderate operating conditions. Studies have investigated the co-pyrolytic behavior of various plastics, including HDPE, LDPE, PE, and PP as well as coal having less volatility, and the different blending ratios with 5 wt% plastic using a thermogravimetric analyzer. The results revealed a synergistic influence in plastic and coal co-pyrolysis at high temperatures. Additionally, a kinetic analysis was conducted, estimating kinetic parameters for coal, plastic, and their mixtures within the range of 35.7–572.8 kJ/mol of activation energy and $27–1.7 \times 10^{38}$/min as pre exponential factor. In one study, co-pyrolysis of palm shell and polystyrene mixtures (1:1 weight ratio) in a fixed bed reactor at 500°C with an N_2 flow rate of 2 L/min and reaction time of 60 minutes was done. The outcomes indicated that co-pyrolysis of palm shell & polystyrene increased the oil yield to around 61.63% (Abnisa et al. 2013).

Various biomasses, including cotton stalk, hazelnut shell, sunflower residue, and so forth, and plastic wastes, including polyvinyl chloride and polyethylene terephthalate were co-pyrolyzed (Çepelioğullar and Pütün 2013). The weight ratio used for blending was 1:1, heating was done at the rate of 10°C/min, and experiments were performed in an inert atmosphere using N_2 gas. Pyrolysis of sugarcane bagasse, LDPE, and their corresponding mixtures at various ratios in a semi-batch reactor, maintaining a temperature range of 350–600°C and heating at a rate of 20°C/min, was studied (Dewangan, Pradhan, and Singh 2016). The pyrolysis of sugarcane bagasse and LDPE yielded the highest oil yields of 47.15% and 74.40%, respectively. Nonetheless, at 500°C and a 1:1 blend ratio, co-pyrolysis produced maximum oil production, of 52.75%. Pyrolysis of a blend including cellulose and LDPE boosted the yield of the liquid product and increased energy recovery by 83% when compared to pyrolysis of the individual components (Gunasee et al. 2017).

The interplay during the co-pyrolysis of polypropylene (PP) and rice bran wax (RBW) was studied. At temperatures of 500°C and 600°C, the maximum liquid yields for PP and RBW were around 76% and 86%, respectively. The highest oil yield from co-pyrolysis was achieved with a 1:3 blend, reaching approximately 81% at 550°C (Akancha et al. 2019). An investigation of co-pyrolysis involving bamboo sawdust (BSD) and linear low-density polyethylene (LLDPE) was conducted through thermogravimetric analysis (TGA) (Alam et al. 2020). This analysis spanned the temperature range of 30–900°C, with heating rates of 5, 10, and 20°C/min. It was observed that a blend containing 25 wt% BSD and 75 wt% LLDPE (BP1:3) displayed the most significant synergistic effect in comparison to the other blends examined.

Residual biomass from agro-based industries was pyrolyzed, including materials like rice husk, peanut shell, and wheat straw (Fermanelli et al. 2020). The impact of operating temperature in the range of 350–650°C was investigated on the yields of biochar, bio-oil, and biogas. The co-pyrolysis of bio-oil yields showed variations ranging from 41 to 46 wt% for all the mixtures, and there was a notable reduction in bio-oil water content, reaching up to 15%, when rice husk was combined with peanut shells or wheat straw.

The co-pyrolysis of cornstalks and high-density polyethylene (HDPE) was explored as an efficient approach for addressing environmental pollution and repurposing problematic plastics to

TABLE 5.1
Plastic Waste: Harmful Effects and Management Techniques

Serial Number	Type of Plastic Waste	Harmful Effects	Management Approach	Reference
1	High-density polyethylene (HDPE)	When discarded improperly, these plastics linger in the environment for hundreds of years, leading to soil infertility and water pollution.	Hot-press molding was used to create composite panels.	Adhikary, Pang, and Staiger, 2008
2	Plasticized polyvinylchlorie	Contains harmful chemical bis phenol-A BPA.	Recycled to compost bin.	Mercea et al., 2018
3	LDPE	Hazardous materials could mix in with the plastic.	Recycled to bin liners, pallet sheets.	Soganciogu, Yel, and Ahmetli, 2017
4	Polypropylene	Aquatic organisms may ingest plastics mistaking them for food leading to increased bioaccumulation potential.	Recycled to pegs, bins, pipes, pallet sheets.	Ranjan and Goel, 2021
5	Commercial plastic mulch films	• Leaks into water systems, contaminating aquatic environments. • Decomposes in the soil to produce microplastics.	Plastic mulch film biodegradation on soil in a lab setting.	Sameshima-Yamashita et al., 2019
6	Poly-butylene succinate and poly-butylene succinate-co-butylene adipate	• Soil contamination by chemicals. • Changes to the physical composition of the soil. • Biological contamination by pathogens.	Microbiological deterioration.	Thirunavukarasu et al., 2016
7	Flax-fiber-reinforced polylactide	Soil health, nutrient cycling, water retention, and crop growth.	Enzymatic degradation.	Stepczyńska and Rytlewski, 2018
8	Polyethylene terephthalate (PET)	Pollutes ocean life, groundwater, and the air.	Depolumerization using I. sakaiensis PETase enzyme.	Knott et al., 2020
9	Polystyrene	Non-biodegradable.	Pyrolyzed to produce pyrolytic oil.	Yan et al., 2014
10	Polyvinyl chloride (PVC) and polyethylene terephthalate	Risk to the biota due to entanglement, suffocation, or internal damage when consumed.	Combined in a certain ratio with various bio-based materials.	Çepelioğullar and Pütün, 2013
11	Low-density polyethylene (LDPE)	Comes into contact with dangerous substances.	Utilized with different bio-wastes in pyrolysis process.	Dewangan, Pradhan, and Singh, 2016
12	LDPE	Polyethylene pollution disrupts ecosystems by altering natural habitats.	Used in pyrolysis process.	Gunasee et al., 2017
13	Polypropylene	Causes skin, digestion. and respiratory irritation and other health effects.	Utilized in co-pyrolysis with bio-waste.	Akancha, Kumari, and Singh, 2019
14	Polystyrene	Pollutes the environment, fills landfills.	Used in co-pyrolysis process with mahua seeds to produce pyrolytic oil.	Kumar Mishra and Mohanty, 2020

generate valuable energy (Kai et al. 2019). A study investigated the effects of various combinations of waste plastics, including waste nitrile gloves PS with mahua seeds, to improve fuel quality and oil yield. Different amounts of plastic waste (10 wt%, 20 wt%, and 30 wt%) were mixed and then burned in a semi-batch reactor set to the best conditions (550°C operating temperature, 80°C/min heating rate, and 100 mL/min N flow rate). The physical and chemical characteristics revealed that it was possible to produce valuable chemicals and renewable fuel (Liu et al. 2013).

5.3.2 Disposal in Landfills

Disposal of plastic waste in landfills is another approach for waste management. The places where plastic waste is disposed of after use are considered landfills (Hidayah and Syafrudin 2018). This management approach requires proper attention in terms of precautions to evade aftereffects including, groundwater contamination and land deprivation. The major aim to dispose waste in landfills is to offer a harmless zone of waste to guard all aspects of the ecosystem. A huge amount of manpower and hard work is required for this purpose (Tamoor et al. 2021). For example, a large hole needs to be dug to dump waste, and a large amount of time is required for the decomposition. Throughout the decomposition process, every molecule degrades and decomposes biologically. Some plastics comprising long-chain polymers require years for degradation (Chidambarampadmavathy et al. 2017). Various factors that influence the degradation include biological and chemical properties of polymers, weather conditions, and so forth. Soil provides carbon dioxide and methane for the degradation to occur biologically. Landfill disposal of plastic waste helps in keeping cities clean and hygienic and segregating toxic waste from other types of waste (Kibria et al. 2023; Pandey et al. 2023). In addition, landfill disposal is an inexpensive approach for plastic waste management. However, some of the drawbacks include contamination of soil and water and climate change, which ultimately influences wildlife (Kaushal et al. 2021; Moharir et al. 2019).

The widespread pollution of the environment with plastic refuse has wide and long-term ramifications for aquatic life as well as for soil quality. In water, plastic litter, from large floating objects to imperceptible microplastic particles, disturbs aquatic and marine life. Aquatic life forms tend to eat plastics as food, which causes internal damage, hunger, and death. Moreover, microplastics are carriers of harmful chemicals and pathogenic bacteria and are responsible for their bioaccumulation in the food web, and posing risks to biodiversity as well as human health. Plastics also physically manipulate environments; they can, for instance, smother coral reefs, block waterways, and block sunlight, thus affecting photosynthesis and aquatic vegetation (Kumar et al. 2021).

Inland, plastic garbage, especially if disposed of carelessly in landfills or farms, undermines soil structure and fertility. Plastic particles can disrupt inherent soil aeration, water holding, and microbeial activity needed for plant establishment and nutrient circulation. Microplastics in soils have been shown to hinder the movement and reproduction of earthworms, interrupt the function of nitrogen-fixing bacteria, and inhibit root development. Furthermore, plastic litter has the potential to leach toxic additives, for example, phthalates, bisphenol A (BPA), and flame retardants into soils, with implications for long-term crop health, groundwater quality, and food safety. Because plastics break down very slowly, their persistence in soils and sediments results in an enduring environmental impact, necessitating action on plastic pollution through better waste management, regulation, and public awareness (Kibria et al. 2023).

Plastic waste, although quite inert in relation to organic waste, has a singular contribution to the formation of landfill gas—both by affecting environments of decomposition and by donating certain chemical ingredients. Although the majority of traditional plastics, such as PE, polypropylene (PP), and PS are not biodegradable at will, additives in their chemistry and incidental degradability with prolonged pressure, heat, and microbial action have the potential to donate landfill gas emissions (Figure 5.3). Although conventional plastics make a negligible contribution to biogenic landfill gas, their breakdown, particularly under thermal and mechanical stress, emits toxic compounds that worsen landfill contamination. In addition, their secondary effects—altering microbial environments and heat trapping—amplify gas production from organic materials. It is essential

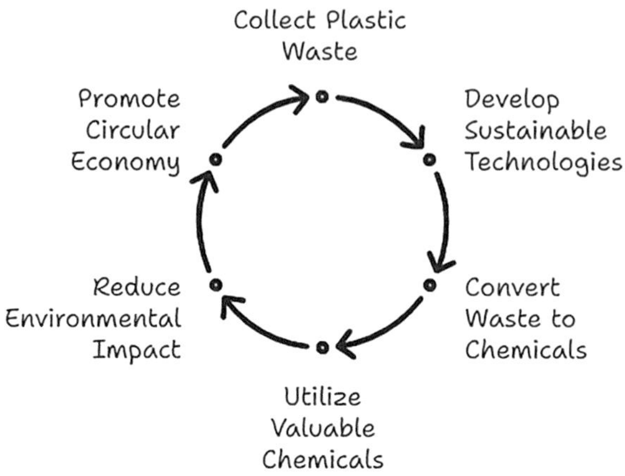

FIGURE 5.3 Circular economy through plastic waste valorization.

to understand these interactions in integrated waste management planning, particularly in landfill design, enhancing gas collection systems and policymaking on biodegradable plastics.

5.3.3 Biodegradation of Plastic Waste Using Enzymes

A method of managing plastic trash is biological breakdown, which involves exposing the garbage to nitrogen and oxygen and creating a carbon double bond around it. To break up polymers, extracellular enzymes are also introduced. Enzymes play a catalytic role in this process, accelerating the reaction. The process employs enzymes such as microbial oxidoreductases, microbial oxygenases, laccases, peroxidases, lignin peroxidases, hydrolases, microbial manganese peroxidases, and microbial lipases. The high cost of using enzymes to break down the polymers is one of the main disadvantages of this technique.

5.3.4 Reuse of Plastic Waste

Recycling of plastic waste comprises the collection of waste and its conversion into other value-added products. The main aim is to protect the ecosystem and lessen the negative impact on society. Plastic wastes are mainly non-biodegradable in nature; however, they can be molded into various other forms like chairs and tables. Recycling of plastic waste mainly comprises of five steps: collection, classification, washing, grating, and compounding. Recycling avoids burning of plastic waste, which leads to the reduction of carbon dioxide release into the environment. Recycling also helps in preserving the landfill area and hence helps in achieving a sustainable economy.

5.3.4.1 Collection

Plastic waste is gathered from households, commercial establishments, industries, and municipal solid waste systems. Efficient collection systems, whether informal (rag-pickers and scrap dealers) or formal (municipal and private recyclers), are critical to ensuring that recyclable plastics do not end up in landfills or natural ecosystems.

5.3.4.2 Classification (Sorting)

Once collected, plastics are sorted into different categories based on resin type (e.g., PET, HDPE, LDPE, PP, PS), color, shape, and contamination level. Advanced sorting technologies, such as infrared scanners, air classifiers, and density separation tanks, are increasingly used to improve accuracy

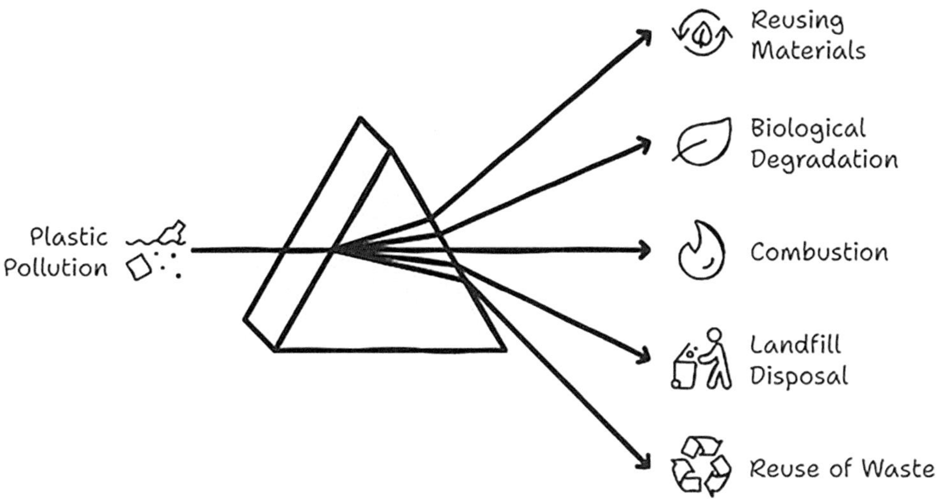

FIGURE 5.4 Plastic waste management approaches.

and reduce contamination. Proper classification enhances the efficiency and quality of the final recycled product.

5.3.4.3 Washing

Washing removes dirt, labels, adhesives, food residues, and other impurities. This is a crucial step because any contamination can affect the quality and mechanical properties of recycled plastics. Washing typically includes hot water baths, detergent treatments, and friction washers.

5.3.4.4 Shredding or Grating

Clean plastic materials are mechanically reduced to smaller pieces, often called flakes or regrind, using shredders or granulators. This increases surface area and uniformity, making subsequent processes such as melting and compounding more efficient.

5.3.4.5 Drying

After washing, the plastic flakes must be thoroughly dried to eliminate moisture, which can interfere with melting and cause hydrolytic degradation during processing. Drying is performed using air dryers or dehumidifying systems.

5.3.4.6 Compounding (Extrusion and Pelletizing)

In the final stage, the dried plastic flakes are melted, filtered, and mixed with stabilizers, colorants, or fillers in an extruder. The resulting molten plastic is extruded into strands, cooled, and cut into uniform pellets or granules, which serve as raw material for the production of new plastic products.

5.4 SUSTAINABLE TECHNOLOGIES FOR VALORIZATION OF PLASTIC WASTE TO CHEMICALS

In spite of having various benefits, recycling techniques have limitations when it comes to plastic waste management (Bazargan and McKay 2012). During the recycling process, various harmful chemicals also get discharged into the environment because of the composition of the plastic waste. These harmful substances destroy the nearby flora and fauna (Nyakuma and Ivase 2021; Chen et al. 2021). Plastic down cycling segregates wastes for ongoing recycling, which typically renders the finished product unsuitable for further recycling. This suggests that rather than being considered for secondary use for non-consumed plastic waste, it ultimately finds its way into a landfill.

Although recyclable plastic is comprehensively touted as a sustainable way of dealing with the burgeoning plastic waste problem, it has its own limitations and environmental implications. The recycling process, especially when dealing with mixed or contaminated streams of plastics, tends to liberate toxic chemicals into the environment (Bazargan and McKay 2012). These emissions occur in the melting, extrusion, and reprocessing processes, when residual additives, plasticizers, flame retardants, and heavy metals volatilize or leach into air, water, or soil. Toxic by-products like dioxins, furans, and volatile organic compounds can harm adjacent flora and fauna, disturbing ecosystems and causing serious health hazards to workers and nearby communities (Nyakuma and Ivase 2021; Chen et al. 2021). Another serious issue is down cycling, or the process of taking recycled plastic and making products that are lower in quality and diminished functionality, like park benches or highway fill. Although this strategy holds off on sending plastics to landfills momentarily, it reduces their recyclability for use in the future because of degradation in polymer properties. This only postpones—not halts—the material's ultimate deposition in a landfill. Furthermore, the availability of composite materials, dyes, and multilayered packaging makes efficient recycling even more challenging and thus frequently results in incineration or disposal of fractions that are not usable. These factors imply that, while recycling has its advantages, it cannot be the exclusive solution for plastic waste treatment. Rather, it needs to be part of a large strategy that involves waste reduction from the source, extended producer responsibility, creation of alternative materials, and investment in next-generation recycling technologies that overcome these innate limitations.

5.4.1 Valorization of Plastic Wastes

Various thermal treatment techniques, including anoxic pyrolysis, catalytic carbonization, pressure carbonization, flash Joule heating, microwave treatment, molten salt methods, and hydrothermal procedures, can be employed to produce carbon compounds from plastic wastes (Table 5.2).

5.4.1.1 Anoxic Pyrolysis

Anoxic pyrolysis, also referred to as thermal cracking in the absence of oxygen, is an advanced thermochemical conversion technique used for the valorization of plastic waste. In contrast to incineration or combustion, which needs oxygen and results in complete oxidation of materials (and emission of substantial quantities of CO_2), anoxic pyrolysis does not need any oxygen and takes place in a low-oxygen environment, heat degrading plastic polymers into useful liquid, gaseous, and solid products with no major emissions of greenhouse gases or harmful by-products.

In anoxic pyrolysis, plastic waste is heated at temperatures usually between 350°C and 700°C in a closed reactor that excludes oxygen entry. Thermal energy results in chain scission of the long polymer molecules, which are broken down into lower hydrocarbons. The main products are as follows:

- *Pyrolysis oil (liquid hydrocarbons)*: This may be utilized as an alternative to diesel or as raw material for production of new plastic.
- *Syngas* H_2, CO, CH_4, *and light hydrocarbons*: This gaseous mixture can be combusted for heat or used to produce electricity.
- *Char (solid carbonaceous residue)*: Can be used in asphalt, construction materials, or activated carbon production.

Olazar et al. (2024) studied the possibilities of plastics valorization by pyrolysis in line catalytic dry reforming to produce syngas. To increase the H_2 / CO ratio of the syngas and enhance the possibility for some uses, integrated dry-steam reforming was investigated. Raising the S/C ratio up to

TABLE 5.2
Sustainable Techniques for Plastic Valorization

Type of Plastic Waste	Product	Synthesis Technique	Reference
Plastic polyethylene terephthalate (PET) bottles	Crushed recycled aggregate	Crushing	Paihte et al., 2019
polymeric, HDPE & PET, wastes	Bricks	Melt compounding technique	Limami et al., 2020
PET bottle wastes	Interlocking bricks	Conventional method for brick preparation	Alaloul, John, and Musarat, 2020
PET plastic waste	Lightweight concrete	Conventional method for concrete preparation	Hameed and Ahmed, 2019
High-impact polystyrene and low-density polyethylene (LDPE) plastic wastes	Lightweight concrete	Conventional method for concrete preparation	Olofinnade et al., 2020
PET waste	High surface area and extremely high electrical conductivity graphene nanostructures	Molten salt conversion	Kamali et al., 2019
PET waste	Graphene-like nanosheets and porous nanostructures	Molten salt preparation	Kamali and Yang, 2020
LDPE plastic waste	Hierarchical porous carbon	Autogenic pressure carbonization followed by potassium hydroxide (KOH) activation	Zhang et al., 2019
Mixed plastic waste	Porous carbon nanosheets	KOH activation process	Gong et al., 2014
Polypropylene waste	Carbon nanosheets	Pressure carbonization followed by KOH activation	Liu et al., 2021

0.6 makes it possible to produce a syngas with an H_2/CO ratio of 1. Due to the lower H content of PS, the quantity of steam needed to bring the H_2/CO ratio to 1 was greater; that is, the S/C ratio must be raised to 0.8. PET, however, contains a complicated composition with low levels of H and high level of O, which restricts its interest for pyrolysis and inline dry or combined reforming process, since it is not feasible to achieve acceptable H_2/CO ratios. Generally, pyrolysis and an inline reforming process has large-scale potential for the simultaneous valorization of CO_2 and waste plastics and to produce a syngas that can be utilized as a feedstock for additional chemical processes.

Barracco et al. (2024) investigated the application as an adsorbent of an actual plastic-derived char as such procured from a pyrolysis pilot plant. This is a new methodology, wherein the char was generally procured as the desired product. The adsorbent was characterized through various techniques, and methylene blue was selected as a representative contaminant of wastewater. The inorganic phase (80 wt%) was primarily related to the presence of rod-type structures, Si-rich, and the organic phase evidenced the presence of aromatic species. From a thermodynamic perspective, the Langmuir isotherm well described the experimental data ($R_2 = 0.98$), whereas kinetic findings were well expressed by a pseudo-second-order law ($R_2 > 0.99$). The negative Gibbs free energy showed that the process of adsorption was spontaneous, and the change in enthalpy (ΔH_0 = 7.18 kJ/mol) demonstrated the endothermic character of the process, in the range typical for physisorption processes.

5.4.1.2 Catalytic Carbonization

Gong et al. (2014) investigated the synthesis of nanosheets using naturally modified montmorillonite in the upcycling of polypropylene (PP) using catalytic carbonization with high yield (Gong et al. 2014). The researchers catalytically carbonized PP at 700°C to investigate the yield, morphological properties, microstructural properties, phase structure, composition of surface elements, and thermal stability of nanosheets. The results demonstrated that the catalyst guaranteed the conversion of plastic into aromatics and lighter hydrocarbons, as well as further conversion of the resultant products. Overall, the study demonstrated that the method has the capability to be a successful means of using the chosen plastic trash as a source of carbon for upcycling.

Plastic was recycled using the environment of vanadium-based photocatalyst that promotes carbon–carbon bonding for the production of useful carbon-based products (Gazi et al. 2019). Moreover, no alkali treatment was needed in the aforementioned process. Pandey et al. (2021) prepared graphene nano-sized structures using multi-step pyrolysis of plastics with the application of catalysts. In addition, graphene oxide and plastic waste were used for the preparation of supercapacitors and color-sensitive solar cells (Pandey et al. 2020). In the experiments, bentonite clay was applied for deprivation at the temperatures of 449.85°C and 944.85°C. The synthesized graphene nanostructures had huge energy density and power density.

5.4.1.3 Pressure Carbonization

Pressure carbonization, referred to as hydrothermal carbonization (HTC) when applied to wet biomass, is a high-pressure thermochemical process that converts carbonaceous material such as plastic wastes to solid, liquid, and gaseous products. In contrast to pyrolysis, which often occurs at atmospheric pressure under an oxygen-free environment, pressure carbonization is performed under high pressures (2–6 MPa) and medium temperatures (180–350°C). This process mimics natural geological coal and hydrocarbon formation but at a significantly shortened timescale, which is an attractive approach to plastic waste valorization.

In pressure carbonization, plastic waste is exposed to high pressure and temperature within a closed reactor vessel, frequently with minimal or no oxygen. The process allows for thermal decomposition, condensation, and polymer rearrangement reactions, yielding the following:

- *Solid char (carbonaceous residue)*—high in carbon content and usable as a solid fuel or precursor to activated carbon.
- *Liquid fraction (oily tar or condensates)*—containing complex hydrocarbons that can be refined or used for chemical feedstocks.
- *Gas phase*—a mix of light hydrocarbons, CO_2, H_2, and other volatile compounds.

This process is particularly effective for plastics that are difficult to recycle through conventional mechanical means, such as multilayer films, composite plastics, or plastics with food or oil contamination.

Zhang et al. (2019) investigated the synthesis of porous carbon structures from low density polymers as the primary material using the pressurized carbonization technique. The sample was activated using potassium hydroxide and analyzed through a suitable characterization technique for understanding the morphological features. It was found that the prepared product had a surface area of around 3000 m^2/g along with the respective functional bonds. These prepared porous carbon structures show commendable electrochemical features along with application in supercapacitors. It showed a specific capacitance value of around 350 Farad/g, current density of 2×10^{-1} Amp/g, and excellent stability when recycled.

5.4.1.4 Flash Joule Heating

Flash Joule heating (FJH) is a novel rapid thermal treatment process that has recently been explored for its potential to convert solid carbonaceous waste materials, such as plastic waste, into valuable products such as graphene, carbon nanotubes, and other frontier carbon materials. First designed for electronic waste and biomass, the process has demonstrated great potential for valorizing plastic waste with low environmental footprint and high-energy efficiency.

5.4.1.4.1 Principle and Process Overview

In FJH, an electric current of high voltage is flow pulsed through a carbonaceous material (here being plastic waste), resulting in instantaneous heat to very high temperatures (2000–3000°C) on a timescale of milliseconds. The reaction occurs within a conducting medium or pressurized sealed quartz or ceramic tube, and the violent thermal spike results in the abrupt breakdown of polymer chains and reorganization into highly structured graphitic molecules or turbostratic graphene.

Key features of the process include the following:

- *No oxygen*: The process occurs in an inert atmosphere (argon or nitrogen), preventing combustion.
- *Rapid transformation*: The entire reaction is completed in less than 1 second.
- *No solvents or catalysts required*: It is a dry process with minimal chemical input.

5.4.1.5 Microwave Treatment

Microwave-assisted treatment is an emerging thermochemical technology that uses microwave radiation (typically 915 MHz or 2.45 GHz) to heat and decompose plastic waste. Unlike conventional heating, which relies on conduction and convection, microwave treatment induces volumetric heating, meaning the material is heated uniformly from within. This approach is especially promising for the rapid and efficient decomposition of plastics into fuels, gases, and valuable chemicals, making it a viable solution in advanced plastic waste recycling and energy recovery.

5.4.1.6 Molten Salt Methods

Molten salt methods are an innovative and high-temperature thermochemical approach to convert plastic waste into fuels, chemicals, or carbonaceous materials. This method involves using molten salts—typically inorganic compounds like sodium carbonate ($NaCO_3$), potassium carbonate (KCO_3), or chlorides/nitrates—as a heat transfer medium and chemical catalyst. The salts act both as a thermal bath to decompose plastics and as a reaction environment that facilitates depolymerization, gasification, or cracking reactions with high efficiency and product purity. Kamali et al. (2019) investigated a fresh salt method comprising oxidative pyrolysis to create nanostructures having high surface area, crystallinity, and conductivity from PET. The scientists established that oxidative conditions could recycle plastic wastes into amorphous carbon, eliminating the need for complex equipment or protective gases like argon and nitrogen commonly used in traditional pyrolysis methods. A study was conducted on converting plastic bottles into nano-sized carbon materials with high conductivity and surface area. The technique involved in the experiments was electrochemical reactions using various salts such as sodium chloride (Kamali et al. 2019; Kamali and Yang 2020). The plastic was heat treated, and product formation occurred at around 1573 K.

5.4.1.7 Hydrothermal Procedures

Hydrothermal procedures for the valorization of plastic waste are innovative approaches that utilize water at elevated temperatures and pressures to break down and convert plastic waste into valuable products, such as fuels, chemicals, or solid carbon-rich materials. These methods are particularly promising for recycling plastics that are difficult to process by conventional means. In another study, PET was applied to recycling for the manufacture of concrete, and the characteristics of the product were examined (Hameed and Ahmed 2019). Various proportions of the plastic, ranging from 1% to 10 % were used to prepare different batches of the cement. Afterward, the mechanical characteristics comprising compressive, flexural, splitting tensile strength, and hardened density for all samples were examined.

In a study, plastic waste was utilized for the preparation of conventional building materials that are environmentally sustainable. In the experiments, plastic bottles were applied by crushing and filling them with the recycled materials and closed (Alaloul, John, and Musarat 2020). Bottles having aggregate particle size less than around 400 microns resulted in more compressive strength as compared to the lesser-sized particles. Paihte et al. (2019) used high-density polymers to prepare bricks through the melt compound method. The product that was prepared was analyzed for thermal conductivity, heat capacity, and other decrement factor features. Results showed that the thermal conductivity and capacity were enhanced by 40% and 50%, respectively. Some researchers successfully converted waste plastic into C2 grade fuels by using distilled water at room temperature (Jiao et al. 2020). The experiments were stimulated for carbon–carbon bonding, having light induction in a proper sequence, without the presence of oxygen. Solar simulators were used for the experiments.

5.4.1.8 Pilot Scale Studies

Continuous operation at pilot scale is essential to gather process data on industrialization and further assess the commercial viability of plastics recycling. Pilot scale studies for plastic valorization to chemicals have been conducted in a variety of ways. A study aimed to assess the pyrolysis process and its resultant products by utilizing a commercially available pyrolysis bench-scale batch reactor in a non-catalytic manner for converting plastic waste into fuel-grade chemicals (Rotta et al. 2022). Four different kinds of plastic fragments were gathered from the printing plant: polyethylene (PE), polypropylene (PP), PS, and biaxial-oriented polypropylene (BOPP) fractions. The feasibility of repurposing them through batch pyrolysis was examined. The pyrolysis oils exhibited a calorific value (reasonable heating value) that is comparable to that of ordinary diesel, falling between 36 and 45 kJ/g.

In a batch pilot scale reactor, pyrolysis of various plastic wastes was investigated using commercial zeolite catalyst (Sivagami et al. 2021). It was observed that PET plastic waste had oil yield of 17.8% with a calorific value of 30 kJ/kg. Metal-based plastic waste gave an oil yield of 65–70% with a calorific value of 45.17 kJ/kg. Osman et al. (2024) studied rapid pyrolysis at 450°C in a fluidized bed reactor to produce 10 tons of diesel per day at a feed rate of 3000 kg of plastic trash per hour. ASPEN HYSYS V11 was used to simulate production to guarantee the intended yield. In addition, a thorough economic analysis is done to calculate the annual cost and return on investment, as well as the payback time and profit for various scenarios.

In order to create a pyrolysis liquid upgraded with a non-thermal plasma reactor, a quick, oxidative pyrolysis method was designed (Lim et al. 2023). In this investigation, plastic waste was pyrolyzed in a pilot scale reactor using flue gas. Pyrolytic liquid produced using flue gas lowers life-cycle carbon emissions and operating costs. The production of liquid was found to rise, adding up a modest percentage of polymers at a temperature of 350°C. This increases the technical feasibility of scaling up such systems by offering operating parameters for the waste-to-energy and pyrolysis processes at the pilot scale.

The screw reactor system on a pilot scale was utilized for conducting the pyrolysis studies. The basic balances for pyrolysis during chemical recycling for plastic wastes were established (Zeller et al. 2021). All the studied plastic wastes have the potential to yield products for use as recycling feedstock. The feedstock plays a significant role in determining the product quality. The energy required for pyrolysis, heating, melting, and evaporation is approximately 5% of the calorific content of the feedstock. It is possible to recover and may be reintroduce between 50% and 75% of the carbon from the feedstock in the condensate through value chains within the chemical sector.

A pilot scale reactor for the pyrolysis technique has been investigated in one of the research works. At 520°C, waste plastics were shredded in a reactor with an hourly feed rate of 9 kg (Miskolczi et al. 2009). Feedstock came from the packing and agriculture industries, where trash was carefully gathered. In order to facilitate the more thorough breaking of the C—C bonds in plastic, a commercially available ZSM-5 catalyst was tried at a 5.0% concentration. By distillation, the yield was divided into gases, gasoline, and light and heavy oil. Depending on the parameters employed, plastic waste could provide 20–48% and 17–36% gasoline and light oil, respectively.

TABLE 5.3
Pilot Scale Studies for the Utilization of Plastic Waste

Type of Plastic Waste	Details of the Study	Reference
Polyolefins and multilayer plastic	• Different types of multilayer plastics along with zeolite catalyst were pyrolyzed batch pilot scale reactor. • Pyrolysis oil yield obtained between 35% and 70%. • Calorific value of oil varied between 30–47 kJ/g.	Sivagami et al., 2021
Organic and plastic waste	• Oxidative pyrolysis of the plastic waste was done in a pilot scale fluidized bed reactor. • Pyrolysis temperature varied between 300°C and 350°C. • Yield of oil increased to 18%.	Lim et al., 2023
Waste high-density polyethylene and polypropylene	• Pyrolysis of plastic waste was investigated in a horizontal gas-heated tube pilot scale reactor system. • Gaseous products mainly consisted of C3 and C4 hydrocarbons.	Miskolczi et al., 2009
Lightweight packaging sorting residues from construction waste	• The pilot scale screw reactor system was used for the pyrolysis studies. • The extraction rate for oil was 40–75%, while the output for pyrolysis solids ranged from 10% to 30%.	Zeller et al., 2021

A model was constructed to calculate the energy density (ED) of thermoplastic mixes by superimposing data on virgin polymers (Netsch et al. 2024). Differential scanning calorimetry was used to determine the material data, including heat capacity, phase transition enthalpy, and reaction enthalpy. One kilogram per hour screw reactor was used for pilot scale studies. Model calculations were contrasted with these experimental results. For pyrolysis, the feedstock-specific ED was independent of the plastic type. It represented 4.6–6% of the net calorific value of the feedstocks. The validation demonstrated excellent concordance between the virgin plastics and the hydrocarbon plastic blends. The systematic underestimation of the modeled ED in mixes containing heteroatoms revealed changes in the degradation mechanism. The model could resolve numerous phenomena that led to the pyrolysis ED. For more accurate process design, optimization, and assessment, the ED could be easily calculated with detailed information about occurrences.

Table 5.3 shows some pilot-scale studies for the utilization of plastic waste.

5.5 CONCLUSIONS AND PERSPECTIVES

This chapter provided a succinct yet comprehensive overview of the latest developments in environmentally friendly waste plastic treatment and monetization methods. The findings revealed that plastic waste constitutes a significant portion of all produced solid trash, highlighting its increasing environmental burden. Currently, plastics are predominantly managed and disposed of through landfilling, indiscriminate dumping, and open-air burning. Unfortunately, these traditional methods are outdated, inefficient, and highly detrimental to the environment. They result in the release of toxic fumes, greenhouse gas emissions, and hazardous chemical leachates, all of which pose serious threats to public health, safety, and ecological systems. Consequently, there is an urgent call for more innovative and sustainable approaches in addressing the short- and long-term effects of plastic waste. The literature review showcased a range of emerging technologies capable of transforming plastic waste into value-added products such as construction materials, alternative fuels, chemicals, and energy. These technologies—pyrolysis, gasification, chemical recycling, and bioconversion—offer promising alternatives to traditional disposal methods and have the potential to be integrated into the circular economy. However, to efficiently collect, transport, separate, process, and treat plastic wastes, appropriate planning, design, and implementation are needed. Thinking about these technologies seriously demands careful planning, investment, infrastructure improvement, and cooperation between stakeholders. Effective collection, transportation, segregation, and processing systems need to be in place to facilitate proper treatment and economic handling of plastic waste. Numerous socioeconomic, environmental, and technological obstacles were found in the study to impede the management, disposal, and valorization of plastic trash at the moment. Additional issues include lack of public awareness, weak policy frameworks, limited funding, and limited access to emerging recycling technologies in most areas. In addressing these, a multidisciplinary and systemic effort must be employed that includes governments, industries, researchers, and communities. In summary, there are promising solutions, but coordinated action and innovation are necessary to allow sustainable plastic waste management and valorization globally.

REFERENCES

Abnisa, Faisal, and Wan Mohd Ashri Wan Daud. 2014. "A Review on Co-Pyrolysis of Biomass: An Optional Technique to Obtain a High-Grade Pyrolysis Oil." *Energy Conversion and Management* 87. Elsevier Ltd: 71–85. https://doi.org/10.1016/j.enconman.2014.07.007.

Abnisa, Faisal, Wan Mohd Ashri Wan Daud, Sujahta Ramalingam, Muhamad Naqiuddin Bin M. Azemi, and J. N. Sahu. 2013. "Co-Pyrolysis of Palm Shell and Polystyrene Waste Mixtures to Synthesis Liquid Fuel." *Fuel* 108. Elsevier Ltd: 311–18. https://doi.org/10.1016/j.fuel.2013.02.013.

Adhikary, Kamal B., Shusheng Pang, and Mark P. Staiger. 2008. "Dimensional Stability and Mechanical Behaviour of Wood-Plastic Composites Based on Recycled and Virgin High-Density Polyethylene (HDPE)." *Composites Part B: Engineering* 39 (5): 807–15. https://doi.org/10.1016/j.compositesb.2007.10.005.

Akancha, Namrata Kumari, and R. K. Singh. 2019. "Co-Pyrolysis of Waste Polypropylene and Rice Bran Wax–Production of Biofuel and Its Characterization." *Journal of the Energy Institute*. https://doi.org/10.1016/j.joei.2018.07.011.

Alaloul, Wesam Salah, Vivekka Olivia John, and Muhammad Ali Musarat. 2020. "Mechanical and Thermal Properties of Interlocking Bricks Utilizing Wasted Polyethylene Terephthalate." *International Journal of Concrete Structures and Materials* 14 (1). Springer Singapore. https://doi.org/10.1186/s40069-020-00399-9.

Alam, Mahboob, Anjireddy Bhavanam, Ashirbad Jana, Jaimin kumar S. Viroja, and Nageswara Rao Peela. 2020. "Co-Pyrolysis of Bamboo Sawdust and Plastic: Synergistic Effects and Kinetics." *Renewable Energy* 149. Elsevier Ltd: 1133–45. https://doi.org/10.1016/j.renene.2019.10.103.

Arena, Umberto, and Maria Laura Mastellone. 2006. "Fluidized Bed Pyrolysis of Plastic Wastes." In *Feedstock Recycling and Pyrolysis of Waste Plastics*. Wiley. https://doi.org/10.1002/0470021543.ch16.

Barracco, F., E. Parisi, G. Pipitone, E. Simone, S. Bensaid, and D. Fino. 2024. "Valorization of Pyrolytic Plastic-derived Char for Adsorption of Wastewater Contaminants: A Kinetic and Thermodynamic Investigation." *International Journal of Environmental Science and Technology* 21: 6513–30. https://doi.org/10.1007/s13762-024-05467-1.

Bazargan, Alireza, and Gordon McKay. 2012. "A Review—Synthesis of Carbon Nanotubes from Plastic Wastes." *Chemical Engineering Journal* 195–96. Elsevier B.V,: 377–91. https://doi.org/10.1016/j.cej.2012.03.077.

Biessey, Philip, Julia Vogel, Mathias Seitz, and Peter Quicker. 2023. "Plastic Waste Utilization via Chemical Recycling: Approaches, Limitations, and the Challenges Ahead." *Chemie-Ingenieur-Technik* 95 (8): 1199–214. https://doi.org/10.1002/cite.202300042.

Brenner, Isaac (Joe). 2010. "ICPMS Applications." *Encyclopedia of Spectroscopy and Spectrometry*, 991–98. https://doi.org/10.1016/B978-0-12-374413-5.00057-9.

Bujak, Janusz Wojciech. 2015. "Thermal Utilization (Treatment) of Plastic Waste." *Energy* 90 (2): 1468–77. https://doi.org/10.1016/j.energy.2015.06.106.

Çepelioğullar, Özge, and Ayşe E. Pütün. 2013. "Thermal and Kinetic Behaviors of Biomass and Plastic Wastes in Co-Pyrolysis." *Energy Conversion and Management* 75: 263–70.

Chen, Aizhu, Min Quan Yang, Sibo Wang, and Qingrong Qian. 2021. "Recent Advancements in Photocatalytic Valorization of Plastic Waste to Chemicals and Fuels." *Frontiers in Nanotechnology* 3 (July): 1–8. https://doi.org/10.3389/fnano.2021.723120.

Chidambarampadmavathy, Karthigeyan, Obulisamy Parthiba Karthikeyan, and Kirsten Heimann. 2017. "Sustainable Bio-Plastic Production through Landfill Methane Recycling." *Renewable and Sustainable Energy Reviews* 71 (October). Elsevier: 555–62. https://doi.org/10.1016/j.rser.2016.12.083.

Damayanti, Damayanti, Desi Riana Saputri, David Septian Sumanto Marpaung, Fauzi Yusupandi, Andri Sanjaya, Yusril Mahendra Simbolon, Wulan Asmarani, Maria Ulfa, and Ho Shing Wu. 2022. "Current Prospects for Plastic Waste Treatment." *Polymers* 14 (15). https://doi.org/10.3390/polym14153133.

Dewangan, Ashish, D. Pradhan, and R. K. Singh. 2016. "Co-Pyrolysis of Sugarcane Bagasse and Low-Density Polyethylene: Influence of Plastic on Pyrolysis Product Yield." *Fuel* 185: 508–16.

Evode, Niyitanga, Sarmad Ahmad Qamar, Muhammad Bilal, Damià Barceló, and Hafiz M. N. Iqbal. 2021. "Plastic Waste and Its Management Strategies for Environmental Sustainability." *Case Studies in Chemical and Environmental Engineering* 4 (August). https://doi.org/10.1016/j.cscee.2021.100142.

Fermanelli, Carla S., Agostina Córdoba, Liliana B. Pierella, and Clara Saux. 2020. "Pyrolysis and Copyrolysis of Three Lignocellulosic Biomass Residues from the Agro-Food Industry: A Comparative Study." *Waste Management* 102. Elsevier Ltd: 362–70. https://doi.org/10.1016/j.wasman.2019.10.057.

Gazi, Sarifuddin, Miloš Đokić, Kek Foo Chin, Pei Rou Ng, and Han Sen Soo. 2019. "Visible Light–Driven Cascade Carbon–Carbon Bond Scission for Organic Transformations and Plastics Recycling." *Advanced Science* 6 (24). https://doi.org/10.1002/advs.201902020.

Gong, Jiang, Beata Michalkiewicz, Xuecheng Chen, Ewa Mijowska, Jie Liu, Zhiwei Jiang, Xin Wen, and Tao Tang. 2014. "Sustainable Conversion of Mixed Plastics into Porous Carbon Nanosheets with High Performances in Uptake of Carbon Dioxide and Storage of Hydrogen." *ACS Sustainable Chemistry and Engineering* 2 (12): 2837–44. https://doi.org/10.1021/sc500603h.

Gunasee, S. D., B. Danon, J. F. Görgens, and R. Mohee. 2017. "Co-Pyrolysis of LDPE and Cellulose: Synergies during Devolatilization and Condensation." *Journal of Analytical and Applied Pyrolysis* 126: 307–14.

Hameed, Awham Mohammed, and Bilal Abdul Fatah Ahmed. 2019. "Employment the Plastic Waste to Produce the Light Weight Concrete." *Energy Procedia* 157 (2018). Elsevier B.V.: 30–38. https://doi.org/10.1016/j.egypro.2018.11.160.

Hidayah, Nurul, and Syafrudin. 2018. "A Review on Landfill Management in the Utilization of Plastic Waste as an Alternative Fuel." *E3S Web of Conferences* 31: 1–6. https://doi.org/10.1051/e3sconf/20183105013.

Jiao, Xingchen, Kai Zheng, Qingxia Chen, Xiaodong Li, Yamin Li, Weiwei Shao, Jiaqi Xu, et al. 2020. "Photocatalytic Conversion of Waste Plastics into C2 Fuels under Simulated Natural Environment Conditions." *Angewandte Chemie—International Edition* 59 (36): 15497–501. https://doi.org/10.1002/anie.201915766.

Kai, Xingping, Tianhua Yang, Shengqiang Shen, and Rundong Li. 2019. "TG-FTIR-MS Study of Synergistic Effects during Co-Pyrolysis of Corn Stalk and High-Density Polyethylene (HDPE)." *Energy Conversion and Management* 181 (2). Elsevier: 202–13. https://doi.org/10.1016/j.enconman.2018.11.065.

Kamali, Ali Reza, and Jigang Yang. 2020. "Effect of Molten Salts on the Structure, Morphology and Electrical Conductivity of PET-Derived Carbon Nanostructures." *Polymer Degradation and Stability* 177. Elsevier Ltd: 109184. https://doi.org/10.1016/j.polymdegradstab.2020.109184.

Kamali, Ali Reza, Jigang Yang, and Qiang Sun. 2019. "Molten Salt Conversion of Polyethylene Terephthalate Waste into Graphene Nanostructures with High Surface Area and Ultra-High Electrical Conductivity." *Applied Surface Science* 476. Elsevier B.V.: 539–51. https://doi.org/10.1016/j.apsusc.2019.01.119.

Kantarelis, E., P. Donaj, W. Yang, and A. Zabaniotou. 2009. "Sustainable Valorization of Plastic Wastes for Energy with Environmental Safety via High-Temperature Pyrolysis (HTP) and High-Temperature Steam Gasification (HTSG)." *Journal of Hazardous Materials* 167 (1–3): 675–84. https://doi.org/10.1016/j.jhazmat.2009.01.036.

Kassab, Ali, Dawood Al Nabhani, Pravansu Mohanty, Christopher Pannier, and Georges Y. Ayoub. 2023. "Advancing Plastic Recycling: Challenges and Opportunities in the Integration of 3D Printing and Distributed Recycling for a Circular Economy." *Polymers* 15 (19). https://doi.org/10.3390/polym15193881.

Kaushal, Jyoti, Madhu Khatri, and Shailendra Kumar Arya. 2021. "Recent Insight into Enzymatic Degradation of Plastics Prevalent in the Environment: A Mini—Review." *Cleaner Engineering and Technology* 2 (March). Elsevier Ltd: 100083. https://doi.org/10.1016/j.clet.2021.100083.

Kibria, Md Golam, Nahid Imtiaz Masuk, Rafat Safayet, Huy Quoc Nguyen, and Monjur Mourshed. 2023. "Plastic Waste: Challenges and Opportunities to Mitigate Pollution and Effective Management." *International Journal of Environmental Research* 17. Springer International Publishing. https://doi.org/10.1007/s41742-023-00507-z.

Knott, Brandon C., Erika Erickson, Mark D. Allen, Japheth E. Gado, Rosie Graham, Fiona L. Kearns, Isabel Pardo, et al. 2020. "Characterization and Engineering of a Two-Enzyme System for Plastics Depolymerization." *Proceedings of the National Academy of Sciences of the United States of America* 117 (41): 25476–85. https://doi.org/10.1073/pnas.2006753117.

Kumar, Rakesh, Anurag Verma, Arkajyoti Shome, Rama Sinha, Srishti Sinha, Prakash Kumar Jha, Ritesh Kumar, et al. 2021. "Impacts of Plastic Pollution on Ecosystem Services, Sustainable Development

Goals, and Need to Focus on Circular Economy and Policy Interventions." *Sustainability (Switzerland)* 13 (17): 1–40. https://doi.org/10.3390/su13179963.

Kumar Mishra, Ranjeet, and Kaustubha Mohanty. 2020. "Co-Pyrolysis of Waste Biomass and Waste Plastics (Polystyrene and Waste Nitrile Gloves) into Renewable Fuel and Value-Added Chemicals." *Carbon Resources Conversion* 3 (October). Elsevier B.V.: 145–55. https://doi.org/10.1016/j.crcon.2020.11.001.

Lange, Jean Paul. 2021. "Managing Plastic Waste-Sorting, Recycling, Disposal, and Product Redesign." *ACS Sustainable Chemistry and Engineering* 9 (47): 15722–38. https://doi.org/10.1021/acssuschemeng.1c05013.

Lim, Mook Tzeng, Ee Sann Tan, Yee Ho Chai, Bridgid Lai Fui Chin, Filbert H. Juwono, Danial Hakim Badrul Hisham, and Muhammad Athir Mohamed Anuar. 2023. "Pilot-Scale Flue Gas Pyrolysis System for Organic and Plastic Wastes with Improved Liquid Properties in a Non-Thermal Plasma Reactor." *Journal of Analytical and Applied Pyrolysis* 173 (August): 1–6. https://doi.org/10.1016/j.jaap.2023.106062.

Limami, Houssame, Imad Manssouri, Khalid Cherkaoui, Malak Saadaoui, and Asmae Khaldoun. 2020. "Thermal Performance of Unfired Lightweight Clay Bricks with HDPE & PET Waste Plastics Additives." *Journal of Building Engineering* 30. Elsevier Ltd: 101251. https://doi.org/10.1016/j.jobe.2020.101251.

Lin, Xiaoqing, Yunfeng Ma, Zhiliang Chen, Xiaodong Li, Shengyong Lu, and Jianhua Yan. 2020. "Effect of Different Air Pollution Control Devices on the Gas/Solid-phase Distribution of PCDD/F in a Full-scale Municipal Solid Waste Incinerator." *Environmental Pollution* 265: 114888. https://doi.org/10.1016/j.envpol.2020.114888.

Liu, Wu Jun, Ke Tian, Hong Jiang, Xue Song Zhang, and Guang Xi Yang. 2013. "Preparation of Liquid Chemical Feedstocks by Co-Pyrolysis of Electronic Waste and Biomass without Formation of Polybrominated Dibenzo-P-Dioxins." *Bioresource Technology* 128. Elsevier Ltd: 1–7. https://doi.org/10.1016/j.biortech.2012.10.160.

Liu, Xiaoguang, Changde Ma, Yanliang Wen, Xuecheng Chen, Xi Zhao, Tao Tang, Rudolf Holze, and Ewa Mijowska. 2021. "Highly Efficient Conversion of Waste Plastic into Thin Carbon Nanosheets for Superior Capacitive Energy Storage." *Carbon* 171. Elsevier Ltd: 819–28. https://doi.org/10.1016/j.carbon.2020.09.057.

Mercea, Peter Viktor, Christoph Losher, Marcus Petrasch, and Valer Toşa. 2018. "Migration of Stabilizers and Plasticizer from Recycled Polyvinylchloride." *Journal of Vinyl and Additive Technology* 24: E112–24. https://doi.org/10.1002/vnl.21609.

Mihai, Florin Constantin, Sedat Gündogdu, Laura A. Markley, Arianna Olivelli, Farhan R. Khan, Claire Gwinnett, Jutta Gutberlet, et al. 2022. "Plastic Pollution, Waste Management Issues, and Circular Economy Opportunities in Rural Communities." *Sustainability (Switzerland)* 14 (1). https://doi.org/10.3390/su14010020.

Miskolczi, N., A. Angyal, L. Bartha, and I. Valkai. 2009. "Fuels by Pyrolysis of Waste Plastics from Agricultural and Packaging Sectors in a Pilot Scale Reactor." *Fuel Processing Technology* 90 (7–8). Elsevier B.V,: 1032–40. https://doi.org/10.1016/j.fuproc.2009.04.019.

Moharir, Rucha V., Pratibha Gautam, and Sunil Kumar. 2019. "Waste Treatment Processes/technologies for Energy Recovery." In *Current Developments in Biotechnology and Bioengineering: Waste Treatment Processes for Energy Generation*. Elsevier B.V. https://doi.org/10.1016/B978-0-444-64083-3.00004-X.

Netsch, Niklas, Michael Zeller, Frank Richter, Britta Bergfeldt, Salar Tavakkol, and Dieter Stapf. 2024. "Energy Demand for Pyrolysis of Mixed Thermoplastics and Waste Plastics in Chemical Recycling: Model Prediction and Pilot-Scale Validation." *ACS Sustainable Resource Management*, 9–13. https://doi.org/10.1021/acssusresmgt.4c00109.

Nyakuma, Bemgba B., and Tertsegha J. P. Ivase. 2021. "Emerging Trends in Sustainable Treatment and Valorisation Technologies for Plastic Wastes in Nigeria: A Concise Review." *Environmental Progress and Sustainable Energy* 40 (5). https://doi.org/10.1002/ep.13660.

Olazar, Leire, Juan Fernando Saldarriaga, Gartzen Lopez, Laura Santamaria, Maider Amutio, Martin Olazar, and Maite Artetxe. 2024. "Insight into the Joint Valorization of CO_2 and Waste Plastics by Pyrolysis and in Line Dry Reforming for Syngas Production." *Fuel Processing Technology* 253: 108024. https://doi.org/10.1016/j.fuproc.2023.108024.

Olofinnade, Oluwarotimi, Sarath Chandra, and Paromita Chakraborty. 2020. "Recycling of High Impact Polystyrene and Low-Density Polyethylene Plastic Wastes in Lightweight Based Concrete for Sustainable Construction." *Materials Today: Proceedings* 38 (Part 5). Elsevier Ltd: 2151–56. https://doi.org/10.1016/j.matpr.2020.05.176.

Osman, Matin Darwisy, Nurizzati Erina Selimi, Abdul Bari Juffri, Nabisab Mujawar Mubarak, and Rama Rao Karri. 2024. "Pilot Scale Pyrolysis and Simulation of Plastic Waste into a Value-Added Product: An Integrated Approach." *Chemical Engineering & Technology* 47 (8): 23–25.

Paihte, P. Lalzarliana, Andy C. Lalngaihawma, and Gaurav Saini. 2019. "Recycled Aggregate Filled Waste Plastic Bottles as a Replacement of Bricks." *Materials Today: Proceedings* 15. Elsevier Ltd.: 663–68. https://doi.org/10.1016/j.matpr.2019.04.135.

Pandey, Prashant, Manisha Dhiman, Ankur Kansal, and Sarada Prasannan Subudhi. 2023. "Plastic Waste Management for Sustainable Environment: Techniques and Approaches." *Waste Disposal and Sustainable Energy* 5 (2). Springer Nature Singapore: 205–22. https://doi.org/10.1007/s42768-023-00134-6.

Pandey, Sandeep, Manoj Karakoti, Karan Surana, Pawan Singh Dhapola, Boddepalli Santhi Bhushan, Swaroop Ganguly, Pramod K. Singh, Ali Abbas, Anurag Srivastava, and Nanda Gopal Sahoo. 2021. "Graphene Nanosheets Derived from Plastic Waste for the Application of DSSCs and Supercapacitors." *Scientific Reports* 11: 3916. https://doi.org/10.1038/s41598-021-83483-8.

Pandey, Umesh, Jan Arild Stormyr, Alireza Hassani, Rajan Jaiswal, Hildegunn H. Haugen, and Britt M. E. Moldestad. 2020. "Pyrolysis of Plastic Waste to Environmentally Friendly Products." *WIT Transactions on Ecology and the Environment* 246: 61–74. https://doi.org/10.2495/EPM200071.

Pol, Vilas Ganpat. 2010. "Upcycling: Converting Waste Plastics into Paramagnetic, Conducting, Solid, Pure Carbon Microspheres." *Environmental Science and Technology* 44 (12): 4753–59. https://doi.org/10.1021/es100243u.

Ranjan, Ved Prakash, and Sudha Goel. 2021. "Recyclability of Polypropylene after Exposure to Four Different Environmental Conditions." *Resources, Conservation and Recycling* 169 (October 2020). Elsevier B.V.: 105494. https://doi.org/10.1016/j.resconrec.2021.105494.

Rotta, Arnella Nechita, Cristina Bota, Balázs Brém, Dan Ioan Porumb, and Emese Gál. 2022. "Analysis of Products From the Pyrolysis of Plastics Waste From a Printing Plant: A Pilot Scale Study." *Studia Universitatis Babes-Bolyai Chemia* 67 (4): 169–85. https://doi.org/10.24193/subbchem.2022.4.11.

Sakao, Tomohiko, Nancy Bocken, Nabil Nasr, and Yasushi Umeda. 2024. "Implementing Circular Economy Activities in Manufacturing for Environmental Sustainability." *CIRP Annals* 73 (2). Elsevier Ltd: 457–81. https://doi.org/10.1016/j.cirp.2024.06.002.

Sameshima-Yamashita, Yuka, Hirokazu Ueda, Motoo Koitabashi, and Hiroko Kitamoto. 2019. "Pretreatment with an Esterase from the Yeast Pseudozyma Antarctica Accelerates Biodegradation of Plastic Mulch Film in Soil under Laboratory Conditions." *Journal of Bioscience and Bioengineering* 127 (1). Elsevier Ltd: 93–98. https://doi.org/10.1016/j.jbiosc.2018.06.011.

Saxena, Saumitra. 2025. "Pyrolysis and beyond: Sustainable Valorization of Plastic Waste." *Applications in Energy and Combustion Science* 21 (November 2024). Elsevier Ltd: 100311. https://doi.org/10.1016/j.jaecs.2024.100311.

Schultz, Felix Carl, and Robert Jaroslav Reinhardt. 2023. "Technological Challenges and Opportunities to Plastics Valorization in the Context of a Circular Economy in Europe." *Sustainability (Switzerland)* 15 (4): 1–10. https://doi.org/10.3390/su15043741.

Sheriff, Salia S., Abdulfatah Abdu Yusuf, Oluwole O. Akiyode, Ezekiel Fayiah Hallie, Saidi Odoma, Rebecca Alicia Yambasu, Kula Thompson-Williams, Charles Asumana, Sylvester Z. Gono, and Mohammed A. Kamara. 2025. "A Comprehensive Review on Exposure to Toxins and Health Risks from Plastic Waste: Challenges, Mitigation Measures, and Policy Interventions." *Waste Management Bulletin* 3 (3). Elsevier B.V.: 100204. https://doi.org/10.1016/j.wmb.2025.100204.

Sivagami, Krishnasamy, Govindaraj Divyapriya, Ramya Selvaraj, P. Madhiyazhagan, N. Sriram, and Indumathi Nambi. 2021. "Catalytic Pyrolysis of Polyolefin and Multilayer Packaging Based Waste Plastics: A Pilot Scale Study." *Process Safety and Environmental Protection* 149. Institution of Chemical Engineers: 497–506. https://doi.org/10.1016/j.psep.2020.10.038.

Sogancioglu, Merve, Esra Yel, and Gulnare Ahmetli. 2017. "Pyrolysis of Waste High Density Polyethylene (HDPE) and Low Density Polyethylene (LDPE) Plastics and Production of Epoxy Composites with Their Pyrolysis Chars." *Journal of Cleaner Production* 165. Elsevier B.V.: 369–81. https://doi.org/10.1016/j.jclepro.2017.07.157.

Stepczyńska, Magdalena, and Piotr Rytlewski. 2018. "Enzymatic Degradation of Flax-Fibers Reinforced Polylactide." *International Biodeterioration and Biodegradation* 126 (November): 160–66. https://doi.org/10.1016/j.ibiod.2017.11.001.

Tamoor, Muhammad, Nadia A. Samak, Yunpu Jia, Muhammad Umar Mushtaq, Hassan Sher, Maryam Bibi, and Jianmin Xing. 2021. "Potential Use of Microbial Enzymes for the Conversion of Plastic Waste Into Value-Added Products: A Viable Solution." *Frontiers in Microbiology* 12 (November). https://doi.org/10.3389/fmicb.2021.777727.

Tang, Kuok Ho Daniel. 2022. "Valorization of Plastic Waste through Incorporation into Construction Materials." *Civil and Sustainable Urban Engineering* 2 (2): 96–109. https://doi.org/10.53623/csue.v2i2.141.

Thirunavukarasu, Kathirvel, Subramanian Purushothaman, Janardhanam Sridevi, Mayilvahanan Aarthy, Marichetti Kuppuswami Gowthaman, Toshiaki Nakajima-Kambe, and Numbi Ramudu Kamini. 2016. "Degradation of Poly(butylene Succinate) and Poly(butylene Succinate-Co-Butylene Adipate) by a Lipase from Yeast Cryptococcus Sp. Grown on Agro-Industrial Residues." *International Biodeterioration and Biodegradation* 110. Elsevier Ltd: 99–107. https://doi.org/10.1016/j.ibiod.2016.03.005.

Verma, Rinku, K. S. Vinoda, M. Papireddy, A. N. S. Gowda, and A. N. S. 2016. "Toxic Pollutants from Plastic Waste: A Review." *Procedia Environmental Sciences* 35: 701–8. https://doi.org/10.1016/j.proenv.2016.07.069.

Vuppaladadiyam, Sai Sree Varsha, Arun K. Vuppaladadiyam, Abhisek Sahoo, Ajay Urgunde, S. Murugavelh, Vít Šrámek, Michael Pohořelý, et al. 2024. "Waste to Energy: Trending Key Challenges and Current Technologies in Waste Plastic Management." *Science of the Total Environment* 913 (December 2023). https://doi.org/10.1016/j.scitotenv.2023.169436.

Yan, W., S. Kent Hoekman, Amber Broch, and Charles J. Coronella. 2014. "Effect of Hydrothermal Carbonization Reaction Parameters on the Properties of Hydrochar and Pellets." *Environmental Progress & Sustainable Energy* 33 (3): 676–80. https://doi.org/10.1002/ep.11974

Zeller, Michael, Niklas Netsch, Frank Richter, Hans Leibold, and Dieter Stapf. 2021. "Chemical Recycling of Mixed Plastic Wastes by Pyrolysis—Pilot Scale Investigations." *Chemie-Ingenieur-Technik* 93 (11): 1763–70. https://doi.org/10.1002/cite.202100102.

Zhang, Hua, Xiao Li Zhou, Li Ming Shao, Fan Lü, and Pin Jing He. 2019. "Hierarchical Porous Carbon Spheres from Low-Density Polyethylene for High-Performance Supercapacitors." *ACS Sustainable Chemistry and Engineering* 7 (4): 3801–10. https://doi.org/10.1021/acssuschemeng.8b04539.

Section III

Sustainable Technologies for Plastic Waste Management and Recycling

6 Plastic Waste Valorization into New Plastic

Khursheed B. Ansari[1], Shakeelur Rahman A.R.[2], M. Yusuf Ansari[3], Mohd Shariq Khan[4], and Abubakr Elkhaleefa[1]*

[1]Department of Chemical Engineering, College of Engineering, King Khalid University, Abha, Saudi Arabia

[2]Department of Applied Science, Shri Vile Parle Kelavani Mandal's Institute of Technology, Dhule, India

[3]Department of Petroleum Studies, Zakir Husain College of Engineering and Technology, Aligarh Muslim University, Aligarh, India

[4]Department of Chemical Engineering, Dhofar University, Salalah, Oman

6.1 INTRODUCTION

Plastic has become an integral part of humankind's life. The short-term use of plastic, such as packaging, single-use plastic, and textiles enabled an escalating generation of global plastic waste, which must be tackled with appropriate technology and minimal environmental impact. Since 1950, global plastic production has amplified 200-fold, from 2 million MT to 400 million MT in 2022 (Abd Karim et al., 2023). Rapidly growing urbanization, population, and economic development are primarily responsible for this (Amasuomo, 2016). Global plastic production is expected to triple by 2060 and hence creates challenges of its proper handling/disposal/conversion. Countries have to develop and implement robust policies and/or rules for plastic waste minimization (or recycling) (Singh and Walker, 2024). If not handled properly, plastic waste can have a severely detrimental effect on the environment and, hence, human health, as mentioned elsewhere (Al-Mosawi et al., 2017; Nayanathara Thathsarani Pilapitiya and Ratnayake, 2024). Conventionally, plastics are produced via polymerization of monomers (such as ethylene or propylene) extracted from petrochemicals. The degradation of such plastics in the environment takes a long time. One of the alternative ways is to produce plastics from biomaterials instead of traditional petroleum-based monomers (Kognole et al., 2019; Tiwari et al., 2023). Biomaterials are often derived from nature (i.e., vegetable oils and plant-based starch, and so forth); hence, they can easily degrade in the environment with less impact and minimal carbon footprint than petroleum-derived plastics. However, bioplastics account for only 1% of the total plastic production and need further attention in research for their large-scale production (Shah et al., 2021; Narancic et al., 2020).

The pollution caused by plastic waste accumulation in the environment becomes a significant threat to ecosystems, wildlife, and human health (Evode et al., 2021). The plastics under solar radiation also break down into micro- (< 5 mm) and nanoplastics (< 100 nm), which can easily percolate into marine life and other water bodies, thus causing a threat to humans upon exposure/consumption of marine species. Plastic waste management has become essential to utilize plastic waste effectively and convert it into a valuable end product. Converting plastic waste into useful end

* Corresponding author's email: khansari@kku.edu.sa

DOI: 10.1201/9781003581000-8

FIGURE 6.1 Traditional plastic recycling. (From Turner et al., 2018.)

products also contribute to environmental protection since it avoids waste disposal and landfills. However, effective valorization of plastics still depends on addressing the difficulties brought on by the complexity and variety of plastic waste (Mong et al., 2024). Plastic recycling is one of the promising ways of plastic waste valorization. It has been pushed in recent years due to increased environmental awareness, regulatory demands, and technological breakthroughs in recycling. Traditional plastic recycling can be seen in Figure 6.1. On a global scale, only 9% of plastic waste is recycled, and almost 80% goes to landfills. This indicates a vast opportunity in plastic waste recycling. Modern developments in plastic recycling techniques (such as mechanical, chemical, and biological) are enabling new routes to handle a variety of plastics with the capability of tackling the feed complexity. For example, contaminated, complex, or multi-material plastics can be handled by chemical recycling techniques.

This chapter presents and discusses the issues related to plastic pollution and discusses the various techniques of plastic waste management, including traditional plastic recycling, plastic conversion into new plastics and chemicals (chemical, biological, and solvent-based recycling), integration of plastics with petroleum refining, monomer recovery from the plastic waste, direct reuse of plastics, and economic aspects. The information assembled and discussed in this chapter opens a new domain for plastic waste management.

6.2 TRADITIONAL PLASTIC WASTE MANAGEMENT

6.2.1 Landfilling

The accumulation of plastic waste has escalated significantly in the last few decades leading to a significant environmental challenge (Gilbert, 2017). In 2021 alone, global plastic production surpassed 390 million tons, with single-use plastics, especially those used for packaging, making up approximately 40% of the total. This trend highlights the unsustainable trajectory of plastic utilization. While plastics have become embedded in modern life, offering convenience, they simultaneously contribute to an ever-growing waste crisis, much of which remains poorly managed.

Despite efforts at recycling, a small percentage (~9%) of plastic waste is effectively recycled, leaving the majority either destined for landfills or incineration. Specifically, about 50% of plastic waste is relegated to landfills, 19% is incinerated, and 22% is left unmanaged, ultimately leaking into the environment (Evode et al., 2021). The heavy reliance on landfilling underscores a critical gap in global waste management systems, particularly in addressing the life cycle of single-use plastics. Various plastic waste management strategies exist, but current practices remain unsustainable, with landfilling still being the predominant method. Landfilling, which involves burying waste underground, has become a widely adopted approach for plastic waste disposal due to several factors.

One of the most significant issues is the inadequate recycling infrastructure, which is unable to keep pace with the growing volume of plastic waste. For instance, in the United States, approximately 85% of municipal plastic waste ended up in landfills in 2021 (Nayanathara Thathsarani Pilapitiya and Ratnayake, 2024; Wojnowska-Baryła et al., 2022). This heavy reliance on landfilling highlights major deficiencies in recycling efforts and policies aimed at reducing plastic waste. Additionally, a lack of public awareness regarding proper waste disposal and recycling further exacerbates the problem. Many people either do not know which types of plastics can be recycled or lack access to appropriate recycling facilities. As a result, landfilling remains the default option for many municipalities, perpetuating the cycle of plastic waste mismanagement (Kibria et al., 2023). The process of selecting and preparing landfill sites is critical for ensuring environmental protection and public health. Landfills are typically located in areas where land is available at a relatively low cost while minimizing environmental risks, particularly the potential for groundwater contamination. Factors considered during site selection include the proximity to waste generation sources, topography, climate, and existing land use. Once a suitable site is identified, it is prepared by lining the base with layers of clay and impermeable geomembranes, which serve to prevent leachate from entering and contaminating groundwater. Preparation is essential for containing the toxic by-products from decomposing waste, especially plastics, which do not easily break down and can leach harmful substances into the surrounding ecosystem throughout their prolonged lifespan in landfills (Arabeyyat et al., 2024).

The degradation of plastics retrieved from the landfills can be evaluated through their physical and chemical characteristics. Zhou et al. (2014) studied that after 24 years of landfilling, plastic waste constituted approximately 10.62% ± 5.12% of the total waste in an aging landfill. Among landfilling, 69.13% were identified as plastic bags, which included 11.34% white polyethylene (PE) bags, 29.77% colored PE bags, and 28.02% other types of plastic bags. Additionally, 30.87% of the plastic waste consisted of different materials such as polypropylene (PP), polyvinyl chloride (PVC), and polystyrene (PS). The characteristics of landfill plastic waste showed notable differences from typical plastic waste. Specifically, landfill plastics had significantly reduced levels of carbon and sulfur while exhibiting higher concentrations of elements like oxygen, silicon, and aluminum; nitrogen and chlorine levels remained comparable between the two types of waste. Moreover, the ash content in landfill-mined plastic waste remained 9.03-fold greater compared to the standard plastic waste (Hopewell et al., 2009; Wojnowska-Baryła et al., 2022).

6.2.2 Plastic Waste Incineration

Incineration has emerged as a common practice for managing plastic waste, primarily because it reduces the volume of waste that occupies landfill space. By burning plastics, the overall footprint of landfill facilities can be considerably minimized, which is crucial as landfill capacity diminishes globally. This method transforms plastic waste into energy, providing an additional benefit for its adoption in waste-to-energy systems.

6.2.2.1 Emission of Harmful Substances during Incineration

The increasing global reliance on plastic has led to a burgeoning demand for plastic waste removal methods. As a consequence, plastic incineration has become a prevalent practice, raising critical public health concerns due to the emissions generated during these processes.

Incineration has long been considered a practical method for managing waste, but it presents significant environmental and health challenges due to the release of harmful pollutants. Studies on the effect of plastic incineration on public health effects are still limited; existing studies suggest that exposure to these emissions can result in significant systemic and respiratory health issues. Research indicates an enhanced risk of neurological disorders and heart issues among populations exposed to the emissions from plastic waste incineration. For example, residents living near plastic reprocessing facilities have reported experiencing elevated respiratory problems and mucocutaneous symptoms (Moharir et al., 2019; Rogers and Jaspers, 2024). Additionally, workers in industries that process fiberglass and plastic have shown higher instances of respiratory complications and significant impairments in lung function due to exposure to volatile organic compounds (VOCs) and various hazardous substances, including particulate matter, carbon monoxide, and heavy metals, released into the atmosphere. These pollutants can severely impact air quality and pose serious risks to public health, particularly for communities located near incineration facilities. To address these concerns, the implementation of effective strategies is crucial (Pathak et al., 2023; Verma et al., 2016). Advanced emission control technologies, designed to capture and reduce harmful outputs, play a key role in minimizing the environmental footprint of incineration.

6.2.2.2 Energy Generation and Recovery during Incineration

Plastic waste incineration is a predominant method, wherein plastics are combusted to release energy. This process not only generates heat, which can be converted into steam to drive turbines for electricity production, but also provides an effective means of reducing the volume of waste sent to landfills. The efficiency of energy recovery through incineration is contingent upon various factors, including the composition of the plastic waste and the technologies employed in the incineration facilities.

6.2.3 Mechanical Processing of Plastic Waste

Mechanical recycling is a crucial process in managing plastic waste, promoting sustainability while minimizing environmental impact. This method enables the recovery of secondary raw materials from plastic waste without significantly altering the chemical structure of these materials (Ragaert et al., 2017).

6.2.3.1 Technologies of Mechanical Processing

Mechanical recycling is widely recognized as the leading method for recycling plastic waste, playing an integral role in advancing towards an environmentally sustainable plastic economy. The mechanical recycling process can be broken down into several stages (Patel et al., 2024; Van Camp et al., 2024). However, the sequence and frequency of each step may vary concerning the plastic types being treated and the final product. The initial phase in mechanical recycling is collection, which involves acquiring discarded plastic products from different sources of waste. Effective collection mechanisms are critical to enhancing the quality of the recycling process. When plastics are collected with minimum contamination and proper segregation, the quality of the recovered material improves significantly. Separated collection systems can result in tenfold increases in recycling rates when compared to mixed collections, highlighting the necessity of effective collecting procedures in producing high-quality recovered products (Anwar et al., 2024; Jung et al., 2023). After sorting, washing is required to remove impurities such as dirt, labels, and residues from food or beverages. Cleanliness is crucial to ensuring the quality of the final recycled product, and advanced washing techniques, including vacuum and spray washing, can be used to improve plastic cleanliness. Thorough washing prepares the plastic for the next step in mechanical recycling: shredding. In the shredding process, clean plastic is broken down into smaller fragments, which allows for easier processing in later stages. Following shredding, secondary sorting is frequently used to remove contaminants, ensuring that only clean and acceptable material is sent to extrusion. The quality of shredding is critical in determining the overall efficiency of the recycling process, and repeated shredding and sorting can increase the purity of the recycled material. During the extrusion phase, the shredded plastic is melted and formed into uniform pellets or granules. These pellets can subsequently be utilized as raw materials to create new products that resemble virgin plastic in many ways. However, it is crucial to note that mechanical recycling can cause material qualities to degrade as a result of temperature and environmental conditions that plastics are exposed to throughout their service life and during processing (Jung et al., 2023).

6.2.3.2 Limitation of Products through Mechanical Processing

The mechanical recycling of plastic waste has several significant limitations that restrict its effectiveness and impact on sustainability. Although mechanical recycling is a popular technology due to its simple physical degrading process (such as grinding and shredding), its drawbacks have received increased attention in recent years. The primary limitation of mechanical recycling is the reduction in molecular weight of the plastics during processing. As the material is repeatedly heated and processed, the molecular chains that give plastic its strength and integrity shorten and weaken, resulting in a finished product that loses the original material's mechanical characteristics. This degradation reduces the use of recycled plastics in the production of high-quality items (Shamsuyeva and Endres, 2021). Consequently, discarded plastics are often downcycled rather than reused in applications that require high performance. Mechanical recycling is less effective with diverse plastic types, as additives and polymers complicate the process. Mixing different plastics often produces secondary materials with inferior properties, such as dull, gray products unsuitable for aesthetic applications. Furthermore, mechanical recycling is energy-intensive, requiring significant energy for heating, shredding, and re-granulating waste. This can result in a notable carbon footprint, potentially offsetting the environmental benefits compared to producing new plastics. As energy use increases, the sustainability of recycling becomes more complex. The efficiency of mechanical recycling is limited, with only 9% of global plastic waste being recycled, while much ends up in landfills or is incinerated, worsening pollution. Low recycling rates, contaminated waste streams, and inadequate sorting hinder its potential, driving interest in alternatives like chemical recycling (Ragaert et al., 2017; Suzuki et al., 2022; Vuppaladadiyam et al., 2024).

6.3 PLASTIC WASTE VALORIZATION INTO PLASTICS AND OTHER CHEMICALS

6.3.1 Chemical Recycling

Compared to mechanical recycling, chemical recycling is a gradually developing route that has high capability even for heterogeneous and contaminated plastic waste material (Ragaert et al., 2017). Based on the American Society for Testing and Materials, chemical recycling is also known as tertiary recycling (D5033) (Narro et al., 2019). Table 6.1 shows various chemical recycling techniques;

TABLE 6.1
Chemical Recycling Techniques for Plastic Waste Valorization

Recycling Technique	Explanation	Type of Plastic	Features	Difficulties
Gasification	Conversion of plastic waste into gases by a partial oxidation process	All types	• No need to separate the plastic into different categories. • Production of heat and electricity.	• Nonapplicable for monomers production. • Produce harmful gases.
Pyrolysis	Polymer chain decomposition in the existence of an inert atmosphere	All polymers	• Fewer harmful gases. • Production of heat and electricity. • Wide range of products in different states (solid, liquid, and gaseous).	• High volumes are required for reasonable efficiency. • Lower product quality due to the hetero atoms.
Solvolysis	Breaking of plastic waste by solvents	PLA, PET, PC	• High efficiency and easy controllable. • Wide range of reactions.	• Separation of feedstock impurities is difficult. • Depolymerization takes a long time.

of these, pyrolysis (between 400°C and 800°C) and gasification (between 700–1000°C) stand out as the most promising options for handling plastic waste that are challenging to recycle through mechanical means and are also hard to break down into monomers (Rahimi and García, 2017; Trinh et al., 2016). Chemical recycling could significantly benefit the economy by providing substantial cost savings. Nevertheless, past efforts in the 1980s and 1990s were unsuccessful due to issues such as impurities, availability of feedstock, logistical challenges, corrosion, and suboptimal economic conditions (Dogu et al., 2021). In contrast to combustion, both pyrolysis and gasification operate at lower temperatures (Czajczyńska et al., 2017).

From an environmental perspective, pyrolysis and gasification offer a viable alternative to landfilling, which could lead to a decrease in greenhouse gases (GHGs) and CO_2 emissions (Al-Salem et al., 2009). The role of chemical recycling in the circular economy is defined by its ability to eliminate waste and produce the raw materials that form the basis of plastics, effectively closing the loop (Singh et al., 2017; Tran et al., 2025). This process assists in maintaining the continuous use of the same products and materials, thereby supporting the restoration of natural systems. To enhance particular product selectivity and efficiency, it is also important to develop polymers that are compatible with chemical recycling (Zhu et al., 2018).

Pyrolysis and gasification offer further economic benefits since they do not need as many feedstock pretreatment processes compared to other waste treatment techniques (Al-Salem et al., 2017; Trinh et al., 2020). Moreover, fuel-grade products with high calorific values or useful chemicals can be easily marketed and potentially be sold with a profit (Rodríguez et al., 2019). Furthermore, process handling for pyrolysis and gasification is easier and more flexible than conventional recycling methods since it does not require an extreme sorting process; it is less labor-intensive as not every step in sorting is fully automated yet (Sharuddin et al., 2016). One more operational advantage is the use of residual output of char, which can be utilized as a fuel or feedstock for other petrochemical processes (Faravelli et al., 2003). The pyrolysis of plastic mixtures, when different polymers are decomposed together, may also have an extra advantage; single polymer chains in a mixture can partially decompose at different temperatures. Some researchers found that the decomposition of mixed polymer feedstock and the separation of the product stream can be effectively combined by applying a stepwise pyrolysis process at reasonable temperatures (Bockhorn et al., 1999).

6.3.2 Biological Recycling

Both biodegradable and non-biodegradable plastics are widely available in the market. Most petroleum-based plastics are highly resistant to microbial degradation and thus will hold out in the environment for extensive time (D'Hooge et al., 2015). On the other hand, bio-based plastics commonly exhibit lower life cycle GHG emissions compared to their fossil fuel-based equivalents. Naturally biodegradable polymers include polylactic acid (PLA), polyhydroxyalkanoate (PHA), thermoplastic starch (TPS), poly(butylene adipate-co-terephthalate) (PBAT), and polycaprolactone (PCL). PBAT and PCL are fossil fuel-based, while TPS, PHA, and PLA are bio-based polymers (Chen and Patel, 2012). Biochemical recycling uses different organisms such as algae, microorganisms, worms, and insect larvae, which are efficient and capable of converting complex polymers into simple molecules like carbon dioxide and water (Krueger et al., 2015; Yang et al., 2015). Mealworms (*Tenebrio molitor*) fed only on polystyrene pieces for around 2 weeks were able to convert up to approximately 50% of the ingested carbon into carbon dioxide, while the remaining carbon was egested as feces (Yang et al., 2015). Another study reported that *Streptomyces* sp. and *Aspergillus niger* degraded 46.7% and 26.17% (w/w) LDPE powder (53–75 μm), respectively, after 6 months (Silva et al., 2018). These processes take place at very low rates and need to keep the organisms alive for long and extensive amounts of time, which is a big challenge for industrial applications.

6.3.3 Solvent-Based Recycling

6.3.3.1 Dissolution–Precipitation Process

Dissolution–precipitation (DP) processes play an important role in several fields such as geology and material science, and their importance in environmental science, namely, plastic waste recovery and recycling, is immense (Aparício et al., 2024; Karanam et al., 2024). A typical process involves dissolving the solid in a solvent due to changes in temperature, pressure, and chemical composition, resulting in ions/molecules of the solid entering the solution. After dissolution, when the solution condition changes due to heating/cooling/evaporation, the dissolved molecules/ions precipitate to form a solid again. Either as the same or different phase material. The solvent DP process is followed by a filtration step (low energy step) to remove any undissolved contaminants. After this, the polymer can be recovered in pure resin form. The solvent DP process has immense potential in successfully recycling "plastic to plastic" and enables a circular economy in one single step in comparison to incineration (associated with producing toxic ash such as dioxins, chlorocompounds (Pappa et al., 2001), depolymerization, pyrolysis, or gasification that require several additional steps and resources to plastic again (Lange, 2021). The dissolution–precipitation technique for polymer recycling is demonstrated in Figure 6.2, which shows that the waste polymer is subjected to cutting, washing (with water), and segregation before use as a feed. Then, it is added to the solvent, which selectively dissolves one of the polymers at specified conditions. Later, the suspension undergoes filtration to eliminate the undissolved polymer. Finally, the addition of an antisolvent to the filtrate enables the precipitation of the dissolved polymer. The precipitated polymer is further exposed to filtration and drying to obtain its final form. The dissolution–precipitation is suitable for the lab and pilot scale recycling of PP, low-density polyethylene (LDPE), and high-density polyethylene (HDPE). Thus, the dissolution–precipitation approach finds uses in several places, and one of the

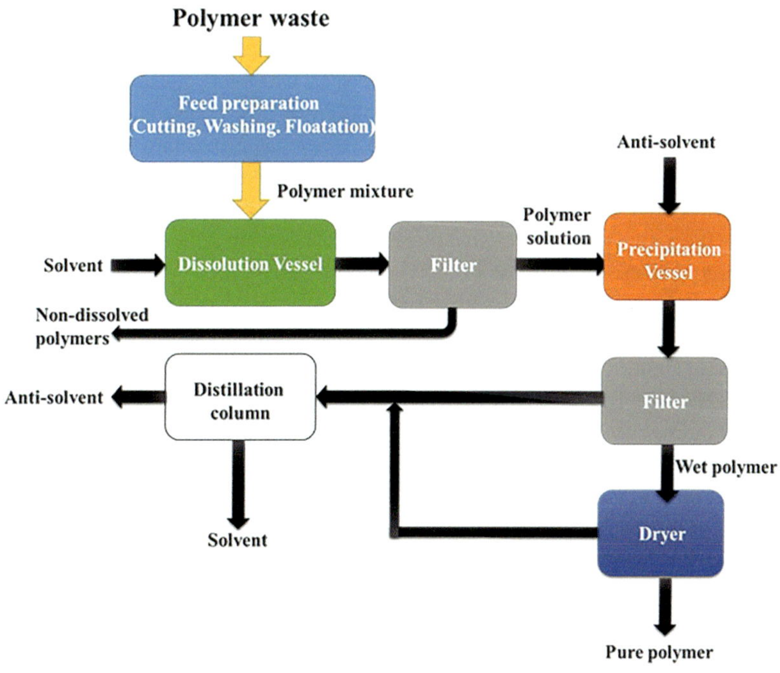

FIGURE 6.2 Dissolution–precipitation technique for the separation of polymers during plastic waste recycling. (From Pappa et al., 2001.)

niche areas is toxicology, where high-quality microsized plastic materials are created for use as reference materials (Balakrishnan et al., 2019; Bhargava et al., 2018; Johnson et al., 2021; Karanam et al., 2024). Some studies reported polystyrene plastic particle synthesis; however, particle size distribution and other property matching of microplastic materials still require clear explications (Yabu et al., 2005; Yip et al., 2022).

6.3.3.1.1 Dissolution–Precipitation-Based Plastic Recycling: Case Study

One good example of a solvent-based DP process was reported by Uekert et al. (2023), in which polyethylene terephthalate (PET) was recycled using benzyl alcohol as a solvent and methanol as an anti-solvent. A solvent DP plant of 150 metric tons per annum (mpta) capacity enabled the recycled PET production. The minimum selling point of the recycled PET was \$0.87/kg, while the PET derived from fossil fuel showed a minimum selling point of \$1.19/kg. The process indicated higher GHG emissions with lower total energy demand compared to fossil fuel-based PET production. Thus, in all the metrics, the solvent DP process is not the winner; to choose between polymer manufacturing and recycling, economic and environmental performance must be compared using life-cycle and techno-economic analysis. For instance, mechanical recycling displays the best economics, PET glycolysis provides the best environmental performance, and solvent DP displays the best recyclate material qualities.

6.3.4 Feedstock Recycling

6.3.4.1 Plastic Conversion to Monomers and Repolymerization

Plastic conversion into monomers aimed to address plastic waste while promoting sustainability, typically involving breaking large polymers conventionally through thermal decomposition (heating from 220–500°C). The process starts with depolymerization, in which the plastic waste (as feed) is subjected to a process of breakdown to obtain its monomers. Once the monomer is recovered, it undergoes further purification to remove any impurities and additives that were part of the initial waste plastic. This is crucial to get a good quality of the monomer. Several processes are used in this step, whereby the polymer is broken down using thermal, chemical, and catalytic processes. When decomposition occurs, radicals form at the site of broken bonds, which leads to de-polymerization. Monomer recovery is a key challenge in the de-polymerization process. For example, in the case of poly(α-methylstyrene), the recovery is excellent at 95%; however, for polymers like polyethylene and polyproline, monomer recovery is as poor as 0.025–2% (2005). In PVC, thermal decomposition becomes more challenging with the release of HCL acid and vinylenes. To mitigate the challenges of thermal decomposition, several methods, like reversible-deactivation radical polymerization (RDRP) (Corrigan et al., 2020), mechanical recovery, chemical recycling, catalytic process, and enzymatic degradation are adopted to break down plastics into monomers (Kristensen et al., 2024). Selection of the best method is guided by several metrics that involve energy, economic viability, and meeting of regulatory frameworks (Coates and Getzler, 2020). For instance, the depolymerization and recovery of the building blocks (or monomers) from polyurethane (PU) and epoxy polymer is demonstrated (Kristensen et al., 2024) (see Figure 6.3). These two polymers are the largest thermoset polymers. The recovery of pyrrole from the PU involved several routes such as solvolysis, acidolysis, aminolysis, and split-phase glycolysis. Further, to break the epoxy polymer's molecular skeleton, chemical catalysis is found to be the best approach and enables bisphenol A (BPA) recovery through C—O bond cleavage (see Figure 6.4). However, efforts are still in progress to evaluate the suitability of monomer recovery strategies at the commercial level.

Another possible route is to produce virgin plastic from plastic waste. The process is completed in a closed loop and can result in a virgin polymer. For example, the PET is reportedly depolymerized into its monomers, which can then be purified (Hopewell et al., 2009). The purified monomers can then be repolymerized through any polymerization method, such as condensation or addition polymerization, to form a repolymerized virgin plastic. Advances in catalysis and reaction

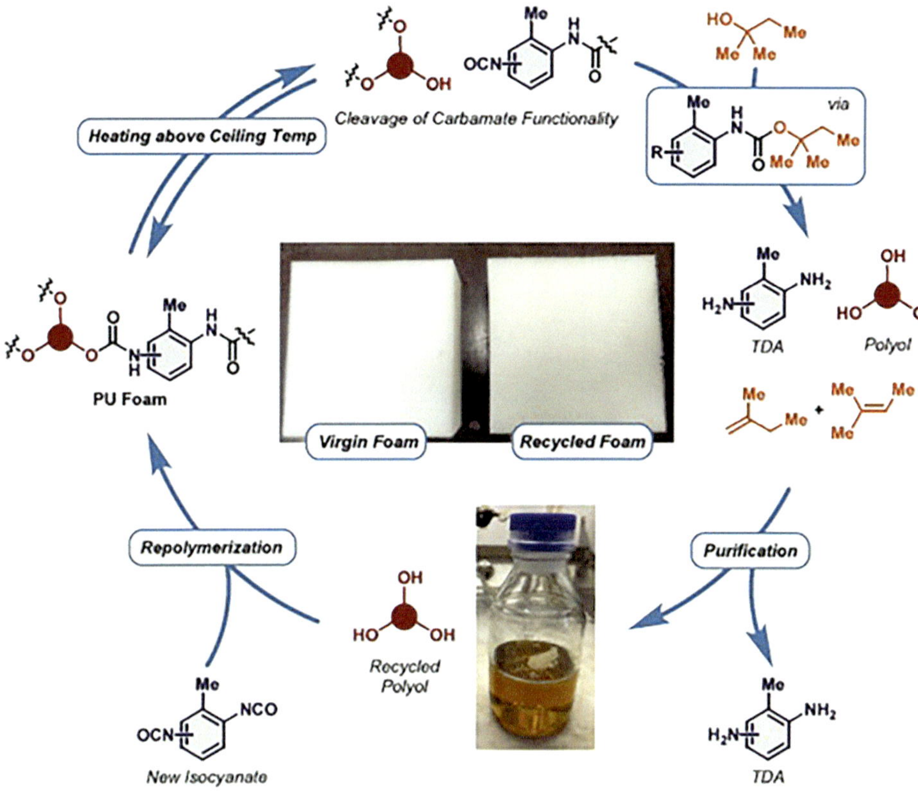

FIGURE 6.3 Recovery of pyrol and bisphenol A from PU and expoxy polymer, respectively.

conditions are making it possible to achieve high yields and purities in the repolymerization process (Wimberger et al., 2024). Since 2005, the United Kingdom (UK) has been utilizing the PET sheet for thermoforming, containing 50–70 % recycled PET (rPET). Table 6.2 assembles information on the formation of virgin plastics from waste plastics (Perugini et al., 2005). Further, food-grade rPET is also available to be utilized. However, it merely exhibits a minor reduction in transparency when compared to virgin PET and is utilized as a 30–50% substitute for virgin PET in numerous applications and 100% of the material in some PET bottles.

It is worthwhile to note that the employment of monomer recovery and repolymerization technologies faces several challenges. The complexity and variability of plastic waste streams can complicate the recycling process, and the presence of additives and contaminants can degrade the quality of recycled materials. Additionally, scaling these technologies to meet the global demand for recycled plastics poses significant logistical and economic hurdles (Hopewell et al., 2009). A continual effort is still in progress to enhance the employability of such techniques and decrease the cost factors associated with monomer recycling and repolymerization. Innovations in sorting and processing technologies, as well as improved catalytic systems for chemical recycling, will be crucial in enhancing the viability of plastic waste valorization on a larger scale (Hopewell et al., 2009; Wimberger et al., 2024).

6.3.4.2 Integration with Petroleum

Integration of plastic production with petroleum refining can enhance plant efficiency and reduce the refining cost by using the by-products of oil processing to produce plastics. This integration can impart great flexibility in the plant to adjust the process for fuel of plastic based on the market need.

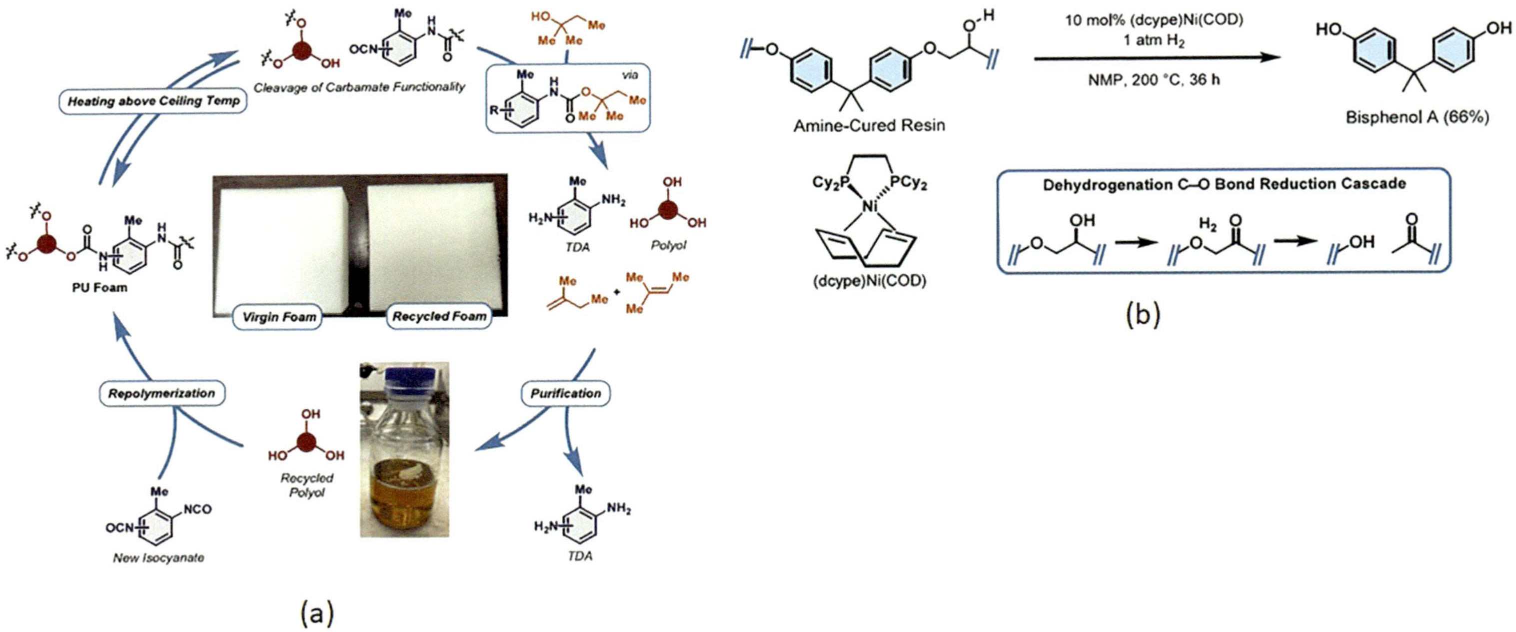

FIGURE 6.4 (a) Recovery of expoxy polymer (polyol) from polyurethane using a closed-loop recycling approach. (b) Formation of bisphenol A through deconstruction of epoxy polymer with amine resin. (From Johansen et al., 2023; Kristensen et al., 2024; open access article.)

TABLE 6.2
Environmental Influence of Polymers and Ability for Plastic Recycling

Polymers	Energy (GJ/ton)	Water (kL/ton)	CO Equivalent (t/ton)	Usage (kton)	Closed-Loop Recycling	Effectiveness in Current Recycling Process
PET	82.7	66	3.4	2160	Yes	High with clear PET from bottles Colored PET is mostly used for fiber Additional issues with CPET trays, PET-G
HDPE	76.7	32	1.9	5468	Some	High with natural HDPE bottles, but more complex for opaque bottles and trays because of wide variety of grades and colors and mixtures with LDPE and PP
PVC	56.7	46	1.9	6509	Some	Poor recovery because of cross-contamination with PET PVC packages and labels present a major issue with PET bottle and mixed plastics recycling
LDPE	78.1	47	2.1	7899	Some	Poor recovery rates, mostly as mixed polyolefins that can have sufficient properties for some applications Most post-consumer flexible packaging not recovered
PP	73.4	43	2	7779	In theory	Not widely recycled yet from post-consumer, but has potential Needs action on sorting and separation, plus development of further outlets for recycled PP
PS	87.4	140	3.1	2600	In theory	Poor, extremely difficult to cost effectively separate from comingled collection, separate collection of industrial packaging and EPS foam can be effective
Recycled plastic	55.08	3.5	1.4	3130	Some	Considerable variability in energy, water, and emissions from recycling processes as it is a developing industry and affected by efficiency of collection, process type, and product mix, etc.

Source: Perugini et al., 2005.

The cracking of plastic waste and its co-cracking with petroleum residue and other used lubricant oil is also an important integration step with refinery operations. Co-cracking brings numerous benefits, like enhanced feedstock utilization and higher hydrocarbon content that provide higher product yields. For instance, co-cracking of the residual fractions of the distillation column with waste plastics can be beneficial for the energy, process economics, and environmental points of view (Sergeyev et al., 2024). The addition of PET to the atmospheric residue showed a positive effect on the yield of light fractions and the destruction of resin-asphaltene substances (RAS) in comparison with the cracking of atmospheric residue (AR) only, impacting both the quality and the quantity of resulting coke. Further, the addition of polypropylene (PP) to the AR reduced the asphaltene content by 9.8-fold. However, PP enabled 1.7 times higher coke formation than PET. Thus, blending plastic with petroleum provides a balanced HC profile with higher-quality fuel besides mitigating otherwise discarded plastic waste. Integration, along with benefits, also brings challenges like complex reaction pathways, which provide difficulty in predicting yields and product distributions. An effective catalyst to work with the varied nature of plastic waste is necessary.

6.3.5 Direct Reuse of Recycled Polymers

Direct reuse of recycled polymers refers to the process of taking post-consumer or post-industrial plastic materials, cleaning and processing them, and then using them directly in the production of new products without extensive chemical alteration (Shen and Worrell, 2024). This method emphasizes the practical and immediate reuse of plastic materials, minimizing the need for new raw materials and reducing the environmental footprint. The direct reuse of recycled polymers involves several key aspects and challenges.

- **Mechanical Recycling**
  Mechanical recycling is the most common method for reusing polymers. This process involves collecting, sorting, and reprocessing plastic waste into new products. However, the quality of the recycled material can be inferior to that of virgin polymers, which limits its applications. Advances in technologies for collection, sorting, and reprocessing are improving the efficiency and quality of mechanically recycled plastics. For instance, improving sorting and separation within recycling plants can enhance both the volume and the quality of recycled resins (Hopewell et al., 2009; Šišková et al., 2021).
- **Chemical Recycling**
  Chemical recycling offers a more promising approach by breaking down polymers into their monomers or smaller molecules, which can then be used to produce new, high-quality materials. This method can overcome some of the limitations of mechanical recycling, such as the degradation of material properties. Light-driven approaches to chemical recycling are being explored to reduce energy consumption and achieve selective transformations that are not possible with heat-driven methods. These methods can produce monomers or valuable small molecules, contributing to a more circular economy (Wimberger et al., 2024).

The direct reuse of recycled polymers faces challenges such as contamination, additives, and the complexity of real-world plastic waste streams. Efficient pre-sorting and removal of additives like plasticizers, flame retardants, and colorants are crucial for producing high-quality recycled materials. Despite these challenges, there are opportunities for significant environmental and economic benefits. Recycling reduces the need for virgin materials, lowers energy consumption, and decreases greenhouse gas emissions. It also offers cost savings compared to using 100% virgin polymers.

Using recycled polymers is economically advantageous as it reduces the cost of raw materials and energy required for production. It also contributes to environmental sustainability by reducing waste, conserving resources, and lowering the carbon footprint associated with producing virgin polymers. Life-cycle assessments are necessary to quantify the overall environmental impacts of different recycling technologies and ensure that the methods used are sustainable (Wimberger et al., 2024).

6.4 CONCLUSIONS AND PERSPECTIVES

This chapter presents various technologies of plastic waste valorization. Traditional plastic waste management methods, namely, landfilling and incineration, continue to dominate but pose significant environmental and public health risks. Landfilling remains prevalent due to inadequate recycling infrastructure and low public awareness, yet it contributes to long-term ecological harm through the leaching of toxic substances and persistence of non-degradable materials. Similarly, incineration emits hazardous pollutants, such as VOCs and heavy metals, which can lead to severe health consequences, especially for nearby populations. On the other hand, mechanical recycling remains attractive in supporting a circular plastic economy because of no or minimal chemical alteration. However, material degradation, unsuitability for mixed plastic streams, and the compensating quality of the recycled limit the employability of mechanical recycling. Hence, plastic valorization using advanced techniques such as chemical and solvent-based recycling, along with feedstock recovery and direct reuse, signifies a crucial step toward achieving a circular economy. Further, technologies like pyrolysis, gasification, and dissolution–precipitation show promise due to their operational ease, minimal sorting needs, and potential for high material recovery. Chemical recycling techniques appeared to be more beneficial than mechanical recycling because of the production of fuels and useful chemicals from plastic waste. Meanwhile, biological recycling and integration with petroleum-refining operations could result in positive economics and also convert to valuable end products. Plastic waste valorization can also be achieved through monomer recovery, re-polymerization, and the use of recycled plastics. Despite the progress, there exist substantial technical, economic, and environmental challenges for the large-scale deployment of such technologies. It is advisable to use an integrated strategy combining multiple recycling routes for desired product quality and economic feasibility. In this context, building pilot-to-industrial scale plastic recycling plants with integrated systems for personalized local plastic waste will be critical. Further, the corporations of the public and private sectors can boost the deployment and technology adoption. More investigations into polymer designs, catalysts, selective solvents, and facile processes may improve the suitability of plastic valorization technologies. Further, the robust enticements for using recycled plastic, restrictions on difficult-to-recycle plastics, and grants for developing technologies can change public perspectives and industrial performance. Moreover, the regulatory framework and policies for the use of plastic products remain essential.

REFERENCES

Abd Karim, S.B., Norman, S., Koting, S., Simarani, K., Loo, S.C., Mohd Rahim, F.A., Ibrahim, M.R., Md Yusoff, N.I., Nagor Mohamed, A.H., 2023. Plastic roads in Asia: Current implementations and should it be considered? *Materials (Basel)* 16.

Al-Mosawi, Ali I., Abdulsada, S.A., Rijab, M.A., 2017. Synthesis effect of natural/chemical blend nanoparticles on impact strength of sustainable composite material. *Journal of Catalyst & Catalysis* 4.

Al-Salem, S.M., Antelava, A., Constantinou, A., Manos, G., Dutta, A., 2017. A review on thermal and catalytic pyrolysis of plastic solid waste (PSW). *Journal of Environmental Management* 197, 177–198.

Al-Salem, S.M., Lettieri, P., Baeyens, J., 2009. Recycling and recovery routes of plastic solid waste (PSW): A review. *Waste Management* 29, 2625–2643.

Amasuomo, E., Baird, J., 2016. The concept of waste and waste management. *Journal of Management and Sustainability* 6, 88–96.

Anwar, M.A., Suprihatin, Sasongko, N.A., Najib, M., Pranoto, B., 2024. Challenges and prospects of multilayer plastic waste management in several countries: A systematic literature review. *Case Studies in Chemical and Environmental Engineering* 10, 100911.

Aparício, S.C., Castro, P.M., Ribeiro, B.D., Marrucho, I.M., 2024. Greening the physical recycling of HDPE: Dissolution precipitation with natural solvents. *Green Chemistry* 26, 6799–6811.

Arabeyyat, O.S., Shatnawi, N., Shbool, M.A., Shraah, A.A., 2024. Landfill site selection for sustainable solid waste management using multiple-criteria decision-making. Case study: Al-Balqa governorate in Jordan. *MethodsX* 12, 102591.

Balakrishnan, G., Déniel, M., Nicolai, T., Chassenieux, C., Lagarde, F., 2019. Towards more realistic reference microplastics and nanoplastics: Preparation of polyethylene micro/nanoparticles with a biosurfactant. *Environmental Science: Nano* 6, 315–324.

Bhargava, S., Chen Lee, S.S., Min Ying, L.S., Neo, M.L., Lay-Ming Teo, S., Valiyaveettil, S., 2018. Fate of nanoplastics in marine larvae: A case study using barnacles, Amphibalanus amphitrite. *ACS Sustainable Chemistry & Engineering* 6, 6932–6940.

Bockhorn, H., Hentschel, J., Hornung, A., Hornung, U., 1999. Environmental engineering: Stepwise pyrolysis of plastic waste. *Chemical Engineering Science* 54, 3043–3051.

Chen, G.-Q., Patel, M.K., 2012. Plastics derived from biological sources: Present and future: A technical and environmental review. *Chemical Reviews* 112, 2082–2099.

Coates, G.W., Getzler, Y.D.Y.L., 2020. Chemical recycling to monomer for an ideal, circular polymer economy. *Nature Reviews Materials* 5, 501–516.

Corrigan, N., Jung, K., Moad, G., Hawker, C.J., Matyjaszewski, K., Boyer, C., 2020. Reversible-deactivation radical polymerization (Controlled/living radical polymerization): From discovery to materials design and applications. *Progress in Polymer Science* 111, 101311.

Czajczyńska, D., Anguilano, L., Ghazal, H., Krzyżyńska, R., Reynolds, A.J., Spencer, N., Jouhara, H., 2017. Potential of pyrolysis processes in the waste management sector. *Thermal Science and Engineering Progress* 3, 171–197.

D'Hooge, D.R., Van Steenberge, P.H.M., Derboven, P., Reyniers, M.-F., Marin, G.B., 2015. Model-based design of the polymer microstructure: Bridging the gap between polymer chemistry and engineering. *Polymer Chemistry* 6, 7081–7096.

Dogu, O., Pelucchi, M., Van de Vijver, R., Van Steenberge, P.H.M., D'Hooge, D.R., Cuoci, A., Mehl, M., Frassoldati, A., Faravelli, T., Van Geem, K.M., 2021. The chemistry of chemical recycling of solid plastic waste via pyrolysis and gasification: State-of-the-art, challenges, and future directions. *Progress in Energy and Combustion Science* 84, 100901.

Evode, N., Qamar, S.A., Bilal, M., Barceló, D., Iqbal, H.M.N., 2021. Plastic waste and its management strategies for environmental sustainability. *Case Studies in Chemical and Environmental Engineering* 4, 100142.

Faravelli, T., Bozzano, G., Colombo, M., Ranzi, E., Dente, M., 2003. Kinetic modeling of the thermal degradation of polyethylene and polystyrene mixtures. *Journal of Analytical and Applied Pyrolysis* 70, 761–777.

Gilbert, M., 2017. Chapter 1—Plastics materials: Introduction and historical development, in: Gilbert, M. (Ed.), *Brydson's Plastics Materials* (Eighth Edition). Butterworth-Heinemann, pp. 1–18.

Hopewell, J., Dvorak, R., Kosior, E., 2009. Plastics recycling: Challenges and opportunities. *Philosophical Transactions of the Royal Society B: Biological Sciences* 364, 2115–2126.

Johansen, M.B., Donslund, B.S., Larsen, E., Olsen, M.B., Pedersen, J.A.L., Boye, M., Smedsgård, J.K.C., Heck, R., Kristensen, S.K., Skrydstrup, T., 2023. Closed-loop recycling of polyols from thermoset polyurethanes by tert-amyl alcohol-mediated depolymerization of flexible foams. *ACS Sustainable Chemistry & Engineering* 11, 10737–10745.

Johnson, L.M., Mecham, J.B., Krovi, S.A., Moreno Caffaro, M.M., Aravamudhan, S., Kovach, A.L., Fennell, T.R., Mortensen, N.P., 2021. Fabrication of polyethylene terephthalate (PET) nanoparticles with fluorescent tracers for studies in mammalian cells. *Nanoscale Advances* 3, 339–346.

Jung, H., Shin, G., Kwak, H., Hao, L.T., Jegal, J., Kim, H.J., Jeon, H., Park, J., Oh, D.X., 2023. Review of polymer technologies for improving the recycling and upcycling efficiency of plastic waste. *Chemosphere* 320, 138089.

Karanam, K., Okoffo, E.D., Shukla, P., Thomas, K.V., Rudolph, V., 2024. Preparation of polystyrene microplastic particles by solvent-dissolution-precipitation. *Sustainable Chemistry for the Environment* 7, 100144.

Kibria, M.G., Masuk, N.I., Safayet, R., Nguyen, H.Q., Mourshed, M., 2023. Plastic waste: Challenges and opportunities to mitigate pollution and effective management. *International Journal of Environmental Research* 17, 20.

Kognole, R.S., Shipkule, K., Patil, M., Patil, L., Survase, U., 2019. Utilization of plastic waste for making plastic bricks. *International Journal of Trend in Scientific Research and Development* 3, 878–880.

Kristensen, S.K., Ahrens, A., Donslund, B.S., Skrydstrup, T., 2024. Perspective on the development of monomer recovery technologies from plastics designed to last. *ACS Organic & Inorganic Au* 4, 373–386.

Krueger, M.C., Harms, H., Schlosser, D., 2015. Prospects for microbiological solutions to environmental pollution with plastics. *Applied Microbiology and Biotechnology* 99, 8857–8874.

Lange, J.-P., 2021. Managing plastic waste—Sorting, recycling, disposal, and product redesign. *ACS Sustainable Chemistry & Engineering* 9, 15722–15738.

Moharir, R.V., Gautam, P., Kumar, S., 2019. Chapter 4—Waste treatment processes/technologies for energy recovery, in: Kumar, S., Kumar, R., Pandey, A. (Eds.), *Current Developments in Biotechnology and Bioengineering*. Elsevier, pp. 53–77.

Mong, G.R., Tan, H., Chin Vui Sheng, D.D., Kek, H.Y., Nyakuma, B.B., Woon, K.S., Othman, M.H.D., Kang, H.S., Goh, P.S., Wong, K.Y., 2024. A review on plastic waste valorisation to advanced materials: Solutions and technologies to curb plastic waste pollution. *Journal of Cleaner Production* 434, 140180.

Narancic, T., Cerrone, F., Beagan, N., O'Connor, K.E., 2020. Recent advances in bioplastics: Application and biodegradation. *Polymers* 12, 920.

Narro, G. Martínez, Pozos Vázquez, C., Mercado González, M.O., 2019. Viscosity reduction of heavy crude oil by dilution with hydrocarbons obtained via chemical recycling of plastic wastes. *Petroleum Science and Technology* 37, 1347–1354.

Nayanathara Thathsarani Pilapitiya, P.G.C., Ratnayake, A.S., 2024. The world of plastic waste: A review. *Cleaner Materials* 11, 100220.

Pappa, G., Boukouvalas, C., Giannaris, C., Ntaras, N., Zografos, V., Magoulas, K., Lygeros, A., Tassios, D., 2001. The selective dissolution/precipitation technique for polymer recycling: A pilot unit application. *Resources, Conservation and Recycling* 34, 33–44.

Patel, A.D., Schyns, Z.O.G., Franklin, T.W., Shaver, M.P., 2024. Defining quality by quantifying degradation in the mechanical recycling of polyethylene. *Nature Communications* 15, 8733.

Pathak, G., Nichter, M., Hardon, A., Moyer, E., Latkar, A., Simbaya, J., Pakasi, D., Taqueban, E., Love, J., 2023. Plastic pollution and the open burning of plastic wastes. *Global Environmental Change* 80, 102648.

Perugini, F., Mastellone, M.L., Arena, U., 2005. A life cycle assessment of mechanical and feedstock recycling options for management of plastic packaging wastes. *Environmental Progress* 24, 137–154.

Ragaert, K., Delva, L., Van Geem, K., 2017. Mechanical and chemical recycling of solid plastic waste. *Waste Management* 69, 24–58.

Rahimi, A., García, J.M., 2017. Chemical recycling of waste plastics for new materials production. *Nature Reviews Chemistry* 1, 0046.

Rodríguez, E., Gutiérrez, A., Palos, R., Vela, F.J., Arandes, J.M., Bilbao, J., 2019. Fuel production by cracking of polyolefins pyrolysis waxes under fluid catalytic cracking (FCC) operating conditions. *Waste Management* 93, 162–172.

Rogers, K., Jaspers, I., 2024. ToxPoint: Waste incineration management of plastic materials—an issue of increasing global public health importance. *Toxicological Sciences*, kfae111.

Sergeyev, N.S., Sviridenko, N.N., Urazov, K.K., 2024. Co-cracking of atmospheric residue and plastic waste. *Journal of Analytical and Applied Pyrolysis* 179, 106422.

Shah, M., Rajhans, S., Pandya, H.A., Mankad, A.U., 2021. Bioplastic for future: A review then and now. *World Journal of Advanced Research and Reviews* 9, 56–67.

Shamsuyeva, M., Endres, H.-J., 2021. Plastics in the context of the circular economy and sustainable plastics recycling: Comprehensive review on research development, standardization and market. *Composites Part C: Open Access* 6, 100168.

Sharuddin, S.D. Anuar, Abnisa, F., Wan Daud, W.M.A., Aroua, M.K., 2016. A review on pyrolysis of plastic wastes. *Energy Conversion and Management* 115, 308–326.

Shen, L., Worrell, E., 2024. Chapter 31—Plastic recycling, in: Meskers, C., Worrell, E., Reuter, M.A. (Eds.), *Handbook of Recycling* (Second Edition). Elsevier, pp. 497–510.

Silva, A.B., Costa, M.F., Duarte, A.C., 2018. Biotechnology advances for dealing with environmental pollution by micro(nano)plastics: Lessons on theory and practices. *Current Opinion in Environmental Science & Health* 1, 30–35.

Singh, N., Hui, D., Singh, R., Ahuja, I.P.S., Feo, L., Fraternali, F., 2017. Recycling of plastic solid waste: A state of art review and future applications. *Composites Part B: Engineering* 115, 409–422.

Singh, N., Walker, T.R., 2024. Plastic recycling: A panacea or environmental pollution problem. *npj Materials Sustainability* 2, 17.

Šišková, A.O., Peer, P., Andicsová, A.E., Jordanov, I., Rychter, P., 2021. Circulatory management of polymer waste: Recycling into fine fibers and their applications. *Materials* 14, 4694.

Suzuki, G., Uchida, N., Tuyen, L.H., Tanaka, K., Matsukami, H., Kunisue, T., Takahashi, S., Viet, P.H., Kuramochi, H., Osako, M., 2022. Mechanical recycling of plastic waste as a point source of microplastic pollution. *Environmental Pollution* 303, 119114.

Tiwari, R., Azad, N., Dutta, D., Yadav, B.R., Kumar, S., 2023. A critical review and future perspective of plastic waste recycling. *Science of the Total Environment* 881, 163433.

Tran, H.-V., Dang, T.T., Nguyen, N.H., Tran, H.T., Nguyen, D.T., Do, D.V., Le, T.S., Ngo, T.H., Late, Y.K.E., Amaniampong, P.N., Fletcher, E., Hung, T.Q., Cheng, Y., Nguyen, T.-K., Tran, T.S., Zhang, J., An, H., Nguyen, N.-T., Trinh, Q.T., 2025. Methanol activation: Strategies for utilization of methanol as C1 building block in sustainable organic synthesis. *ChemSusChem* 18, e202401974.

Trinh, Q.T., Banerjee, A., Ansari, K.B., Dao, D.Q., Drif, A., Binh, N.T., Tung, D.T., Binh, P.M.Q., Amaniampong, P.N., Huyen, P.T., Le, M.T., 2020. Upgrading of bio-oil from biomass pyrolysis: Current status and future development, in: Nanda, S., N. Vo, D.-V., Sarangi, P.K. (Eds.), *Biorefinery of Alternative Resources: Targeting Green Fuels and Platform Chemicals*. Springer Singapore, Singapore, pp. 317–353.

Trinh, Q.T., Nguyen, A.V., Huynh, D.C., Pham, T.H., Mushrif, S.H., 2016. Mechanistic insights into the catalytic elimination of tar and the promotional effect of boron on it: First-principles study using toluene as a model compound. *Catalysis Science & Technology* 6, 5871–5883.

Turner, R.P., Kelly, C.A., Fox, R., Hopkins, B., 2018. Re-formative polymer composites from plastic waste: Novel infrastructural product application. *Recycling* 3, 54.

Uekert, T., Singh, A., DesVeaux, J.S., Ghosh, T., Bhatt, A., Yadav, G., Afzal, S., Walzberg, J., Knauer, K.M., Nicholson, S.R., Beckham, G.T., Carpenter, A.C., 2023. Technical, economic, and environmental comparison of closed-loop recycling technologies for common plastics. *ACS Sustainable Chemistry & Engineering* 11, 965–978.

Van Camp, N., Lase, I.S., De Meester, S., Hoozée, S., Ragaert, K., 2024. Exposing the pitfalls of plastics mechanical recycling through cost calculation. *Waste Management* 189, 300–313.

Verma, R., Vinoda, K.S., Papireddy, M., Gowda, A.N.S., 2016. Toxic pollutants from plastic waste- A review. *Procedia Environmental Sciences* 35, 701–708.

Vuppaladadiyam, S.S.V., Vuppaladadiyam, A.K., Sahoo, A., Urgunde, A., Murugavelh, S., Šrámek, V., Pohořelý, M., Trakal, L., Bhattacharya, S., Sarmah, A.K., Shah, K., Pant, K.K., 2024. Waste to energy: Trending key challenges and current technologies in waste plastic management. *Science of the Total Environment* 913, 169436.

Wimberger, L., Ng, G., Boyer, C., 2024. Light-driven polymer recycling to monomers and small molecules. *Nature Communications* 15, 2510.

Wojnowska-Baryła, I., Bernat, K., Zaborowska, M., 2022. Plastic waste degradation in landfill conditions: The problem with microplastics, and their direct and indirect environmental effects. *International Journal of Environmental Research and Public Health* 19.

Yabu, H., Higuchi, T., Ijiro, K., Shimomura, M., 2005. Spontaneous formation of polymer nanoparticles by good-solvent evaporation as a nonequilibrium process. *Chaos: An Interdisciplinary Journal of Nonlinear Science* 15.

Yang, Y., Yang, J., Wu, W.-M., Zhao, J., Song, Y., Gao, L., Yang, R., Jiang, L., 2015. Biodegradation and mineralization of polystyrene by plastic-eating mealworms: Part 1. Chemical and physical characterization and isotopic tests. *Environmental Science & Technology* 49, 12080–12086.

Yip, Y.J., Lee, S.S.C., Neo, M.L., Teo, S.L.-M., Valiyaveettil, S., 2022. A comparative investigation of toxicity of three polymer nanoparticles on acorn barnacle (Amphibalanus amphitrite). *Science of the Total Environment* 806, 150965.

Zhou, C., Fang, W., Xu, W., Cao, A., Wang, R., 2014. Characteristics and the recovery potential of plastic wastes obtained from landfill mining. *Journal of Cleaner Production* 80, 80–86.

Zhu, J., Ruan, B., You, Q., Guo, J., Dai, X., Xiang, Y., 2018. Terahertz imaging sensor based on the strong coupling of surface plasmon polaritons between PVDF and graphene. *Sensors and Actuators B: Chemical* 264, 398–403.

7 Sustainable Treatment and Management of Plastic Waste

*Natarajan Baskar[1], Pragya Rai[1], Shivanand Pai[1], and Chiranjeevi Thota[1]**
[1]Bharat Petroleum Corporation Limited, Udyog Kendra, Greater Noida, India

7.1 INTRODUCTION

Plastic waste pollution has emerged as one of the most pressing environmental issues of the modern era (Awasthi et al., 2017). With global plastic production exceeding 390 million tons annually (Sahin & Kirim, 2018), the accumulation of non-biodegradable plastics in landfills and natural ecosystems poses significant ecological and health risks. These synthetic materials, valued for their durability, lightweight nature, and versatility, have become a cornerstone of human convenience. Plastic pollution originates from various sources, including single-use plastics like bags, bottles, and packaging. These items often end up in the environment due to improper disposal. Plastics are non-biodegradable and can continue to stay in the environment for centuries (Hopewell et al., 2009). This long lifespan means that plastic waste accumulates over time, creating significant pollution problems. Plastics break down into tiny particles called microplastics, which are found in all types of ecosystems such as land, oceans, rivers, and even the air atmosphere. These particles can enter the food chain and pose risks to both wildlife and humans. Chemicals in plastics can leach into food and water, potentially causing health issues. Microplastics have been found in human organs, raising concerns about their long-term effects (Verma et al., 2016). Further, increased greenhouse gas emission is observed due to the production and improper disposal of plastics. For example, plastic products were responsible for 3.4% of global greenhouse gas emissions in 2019, as per a UN report. Therefore, it is important to find solutions for the treatment and management of waste plastics in a sustainable manner.

Sustainable management and treatment of waste plastics is critical to reduce the environmental impacts of plastic pollution and ensure a circular economy. With plastics accumulating in landfills, oceans, and natural ecosystems, managing and treating them sustainably is essential to mitigate the environmental damage caused by their disposal. Sustainable management involves not just handling plastic waste but also preventing its generation, optimization recycling systems, and finding innovative solutions for its reuse and recovery.

Sustainable treatment of plastic waste focuses on minimizing environmental impacts, conserving resources, and promoting long-term ecological health. The goal is to minimize the extent of plastic waste ending up in landfills, incinerators, or the natural environment while also recovering valuable resources for reuse. The sustainable treatment methods include reducing, reusing, recycling, upcycling, and exploring alternative materials. Sustainable management of plastic waste involves a combination of reducing plastic production, improving recycling and upcycling technologies, adopting biodegradable alternatives, and implementing robust waste collection and sorting systems. A circular economy approach, supported by policy changes, consumer behaviour shifts, and technological innovations, is essential to mitigate the global plastic waste challenge effectively.

* Corresponding author's email: chiranjeevit@bharatpetroleum.in

DOI: 10.1201/9781003581000-10

Reducing plastic waste, reusing materials, and ensuring that plastics are recycled or repurposed help to minimize the impact on the environment and preserve resources. These strategies aim to shift away from single-use plastics and to encourage the use of recycled or repurposed plastic products to achieve the circular economy. Sustainable treatment of plastic waste primarily focuses on the "reduce, reuse, recycle" approach, which involves minimizing plastic consumption, finding ways to reuse plastic items instead of discarding them, and recycling plastic waste into new products whereever possible. Additionally, exploring biodegradable plastic options and advanced chemical recycling techniques can contribute to sustainable plastic waste management.

This chapter discusses the sustainable treatment and management methods for plastic waste.

7.2 CONVENTIONAL METHODS FOR PLASTIC WASTE MANAGEMENT

7.2.1 Landfilling of Plastic Waste

The predominant conventional approach for management of plastic waste is landfilling, as 79% of waste plastics end up in landfills (Geyer et al., 2017). It comprises burying plastics in designated areas called landfills, where the plastic waste is managed under controlled conditions. In the landfilling process, plastic waste is collected from residential, commercial, and industrial sources and transported to landfill sites. In some cases, recyclable materials are separated before the remaining waste is sent to landfills. Further, to prevent contamination of the ecosystem, such as soil and water, landfill sites are constructed with liners and drainage systems.

The main advantage of landfilling is lower expense than other waste management methods like recycling or the incineration process; it requires basic infrastructure and is easy to set up and operate. While it is a widely used practice due to its simplicity and cost-effectiveness, the landfilling of plastic waste has significant environmental and societal implications. Plastics do not decompose naturally, leading to long-term environmental challenges; waste plastics can remain intact for hundreds to thousands of years (Webb et al., 2013). Toxic substances from plastics, including additives and chemicals, can leach into the surrounding soil and groundwater. This contamination poses risks to human health and ecosystems. Further, landfills require large areas, which can disrupt natural habitats and compete with other land uses. Over time, the availability of suitable landfill sites diminishes. Organic materials in mixed waste decompose anaerobically in landfills, producing methane, a potent greenhouse gas. Although plastics themselves do not produce methane, their presence contributes to waste pile longevity. Landfills can become eyesores and emit unpleasant odours if not properly managed. Plastics in landfills can be consumed by animals, leading to health problems or fatalities. Over time, plastics break into tiny particles called microplastics, which can infiltrate soil and water systems. Proximity to poorly managed landfills can expose communities to harmful chemicals and pollutants. While landfilling remains a widely practiced method for managing plastic waste, it is not an environmentally friendly long-term solution. The persistence of plastics in landfills causes serious environmental and health challenges. Efforts to reduce, recycle, and find alternatives to plastic are essential to minimize reliance on landfilling and create a more sustainable waste management system. Considering the hazards posed by landfilling, the incineration process is being widely adopted (Evode et al., 2021), which not only reduces the burden on existing landfills and but also recovers energy from plastic waste.

7.2.2 Incineration of Plastic Waste

Incineration is a widely adopted method for managing plastic waste, by burning it at elevated temperatures (between 870°C and 1200°C) in controlled environments (Evode et al., 2021). This process is often used as an alternative to landfilling, particularly when space is limited or energy recovery is desired. While the incineration process reduces the amount of plastic waste, it also raises environmental and health concerns. In a typical process, plastic waste is collected and may undergo basic sorting to remove recyclable or hazardous materials. The plastic waste is shredded into smaller pieces to improve combustion efficiency. The shredded plastics are burned in specially designed furnaces called incinerators at higher temperatures. Heat generated during this process can be used to generate steam, which runs turbines to produce electricity. This is known as waste-to-energy (WTE) process. While it can lower the volume of plastic waste and produce electricity, it has several setbacks. The use of incineration for managing plastic waste has adverse environmental impacts due to the emission of highly toxic substances, primarily CO_2, which contribute to atmospheric pollution through air contamination (Jeswani et al., 2021). Modern incinerators are equipped with scrubbers and filters to reduce the release of harmful gases and particulates. Residual ash, containing non-combustible materials, is collected and typically disposed of in landfills or repurposed for construction. Incineration saves space and provides energy but has higher environmental risks due to emissions; further, incineration is a better option for non-recyclable or contaminated plastics. Incineration is a valuable tool for managing plastic waste, particularly in waste-to-energy systems. However, it must be implemented with advanced technologies and strict regulations to minimize its environmental and health impacts. Long-term solutions should focus on reducing plastic consumption, improving recycling rates, and investing in sustainable alternatives to plastics. The typical incineration process is explained in Figure 7.1.

While landfilling and incineration can help manage plastic waste, it is not a sustainable long-term solution. Landfills occupy vast spaces and release microplastics and toxic leachate into soil and water, while incineration adds greenhouse gases and air pollution to environments. As the world grapples with this mounting challenge, sustainable management and treatment of waste plastic have emerged as a critical pathway to mitigate pollution, conserve resources, and transition towards a circular economy. Emphasizing reduction, reuse, and recycling, along with developing a circular economy, are more effective and sustainable strategies for addressing plastic pollution. There are several effective and sustainable alternatives to conventional management of plastic waste. Some of the processes that can be adopted for sustainable management and treatment of waste plastics are discussed below.

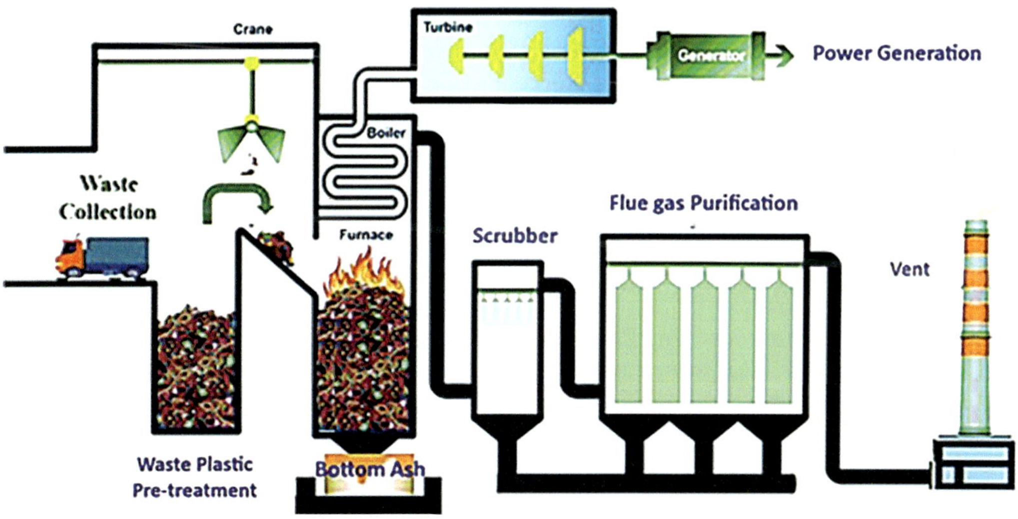

FIGURE 7.1 Schematic representation of the incineration process for waste plastic management.

7.3 SUSTAINABLE MANAGEMENT AND TREATMENT OF PLASTIC WASTE

Sustainable management and treatment of waste plastics is critical to reduce the environmental impacts of plastic pollution and ensure a circular economy (Evode et al., 2021). With plastics accumulating in landfills, oceans, and natural ecosystems, managing and treating them sustainably is essential to mitigate the environmental damage caused by their disposal. Sustainable management involves not just handling plastic waste but also preventing its generation, optimizing recycling systems, and finding innovative solutions for its reuse and recovery.

7.3.1 Recycling

Recycling is a process where plastic waste can be transformed into valuable and usable items such as toys, furniture, and bags (Ángeles-Hurtado et al., 2021). Recycling plastic waste offers numerous environmental, economic, and social benefits. Firstly, it helps reduce the burden on landfills, where plastics take hundreds of years to decompose. Secondly, it decreases pollution in the ecosystem, preventing harm to wildlife and marine life. Thirdly, recycling reduces the demand for new plastic production, which implies the reduction in carbon footprint and greenhouse gas emissions. By recycling, industries can decrease their dependency on non-renewable resources and lower their carbon emissions. Implementing circular economy policies through the recycling of plastic waste is crucial to mitigate the adverse effects of pollution (Sahin & Kirim, 2018).

Furthermore, recycling plastic waste supports the economy by creating job opportunities in waste collection, sorting, processing, and manufacturing. Governments and businesses investing in efficient recycling systems can boost economic growth while promoting environmental sustainability. Additionally, recycling helps conserve energy since producing new plastic from recycled materials requires significantly less energy compared to manufacturing virgin plastic.

The plastic recycling process consists of several essential steps:

Collection and sorting: Plastic waste is collected from different sources, like households, businesses, and recycling centres. It is then sorted based on type, such as polyethylene (PE), polystyrene (PS), polypropylene (PP), polyethylene terephthalate (PET), and polyvinyl chloride (PVC). Proper segregation is critical to maintain the quality of recycled plastics.

Cleaning and shredding: The segregated plastics undergo cleaning to remove contamination, through washing with solvents or water and drying. The cleaned plastic is then shredded into small pieces to facilitate further processing.

Melting and reprocessing: The shredded plastic is melted and reformed into raw materials, such as plastic pellets or fibres. These materials are then used to manufacture new plastic products, such as bottles, containers, and textiles.

Manufacturing new products: The processed plastic is integrated into various industries to create recycled sustainable goods, hence reducing the need for virgin plastic, and promoting a circular economy.

Despite its benefits, plastic recycling faces several challenges. One major issue is contamination, as mixed plastics and residual waste can lower the quality of recycled materials. Additionally, not all plastics are recyclable; certain types, such as multilayered packaging and polystyrene, require advanced recycling technologies. The substantial expenses associated with recycling infrastructure, coupled with inadequate waste management systems in certain areas, significantly impede the effectiveness of recycling initiatives. However, the contaminated or mixed plastic waste can be processed into new products, including mechanical recycling for common plastic types and advanced chemical recycling for more complex plastics. A schematic representation of a typical plastic waste recycling process is shown in Figure 7.2.

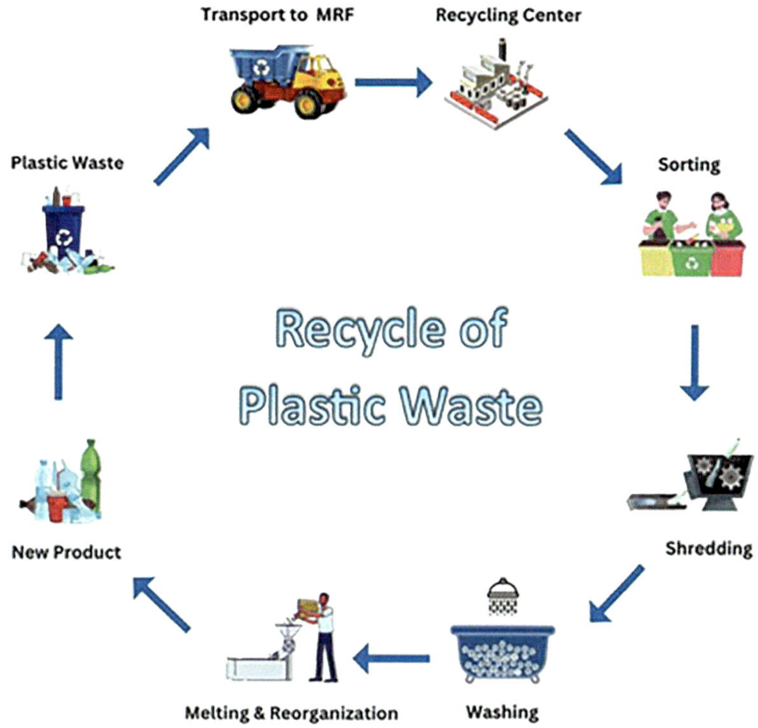

FIGURE 7.2 Schematic of a typical plastic waste recycling process.

7.3.1.1 Mechanical Recycling

Mechanical recycling of waste plastics is one of the most widely used methods for recycling plastic waste. Mechanical recycling, often known as secondary recycling, is a process for recovering plastic waste that allows for its reuse through mechanical methods (Al-Salem et al., 2009). It involves the physical processes of collecting, sorting, separating, cleaning, and reprocessing used plastics to produce new plastic products (Vilaplana & Karlsson, 2008). Mechanical recycling preserves the original polymer structure, allowing plastics to be reused directly in the manufacturing of new items (Hamad et al., 2013). This involves breaking down plastic waste into smaller pieces, which are then melted and remoulded into new products. It's a straightforward and effective way to recycle certain types of plastics. However, a decline in the properties of recycled plastics is a widespread issue across all plastic types, and it continues to be a significant challenge for the effective mechanical recycling process (Schyns & Shaver, 2020).

Plastic waste is collected from different sources and transported to a recycling facility. Here, it is sorted by type and colour to ensure proper processing and to produce high-quality recycled plastic. The sorted plastics are thoroughly cleaned with water and/or solvents to take off contaminants like food residue, labels, and dirt. The cleaned plastics are shredded into small flakes or pellets. This reduces the size of the plastic pieces, making them easier to manage and transport. The advantage of mechanical recycling is it diverts plastic waste from landfills and incinerators, reducing environmental pollution and preserving resources by reusing existing plastics instead of producing new ones. It uses less energy compared to producing new plastic from raw materials and creates, jobs and economic growth, and opportunities in the recycling industry. The limitation of the process is that plastics can only be mechanically recycled a limited number of times due to quality degradation because of the continuous increase of contaminants in recycled plastics, which can reduce, their performance and therefore requires clean sources of materials. Mechanical recycling is a basis for sustainable waste management, offering an effective and practical way to recycle plastic waste into valuable new products.

7.3.1.1.1 Types of Plastics Suitable for Mechanical Recycling

Mechanical recycling works best with thermoplastic plastics, which can be melted and reformed multiple times without substantial degradation of their properties. Examples of thermoplastics suitable for mechanical recycling include the following:

- *Polyethylene (PE)*: Used in plastic bags, bottles, and packaging.
- *Polypropylene (PP)*: Found in food containers, automotive parts, and household goods.
- *Polyethylene terephthalate (PET)*: Common in beverage bottles, textiles, and food containers.
- *Polystyrene (PS)*: Used in disposable cups, cutlery, and packaging.
- *Polyvinyl chloride (PVC)*: Found in pipes, flooring, and window profiles.

However, thermosetting plastics (such as epoxy resins and certain composite materials) are generally not suitable for mechanical recycling because they cannot be remelted and reshaped after being hardened. The mechanically recycled plastics are commonly used to make new packaging materials, including bottles, containers, and films. Many automotive manufacturers use recycled plastics to make parts like dashboards, bumpers, and interior components.

7.3.1.1.2 Innovations in Mechanical Recycling

- *Improved sorting technologies*: Advances in sorting technology, such as AI-powered optical sorters and infrared spectroscopy, help improve the purity and efficiency of the sorting process, ensuring higher-quality recycled plastics.
- *Closed-loop recycling systems*: Some companies are developing closed-loop systems for mechanical recycling, where the plastic is continually recycled into the same product type, improving the overall efficiency and sustainability of the process.
- *Recycling of mixed plastics*: Researchers are working on improving technologies that can process mixed or contaminated plastic waste, which would significantly expand the range of plastics suitable for mechanical recycling.

Mechanical recycling is a critical process for managing plastic waste, offering an environmentally friendly way to recycle plastics, reduce waste, and conserve resources. However, challenges related to quality degradation, contamination, and limited plastic types remain. Advances in sorting technologies, closed-loop systems, and innovations in recycling practices hold the potential to make mechanical recycling more efficient and effective, contributing to a more sustainable plastic economy.

7.3.1.2 Chemical Recycling

Chemical recycling presents a promising avenue for sustainable plastic management by converting waste plastic into valuable chemicals and fuels, offering a potential solution to plastic pollution while creating a circular economy for plastic materials. More valuable products like gaseous hydrocarbon, liquid oils, carbon black, hydrogen, and syngas are recovered through chemical recycling of plastics (Huang et al., 2022). Chemical recycling of waste plastic is an innovative approach that splits plastic waste into its basic molecular components, allowing it to be transformed into new, high-quality plastics or other valuable chemicals. This method presents a viable alternative to the environmental challenges posed by traditional plastic waste management techniques like landfilling and incineration.

TABLE 7.1
Summary of Various Processes for the Chemical Recycling of Waste Plastics

Process	Process Description	Typical Products
Pyrolysis	Thermal decomposition of plastics at elevated temperatures.	Pyrolysis oil, gas, wax
Catalytic hydrocracking	Usage of catalysis and hydrogen to reduce processing temperature and time.	Liquid oil, gas, aromatics
Co-pyrolysis	Involves more than a type of feedstock (e. g., plastics and biomass) in the pyrolysis process.	Bio-oils, organics
Gasification	Conversion of plastic wastes into gases through partial oxidation process.	Hydrogen-rich syngas
Depolymerization or solvolysis	Depolymerization of plastics using solvents. Typical reactions include alcoholysis, hydrolysis, glycolysis.	Monomers, low molecular weight products
Catalytic reforming	Production of syngas from plastic wastes using steam or CO.	Hydrogen-rich syngas, carbon nanotubes

Sources: Huang et al., 2022; Jiang et al., 2022. Adapted from Thew et al., 2023.

It can process a variety of mixed, contaminated, and multilayered plastics that pose challenges for mechanical recycling. The product from the chemical recycling process can be used in high-quality applications, such as food packaging. However, the high cost of these processes may limit their broader adoption. Also, some chemical recycling processes require significant energy input. Overall, chemical recycling represents a significant step towards a more sustainable and circular economy for waste plastics. Key processes for chemical recycling include pyrolysis, gasification, hydrocracking, and depolymerization. Table 7.1 shows various chemical recycling processes for waste plastic management.

7.3.2 Pyrolysis

Pyrolysis of waste plastics (Maqsood et al., 2021) is a process that involves the thermal decomposition of plastic materials in the absence of oxygen to break them down into useful products like fuels, gases, and chemicals (Sharuddin et al., 2016). This method is increasingly being explored to manage plastic waste and recover valuable resources, as well as reduce the environmental impact of plastic pollution (Maqsood et al., 2021). The pyrolysis process offers several benefits, including a reduction in CO_2 emissions, decreased reliance on landfills, accelerated commercialization of products, and the generation of materials that can be utilized for electricity production and heat generation.

In a typical pyrolysis process, plastic waste is heated to temperatures typically ranging from 300°C to 900°C (Sonawane et al., 2014). The heat causes the breaking down of long polymer chains in the plastics into smaller molecules. In the absence of oxygen, the plastic materials do not burn but instead undergo thermal decomposition. This results in the breaking of chemical bonds and the formation of various smaller molecules, which can be condensed into liquids or gases. A flow diagram of pyrolysis is shown in Figure 7.3 (Maqsood et al., 2021).

The key products of plastic pyrolysis include the following:

- *Liquid fuel*: Also known as pyrolysis oil or synthetic crude oil, it can be further refined into diesel or gasoline or used as a chemical feedstock.
- *Gaseous products*: These include methane, ethane, propane, butane, and other gases, which can be used as fuel or processed further.

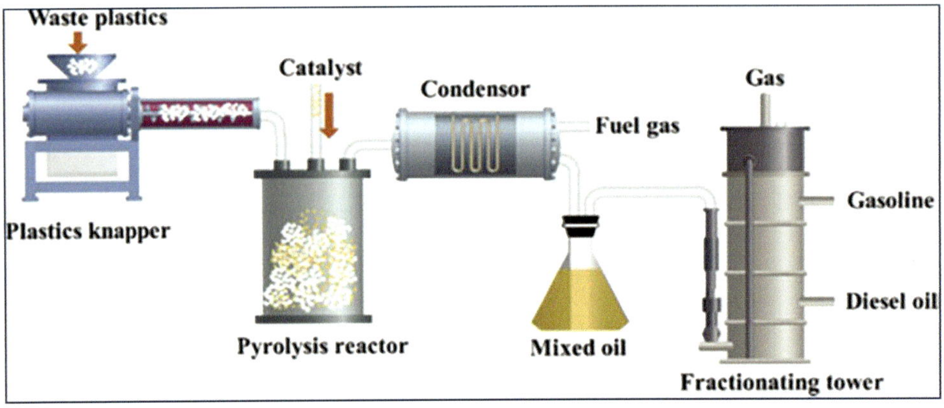

FIGURE 7.3 Schematic of a pyrolysis process for plastic waste. (Reprinted with permission, Maqsood et al., 2021.)

- *Solid residue*: Known as carbon black, it can be used in industrial applications or processed further for other uses, such as in the manufacturing of rubber or as an adsorbent material.

The pyrolysis process can also be applied to mixed plastics, which is especially useful for managing plastic waste that cannot be easily separated for recycling. The major advantage of waste plastic pyrolysis is that it helps divert plastic waste from landfills, which is a major environmental challenge. It brings down the volume of plastic waste and prevents it from entering oceans and ecosystems. The liquid and gaseous products from pyrolysis can be used as alternative fuels, reducing the dependence on fossil fuels. The process also generates valuable chemicals and carbon black for use in various industries. The product distribution of liquid oil, gases, wax, and char materials with respect to various plastic types in the pyrolysis process is shown in Figure 7.4 (Maqsood et al., 2021).

The gases produced during pyrolysis can be used as a source of energy to power the pyrolysis plant itself, making the process potentially self-sustaining. Unlike incineration, pyrolysis typically produces fewer harmful emissions, as the process occurs in a controlled environment with limited oxygen. This can result in lower levels of toxic substances like dioxins and furans. Pyrolysis of waste plastics fits into the concept of a circular economy by transforming waste materials back into usable products and energy. However, pyrolysis of waste plastic poses many challenges, such as the following:

- *Feedstock quality*: The quality and consistency of the plastic waste feedstock can vary, which may affect the efficiency and quality of the products. Mixed or contaminated plastics might require additional pre-processing to remove impurities.
- *Energy requirements*: Pyrolysis is an energy-intensive process, and while the gases produced can be used to generate heat for the system, the initial energy input required to start the process can be significant.
- *Economic viability*: The costs associated with setting up pyrolysis plants, including the collection, sorting, and processing of waste plastics, can be high. The economic success of the process often depends on the price of oil and the demand for the end products.
- *Environmental impact*: While pyrolysis produces fewer emissions compared to traditional incineration, the process still generates some air pollutants (e.g., VOCs, particulate matter) that need to be managed with appropriate pollution control systems.
- *Regulatory challenges*: Pyrolysis plants may face regulatory hurdles related to emissions, safety, and environmental impact, especially in regions with strict environmental standards.

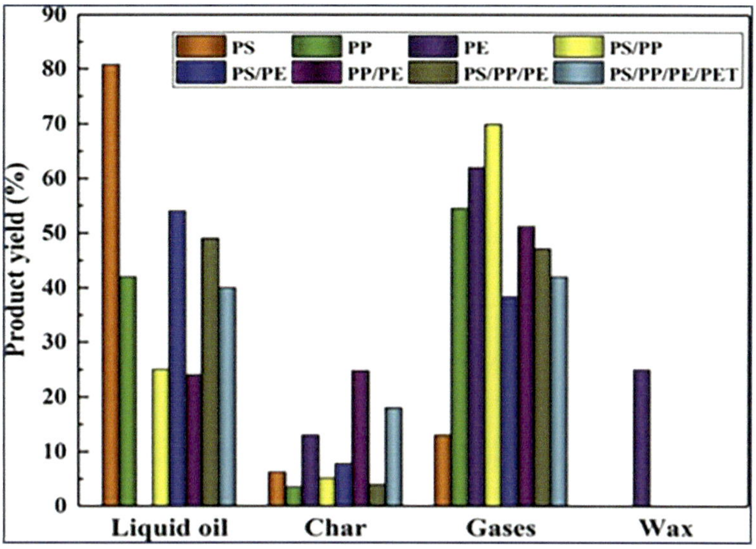

FIGURE 7.4 Product yield with respect to various plastic types. (Reprinted with permission, Maqsood et al., 2021.)

7.3.2.1 Applications of Pyrolysis Products

- *Fuel production*: The pyrolysis oil produced from waste plastic can be used as an alternative fuel for industrial applications, including power plants, or further refined into gasoline and diesel.
- *Chemical feedstock*: The oil can also be used as a raw material for producing valuable chemicals like aromatic compounds, which are used in the production of plastics, synthetic rubber, and pharmaceuticals.
- *Carbon black*: The solid residue, carbon black, has a range of industrial applications including production of tires, inks, coatings, and as a reinforcing agent in rubber manufacturing.
- *Sustainable plastics recycling*: The process can contribute to more sustainable plastic recycling practices, particularly for plastics that are difficult to recycle through traditional mechanical methods.

Research and development in pyrolysis technologies is rapidly advancing. Some notable trends include the following:

- *Integration with renewable energy*: Pyrolysis systems that are powered by renewable energy sources, such as solar or wind, are being explored to reduce the reliance on fossil fuel.
- *Advanced catalytic pyrolysis*: This method involves using catalysts to enhance the pyrolysis reaction, producing higher-quality fuels and chemicals with greater efficiency.
- *Small-scale, decentralized systems*: These systems aim to process plastic waste at local levels, reducing transportation costs and providing a solution for plastic waste management in specific regions.

The pyrolysis of waste plastics is a promising technology for managing plastic waste and recovering energy and valuable materials from it. While the process holds significant potential for waste reduction and resource recovery, it also faces challenges in terms of energy consumption, feedstock quality, and economic feasibility. Continued research and development, along with improvements in efficiency and pollution control technologies, can make pyrolysis a more viable and sustainable solution for addressing global plastic waste issues.

7.3.3 Gasification

Gasification is a process that decomposes plastic waste at high temperatures in the presence of oxygen. This method produces valuable by-products like syngas (mixture of CO and hydrogen), which is further used as raw materials for producing sustainable fuels and chemicals (Locaspi et al., 2023). Gasification of waste plastics is a process where it is converted into gaseous products through a high-temperature, oxygen-limited environment. Unlike pyrolysis, which produces liquid fuels and solid residues, gasification primarily focuses on producing carbon monoxide (CO), hydrogen (H_2), methane (CH_4), and other gases. These gases can be used further to produce chemicals, generate electricity, or processed into fuels.

In a typical gasification process, waste plastics, such as polyethylene (PE), polypropylene (PP), or polystyrene (PS), are sorted and sometimes broken into smaller pieces for efficient gasification. They are then fed into a gasifier, a high-temperature, oxygen-controlled chamber. In the gasifier, plastic materials are heated to temperatures ranging from 700°C to 1200°C in a controlled, limited-oxygen environment. The plastics undergo partial oxidation, a process where only a small amount of oxygen is used to break down the materials, instead of complete combustion, which would occur in the presence of more oxygen. At these high temperatures, the plastics undergo chemical reactions, breaking down into smaller molecules, forming gases like carbon monoxide (CO), hydrogen (H_2), methane (CH_4), and carbon dioxide (CO_2). Hydrogen, a clean fuel, can be separated from the gas mixture and used in fuel cells or as a feedstock in various industries. Solid residues, such as ash and char, are left behind. The syngas generated by gasification can be used in gas turbines or internal combustion engines to generate electricity. It can also be processed into chemicals (e.g., methanol, ammonia) or used to synthesize liquid fuels (like synthetic gasoline or diesel) through the FT process. Before the syngas is used, it often undergoes a cleaning process to remove impurities, such as particulates, sulfur, and nitrogen compounds, which can be harmful to engines or turbines. In some systems, the heat generated during the gasification process is captured and used to produce steam or electricity, improving the overall efficiency of the system. Gasification helps to divert plastic waste from landfills and incineration, reducing environmental pollution and the pressure on waste disposal systems. The process converts non-recyclable plastics into valuable gases (syngas), which can be used for energy production or chemical synthesis, providing a sustainable way to recover energy and materials from waste. Gasification, compared to traditional incineration, typically produces lower levels of harmful pollutants such as dioxins and furans. However, emission control is still needed to manage gases like CO and CO_2. The syngas can be converted into valuable chemicals or fuels, promoting a circular economy by transforming waste into usable products instead of letting it accumulate in landfills. Gasification requires significant energy input to reach the high temperatures needed for the reaction. The type and composition of plastic waste can vary, affecting the efficiency and consistency of the gasification process. Some plastics may require pretreatment, such as shredding or drying. The capital investment for a gasification plant can be high due to the specialized equipment and infrastructure needed. The economic feasibility is influenced by various factors, including the availability of plastic waste, energy costs, and the demand for syngas. Although gasification produces less emissions than incineration, the process still generates gases like carbon monoxide and carbon dioxide, which require careful management and monitoring.

Gasification of waste plastics appears to be a promising technology for addressing the growing plastic waste problem while also generating valuable resources like energy and chemicals. By converting non-recyclable plastics into syngas, this process can contribute to sustainable waste management and resource recovery. However, challenges related to energy input, feedstock variability, and cost remain, and further technological advancements and economic optimization are needed for its widespread adoption.

7.3.4 Hydrocracking

Hydrocracking of waste plastics (Munir et al., 2018) is a process that involves breaking down large polymer chains in plastic waste into smaller, more valuable hydrocarbons (such as naphtha, diesel, or gasoline) using hydrogen gas and a catalyst, typically at high temperatures and pressures (Venkatesh et al., 1995). This process is considered a type of hydrothermal treatment and is an important technique in the recycling and conversion of plastic waste into useful fuels and chemicals.

Hydrocracking is similar to the well-established process used in the petroleum industry to upgrade heavier crude oils into lighter, more desirable products. When applied to plastic waste, hydrocracking can produce fuels that can be used in internal combustion engines or in chemical industries, contributing to the circular economy of plastics.

7.3.4.1 Waste Plastic Hydrocracking

The waste plastics, such as polyethylene (PE), polypropylene (PP), polystyrene (PS), and polyethylene terephthalate (PET), are fed into a hydrocracking reactor and heated in the presence of high hydrogen pressure and temperature conditions (typically between 350°C and 450°C and at pressures up to 1500 psi) (Mosio-Mosiewski et al., 2007) and a catalyst (often a metal-based catalyst like nickel, molybdenum, or palladium). The plastics may need to be pretreated by cleaning and shredding to facilitate the process. During hydrocracking, the polymer chains in the plastics undergo hydrogenation (where hydrogen atoms are added to the molecules) and cracking (where the large polymer chains are broken into smaller hydrocarbons). This process results in the formation of lighter products, such as alkanes (saturated hydrocarbons), which can be further refined into liquid fuels or chemical feedstocks. After the hydrocracking process, the products are separated into various fractions, including light hydrocarbons (e.g., methane, ethane, propane); these can be used as gases or chemical feedstocks. Liquids (e.g., naphtha, diesel, gasoline) can be further refined into useful fuels. The residual solid (e.g., carbon) may be carbon-rich material, which can be used in other industrial applications such as producing needle coke or graphene. The liquid products from hydrocracking, such as diesel or gasoline, may need additional refining to meet quality standards for commercial use. This can include distillation, further hydrogenation, or other upgrading processes.

Hydrocracking produces high-quality fuels, such as diesel and gasoline, which are in high demand and can be used directly in engines and industrial applications. This makes it a valuable method for converting plastic waste into usable energy. Hydrocracking can process a wide range of mixed or contaminated plastics, including those that are difficult to recycle by conventional methods (such as mixed plastics, PVC, and multilayer packaging). The process is not as selective as some other methods, making it versatile for dealing with various waste plastics. The process can contribute to a circular economy by recycling plastic waste back into fuel products, reducing reliance on fossil fuels, and minimizing plastic waste. Further, the hydrogen used in the process can be produced from renewable sources, making hydrocracking potentially more sustainable compared to other methods that rely on fossil fuels.

The hydrocracking process requires high temperatures and pressures, making it energy-intensive. This can result in high operational costs, especially when hydrogen needs to be produced and supplied in massive quantities. The catalysts used in hydrocracking can degrade over time and need to be replaced, adding to the overall cost of the process. Additionally, catalysts can be sensitive to impurities in the plastic feedstock. Producing sufficient hydrogen at a reasonable cost can be a challenge, especially if the hydrogen is produced from non-renewable sources. The sustainability of the process depends on the ability to use green hydrogen (produced from renewable energy) to make it truly eco-friendly. The quality of the products can vary depending on the type of plastic used, the operating conditions, and the catalyst employed. Ensuring consistent quality in the final products can be a challenge. The presence of additives, dyes, or other contaminants in plastic waste can affect the efficiency of hydrocracking and the quality of the final products. Proper sorting and cleaning of the plastics may be required.

The liquid hydrocarbons produced during hydrocracking can be refined into fuels like diesel, gasoline, kerosene, and jet fuel, which can be used for transportation and industrial purposes. Some of the lighter hydrocarbons (such as naphtha) can be used as feedstocks in the petrochemical industry for producing a variety of chemicals (e.g., ethylene, propylene, and other intermediates). The syngas (gaseous hydrocarbons) generated in the process can be used to produce electricity or heat, providing energy for the process itself or other applications.

Researchers are working to develop more efficient, long-lasting, and less expensive catalysts that can improve the hydrocracking process, making it more economically viable. There is an increasing focus on using renewable hydrogen in hydrocracking to make the process more sustainable and reduce its carbon footprint. Several companies and research organizations are running pilot projects to demonstrate the feasibility of hydrocracking waste plastics at commercial scales. Some have integrated the process into waste-to-energy facilities, providing an efficient way to handle mixed plastic waste and generate value from it.

Hydrocracking of waste plastics offers a promising method for converting plastic waste into high-quality fuels and valuable chemical feedstocks (Munir et al., 2018). While it provides advantages in terms of versatility, product quality, and waste reduction, challenges such as high energy requirements, hydrogen supply, and catalyst management must be addressed to make it commercially feasible. With ongoing advancements in technology, hydrocracking could play a significant role in the recycling and upcycling of plastic waste, contributing to the global effort to reduce plastic pollution and dependence on fossil fuels.

7.3.5 Chemical Depolymerization

Depolymerization of waste plastic refers to the process of breaking off long polymer chains found in the plastics into its original monomers or oligomers (Kumar et al., 2011). This process aims to recycle plastic waste by converting it into useful raw materials, which can then be repolymerized into new plastics or used in other applications. Depolymerization is considered a key step in advancing the circular economy of plastics, as it offers a potential method for reducing plastic waste and producing higher-quality recycled products.

By converting plastic waste into reusable monomers or chemicals, depolymerization helps to address the growing plastic waste problem. The products of depolymerization can often be used to make new, high-quality polymers or plastics, unlike traditional recycling methods, which may degrade the quality of plastics. In this process, plastics are chemically depolymerized in the presence of some solvents at elevated temperature (Payne et al., 2019). Several chemical reactions are utilized in the depolymerization of plastics, the most widely commercialized methods being hydrogenolysis, methanolysis, and glycolysis. Typical depolymerization representation of PET via hydrolysis or trans-esterification is shown in Figure 7.5 (George & Kurian, 2014).

Chemical recycling techniques, including pyrolysis, gasification, hydrocracking, and chemical depolymerization, are viewed as effective strategies for managing plastic waste. Nonetheless, several challenges remain that must be addressed. Additional research is essential to improve the recovery, reusability, and longevity of catalysts in large-scale implementations (Jiang et al., 2022). Furthermore, the extreme conditions used in these chemical processes result in the release of harmful pollutants, such as sulfur, carbon, and various volatile organic compounds, into the environment (Evode et al., 2021).

7.3.6 Enzymatic Depolymerization

Enzymatic depolymerization of waste plastics is an emerging and sustainable method for addressing plastic pollution (Lu et al., 2024). This process uses enzymes, typically derived from

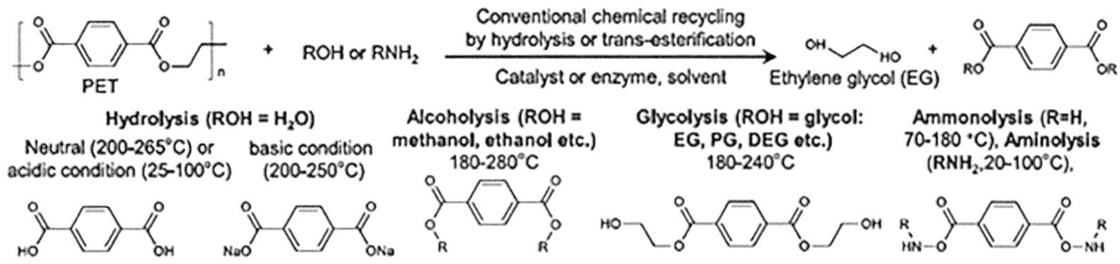

FIGURE 7.5 Depolymerization of PET via hydrolysis or trans-esterification. (From George & Kurian, 2014.)

microorganisms, to break down polymers in plastics into their monomeric building blocks, which can then be repurposed for producing new materials. Unlike traditional chemical or thermal recycling, enzymatic depolymerization operates at moderate conditions (lower temperatures and pressures), which subsequently reduces the energy consumption and minimizes production of harmful by-products.

Enzymatic depolymerization uses specific enzymes to break down the plastic polymer into its constituent monomers (Sun et al., 2024). Specific enzymes, often sourced from microbes naturally capable of degrading certain plastics, are introduced to plastic waste, where they catalyse within the polymer chains, effectively breaking down the plastic into smaller molecules like monomers or oligomers. For instance, enzymes such as hydrolases, esterases, and cutinases are employed to cleave ester bonds in synthetic polymers like polyethylene terephthalate (PET) and polyurethanes (PU) (Lee et al., 2023). Finally, the purified monomers can subsequently be utilized for the creation of new plastic or materials, thereby establishing a closed-loop recycling system.

One notable example is PETase, an enzyme discovered in *Ideonella sakaiensis*, a bacterium capable of degrading PET. PETase hydrolyses the ester bonds in PET, producing its monomer terephthalic acid and ethylene glycol, which can be directly reused for manufacturing new PET products (Tournier et al., 2020). Similarly, enzymes like urethanases and proteinases have been studied for their ability to degrade polyurethanes and polylactic acid (PLA), respectively.

Enzymatic depolymerization offers several advantages over traditional recycling methods: the enzymatic approach operates at moderate conditions, such as room temperature and neutral pH, thus reducing energy consumption and minimizing greenhouse gas emissions. Enzymes are highly selective, targeting specific bonds in the polymer chain without affecting other components. This reduces the formation of unwanted by-products and ensures high-purity outputs. The monomers obtained through enzymatic depolymerization can be re-polymerized to produce new plastics, enabling a circular economy. Enzymes have potential applications in breaking down microplastics in natural environments, addressing a critical aspect of plastic pollution (Lu et al., 2024). Depolymerization of plastic waste with various types of enzymes is shown in Figure 7.6.

Innovations in enzyme engineering leads the development of efficient and robust enzymes for plastic waste depolymerization process. Techniques like protein engineering and directed evolution have been employed to enhance enzyme stability, activity, and specificity. Additionally, integrating enzymatic depolymerization with chemical or physical pretreatments can improve substrate accessibility and overall process efficiency.

Despite its promise, enzymatic depolymerization faces several challenges: many natural enzymes exhibit slow reaction rates and low yields when degrading plastics, particularly those with high crystallinity or complex structures. Large-scale production of enzymes and their application in industrial processes remains costly compared to conventional recycling methods. The additives and fillers used in plastic manufacturing can hinder enzyme binding and activity, reducing the efficiency of depolymerization. While enzymatic depolymerization has shown success at laboratory scales, translating these results to industrial scales presents logistical and technical challenges. Enzymatic depolymerization is mostly in the research stage, but it shows promise for biodegradable plastics,

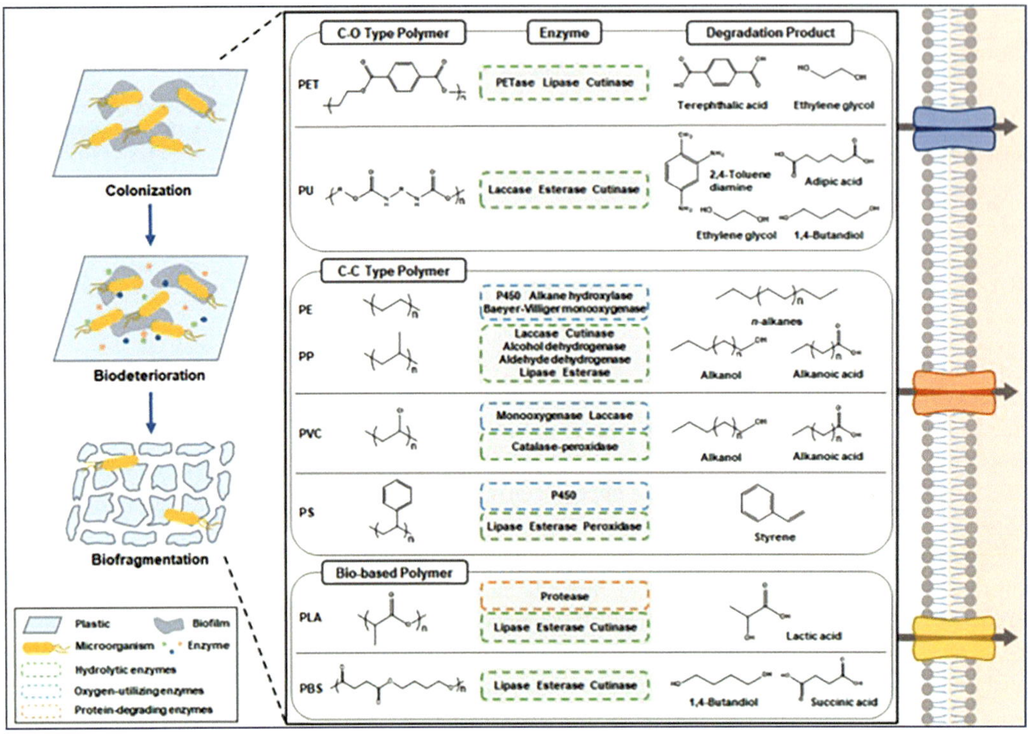

FIGURE 7.6 Depolymerization plastic waste with various types of enzymes. (Reprinted with permission, Lee et al., 2023.)

such as PLA, and even certain types of PET, where enzymes can catalyse the breakdown of plastic polymers.

Depolymerization processes, especially chemical and enzymatic methods, can be expensive and energy-intensive. Not all plastics are easily depolymerized. Some polymers, like PVC, are more resistant to depolymerization. Many depolymerization technologies are still in the research or pilot stages, and scaling them up for industrial applications is a significant challenge. Waste plastic can be contaminated with dyes, additives, or other materials, which can affect the efficiency and quality of depolymerization. Overall, depolymerization is a promising technology for reducing plastic waste, and ongoing research aims to make it more efficient, cost effective, and applicable to a wider variety of plastics.

Enzymatic depolymerization represents a transformative approach to tackle plastic waste, advancing a sustainable alternative to conventional recycling methods. By leveraging the power of enzymes, it is possible to break down plastics into their fundamental components, enabling circularity and reducing environmental impacts. While challenges remain in terms of efficiency, cost, and scalability, continued advancements in biotechnology and interdisciplinary collaboration hold the potential to make enzymatic depolymerization a cornerstone of future waste management and recycling strategies.

7.3.7 Bioremediation of Plastic Wastes

Bioremediation of waste plastics refers to the use of biological organisms, such as microorganisms (bacteria, fungi) or enzymes to break down or degrade plastic waste into less harmful or non-toxic substances (Asgher et al., 2020). This approach is an emerging area of research that seeks to provide a sustainable solution to the growing environmental problem of plastic pollution, especially the challenge of plastic waste that does not easily biodegrade in nature. Unlike traditional methods of plastic disposal, such as incineration or landfilling, bioremediation aims to harness the natural ability of certain organisms to break down synthetic polymers into simpler, biodegradable components, which could eventually be integrated into natural ecosystems.

7.3.7.1 Mechanisms of Bioremediation of Plastic Waste

Certain bacteria and fungi have evolved the ability to break down specific types of plastic. These microorganisms secrete enzymes that can degrade polymers like polyethylene, polystyrene, and PET, which are typically very resistant to degradation in natural environments. *Ideonella sakaiensis*, a bacterium, has been shown to break down PET plastic by producing enzymes that can degrade the polymer into its monomers, which the bacteria then metabolize. Fungi like *Aspergillus* and *Penicillium* are also known to degrade plastic, especially in environments where plastic materials have been exposed to moisture and air.

Certain enzymes, such as PETase (produced by *Ideonella sakaiensis*) and cutinase, have been isolated and studied for their ability to break down plastic polymers. These enzymes work by breaking the long chains of polymers into smaller, more manageable components (monomers or oligomers), which can then be further degraded by microorganisms or absorbed by plants. Fungi are known for their ability to decompose organic materials, and certain species have been found to degrade plastics by releasing various enzymes. For example, *Phanerochaete chrysosporium* is a white-rot fungus that produces lignin-degrading enzymes (Singh & Chen, 2008), which can also break down plastic polymers like PET and PVC.

Some bacteria (e.g., *Pseudomonas* and *Bacillus* species) can also metabolize hydrocarbons, breaking down plastic materials by using them as carbon sources for growth. Some others form biofilms on the surface of plastic materials, which helps them remain attached to the plastic and effectively degrade it. Biofilms consist of groups of microorganisms that are encased in a matrix of extracellular polymeric substances (EPS) that they produce themselves. This matrix not only offers protection to the microorganisms but also improves their capacity to degrade plastics.

7.3.7.2 Types of Plastics That Can Be Bioremediated

Polyethylene (PE): One of the most common plastics in the world, polyethylene is found in products like plastic bags, packaging materials, and plastic bottles. Certain bacteria and fungi have shown the ability to degrade low- and high-density polyethylene.

- *Polypropylene (PP)*: Polypropylene, used in items like food containers and automotive parts, is also a target for bioremediation. Some bacterial strains are known to break down polypropylene in landfills.
- *Polyethylene terephthalate (PET)*: PET, used in bottles, textiles, and food packaging, has received attention due to the discovery of bacteria like *Ideonella sakaiensis*, which that can degrade it. Research into enzymatic degradation of PET is also advancing.
- *Polystyrene (PS)*: Polystyrene, commonly found in disposable cups, plates, and packaging, can be degraded by certain microorganisms and fungi, though it is often more challenging to break down than other plastics.
- *Polyurethane (PU)*: Polyurethane plastics, used in insulation, foams, and coatings, can be decomposed by certain fungi, which can utilize them as a carbon and nitrogen source.

Bioremediation provides a potential solution for addressing plastic pollution without the need for landfilling or incineration, both of which contribute to environmental harm. Biodegradable products can return to the natural cycle and prevent long-term plastic accumulation in ecosystems. Compared to more traditional methods of waste disposal, bioremediation can be more cost effective, as it requires less energy and infrastructure. Microbial or enzymatic degradation can be done at a lower temperature and pressure, making the process potentially more affordable. Unlike incineration, which can release toxic chemicals and greenhouse gases, bioremediation avoids harmful emissions.

The products of plastic degradation through bioremediation process are typically non-toxic and can be safely absorbed into the environment. Bioremediation can reduce the accumulation of plastic waste in oceans, rivers, and landfills, contributing to cleaner environments and reducing the harmful effects of plastic on wildlife.

The natural degradation of plastics *via* microbes or enzymes can be a slow process, especially for highly durable plastics like polyethylene and polypropylene. While significant advances are being made, the scale of bioremediation needed to address global plastic pollution remains a challenge. Not all plastics are amenable to microbial or enzymatic degradation. Some plastics, such as PVC and polytetrafluoroethylene (PTFE), are extremely resistant to biodegradation. Additionally, some bioremediation processes may only work under specific environmental conditions (e.g., suitable temperature, moisture). Plastic waste is often made of composite materials, mixed plastics, or contains additives (e.g., dyes, stabilizers, plasticizers), which can complicate the degradation process. Microorganisms may need to be tailored or engineered to break down these complex materials. While laboratory-scale bioremediation studies are promising, scaling up the process to handle the vast quantities of plastic waste produced globally remains a challenge. The development of efficient methods for large-scale plastic degradation is still a key area of research.

Genetically engineered bacteria or fungi could degrade plastics more efficiently. This involves modifying microbial strains to enhance their plastic-degrading abilities or create entirely new strains capable of targeting a broader range of plastics. Advances in enzyme research, particularly for the enzymes such as PETase, may offer faster and more effective ways to break down plastics, especially PET, which is widely used in beverage bottles and textiles.

Bioremediation could be integrated into broader waste management systems, where microorganisms or enzymes break down plastics alongside other waste recycling processes. This could be especially useful for dealing with plastics that are difficult to recycle using conventional methods. However, challenges such as slow degradation rates, scalability, and the complexity of plastic materials need to be addressed. Ongoing research into genetic engineering, enzyme technology, and microbial processes holds the potential to improve the efficacy for bioremediation process and make it viable for large-scale solution to the plastic pollution crisis.

7.3.8 Biodegradable Plastics

Biodegradable plastics derived from renewable materials such as corn starch or sugarcane can help to mitigate the effects of plastic pollution. Plastics produced from natural resources such as corn and potato starch, cellulose, or PLA are commonly named as bio-based plastics (Moshood et al., 2022). These plastics break down more easily in natural environments. Biodegradable plastics are designed to decompose naturally in the environment through the action of microorganisms such as bacteria and fungi. Biodegradable plastics differ significantly from traditional plastics, which can remain in the environment for centuries. These eco-friendly alternatives decompose into natural elements such as water, carbon dioxide, and biomass in a considerably shorter period (Moshood et al., 2022). It is important to note that not all bioplastics are capable of degrading in natural environments. Plastics made from PLA, polyhydroxyalkanoates (PHA), cellulose, and starch are bio degradable. Additionally, certain plastics derived from petroleum-based monomers, including polycaprolactone (PCL), polybutylene succinate (PBS), and polybutylene adipate (PBA), show biodegradability (Narancic et al., 2020; Hosni et al., 2019).

One of the primary advantages of biodegradable plastics is their ability to reduce environmental pollution. Biodegradable plastics also contribute to reduction of greenhouse gas emissions. Many biodegradable plastics are produced from plant-based materials, which capture carbon dioxide during their growth, offsetting some of the emissions generated during production. Furthermore, their decomposition process often releases fewer harmful by-products compared to conventional plastics. However, the degradation rates of biodegradable plastics differ with respect to environmental condition, leading to the possible accumulation in the environment (Courtene-Jones et al., 2022).

One major issue is the need for specific environmental conditions for proper decomposition. Many biodegradable plastics require high temperatures and controlled conditions, which are typically found in industrial composting facilities but not in natural environments or landfills. This means that improper disposal can still lead to environmental persistence.

Recent advancements in biotechnology and materials science are driving innovations in biodegradable plastics. Scientists are exploring the use of novel feedstocks, such as algae, seaweed, and waste products, to produce biodegradable plastics with lower environmental footprints. Innovations in microbial engineering are also enabling the development of plastics that can degrade under a wider range of environmental conditions, including in marine environments. In this regard, one of the important aspects is to enhance the mechanical properties of biodegradable plastics to make them more durable and versatile. This progress could expand their applications in industries such as packaging, agriculture, and medical devices, making biodegradable plastics a viable alternative in a broader range of contexts. While biodegradable plastics alone cannot solve the plastic pollution crisis, they can play an important role in a broader strategy to achieve sustainability. Combined with efforts to reduce plastic consumption, improve recycling infrastructure, and raise public awareness about waste management, biodegradable plastics can contribute to a cleaner and healthier planet.

Governments, industries, and consumers all have a role to play in promoting the adoption of biodegradable plastics. Policymakers can implement regulations and incentives to encourage the use of sustainable materials, while industries can invest in research and development to improve the efficiency and affordability of biodegradable plastics. Consumers, on the other hand, can make informed choices by opting for eco-friendly products and properly disposing of biodegradable plastics. Educational campaigns and public awareness initiatives can further enhance the understanding and acceptance of biodegradable plastics.

7.3.9 Upcycling of Plastic Waste in Different Sectors

Upcycling of waste plastics is a creative and environmentally beneficial process where plastic waste is repurposed into higher-value materials or products with enriched functionality, aesthetic appeal, or utility. Unlike traditional recycling, which breaks down plastics into raw materials for creating similar products, the upcycling process transforms waste plastics into new, innovative products which often have a longer lifespan or additional value. This approach helps to reduce plastic waste landfills and reduces the demand for virgin plastic and fosters more sustainable consumption.

7.3.9.1 Plastic-Modified Bitumen for Road Construction

Plastic-modified bitumen (PMB) has emerged as a promising technology that addresses both plastic waste and sustainable road infrastructure challenges by utilizing waste plastic in road construction. This method not only enhances the performance of roads but also contributes to waste management, making it an eco-friendly and practical solution. The use of plastics in construction is a sustainable alternative to conventional waste disposal, minimizing environmental hazards and reducing dependence on virgin materials for road construction. PMB involves blending of processed waste plastic with bitumen. In this process, plastic waste is shredded into small pieces (2–4 mm) and mixed with a hot aggregate at high temperatures. The coated aggregate is then mixed with hot bitumen to form PMB. This modified material exhibits enhanced properties, making it superior to conventional bitumen.

The advantages of PMB include the following. It provides better resistance to deformation, cracking, and wear due to its enhanced binding properties. Roads constructed with PMB can withstand heavy traffic and extreme weather conditions. PMB increases the water-resistance capability of roads, reducing the risk of potholes caused by water infiltration. The process helps recycle non-biodegradable plastic waste, reducing environmental pollution and conserving landfill space. The use of plastic waste minimizes the amount of bitumen required for road construction, resulting in cost savings. Roads made with PMB have a higher softening point, allowing them to withstand high temperatures without melting or deforming.

PMB has been successfully implemented in various road construction projects worldwide. For example, India has pioneered this technology by constructing several kilometres of roads using waste plastics. Countries like the UK and South Africa are also exploring similar innovations to improve infrastructure sustainability. PMB is a revolutionary approach to sustainable infrastructure development and waste management. By integrating waste plastic into road construction, it offers a dual solution to two pressing issues: one is plastic pollution and other one is road durability. While challenges exist, the benefits of PMB far outweigh its limitations, making it an essential tool for building resilient, eco-friendly, and cost-effective road networks for the future.

India has pioneered the use of plastic waste in road construction. The first plastic road in India was constructed in 2002, which used waste plastics mixed with bitumen. Since then, many roads have been constructed using plastic waste. In India, guidelines have been developed by Indian Road Congress (IRC) for the use of plastic waste in road construction, and the practice has been adopted widely.

Using waste plastics in road construction is a promising solution to the twin challenges of plastic waste pollution and the need for sustainable infrastructure. By recycling waste plastics into durable, long-lasting roads, we can minimize the ecological impact of waste plastic, conserve natural resources, and improve road quality. However, challenges such as ensuring plastic quality, maintaining long-term durability, and meeting regulatory standards need to be addressed to make this technology more widespread. As more research is conducted and successful case studies emerge, plastic roads may become a standard in the future of road construction, offering a practical solution to a global problem.

7.3.9.2 Pillar Casting and Brick Moulding

Innovative solutions such as column casting and brick moulding using waste plastics have emerged, offering an eco-friendly and sustainable alternative for construction materials. This section explores the processes, benefits, and challenges of utilizing plastic waste in column casting and brick moulding. Plastic bottles have been used in construction projects, such as columns, bricks, fencing, warehouses, residential structures, and many more using waste plastics (Gu & Ozbakkaloglu, 2016). Column casting is a vital process in construction, where columns provide structural support to buildings and other structures. Traditional column casting involves materials like concrete, steel, and aggregates. However, incorporating waste plastics into this process presents a sustainable alternative.

The process starts with the collection and sorting of waste plastic, which is then shredded or melted into small particles. These plastic particles can replace fine or coarse aggregates in the concrete mix. Alternatively, melted plastic can be blended with binders to create composite materials that are poured into column moulds. After proper compaction and curing, the resulting columns are durable and sustainable. The integration of waste plastics in column casting offers numerous benefits. It reduces the dependency on natural aggregates like sand and gravel, conserving natural resources. Additionally, plastics enhance the flexibility and tensile strength of the concrete mix, resulting in more resilient structures. Moreover, the production costs for such columns are often lower, making them an economically viable option, especially in developing regions.

Brick moulding using waste plastics is yet another innovative approach to tackling the global plastic waste crisis. Unlike traditional bricks, which are made from clay and require energy-intensive processes such as firing, plastic bricks are produced through a simpler and more eco-friendly method. The process starts with collecting, cleaning, and shredding waste plastics. These shredded plastics are then blended with additives like sand, fly ash, or cement to improve strength and texture. The mixture is poured into moulds and compressed using hydraulic presses to achieve uniformity. Once cooled and set, the bricks are ready for use in construction.

Plastic bricks have numerous advantages over traditional bricks. They are lightweight, making them easier to transport and handle. They exhibit excellent resistance to water, chemicals, and cracking, which enhances their durability. The manufacturing process requires less energy, significantly reducing carbon emissions. Furthermore, plastic bricks are cost effective, as they utilize inexpensive waste materials. This innovation provides a sustainable solution for both housing and waste management, particularly in areas with high levels of plastic pollution.

Column casting and brick moulding using waste plastics represent a significant step towards sustainable construction and environmental conservation. By repurposing plastic waste, these methods provide a dual solution: addressing the global plastic waste crisis and creating cost-effective, durable construction materials. While challenges remain, continued innovation, standardization, and awareness can pave the way for the broader adoption of these eco-friendly practices.

7.3.9.3 Plastic Lumber

Plastic lumber, a versatile material made from recycled waste plastic, has emerged as a sustainable substitute for traditional wood. It is widely used in construction, landscaping, and other applications due to its durability, weather resistance, and eco-friendliness (Breslin et al., 1998; Herrera et al., 2023). By addressing environmental challenges such as plastic waste and deforestation, plastic lumber represents a forward-thinking solution for modern construction and resource management. Plastic lumber is a material manufactured from recycled plastics such as polyethylene, polypropylene, and polyvinyl chloride. Unlike traditional wood, plastic lumber is entirely synthetic, with no need for cutting down trees. Further, plastic lumber can be made into several shapes and sizes, mimicking the appearance of natural wood while offering enhanced performance properties.

Plastic lumber is highly durable, resistant to moisture, insects, and rot. It does not crack, splinter, or degrade easily, making it suitable for long-term use in outdoor applications.

Additionally, it eliminates the need for cutting trees, aiding in forest conservation. Unlike wood, plastic lumber requires minimal maintenance. It does not need painting, staining, or regular treatments to prevent decay or insect infestation. Plastic lumber can withstand any weather conditions, including intense sunlight, heavy rain, and freezing temperatures, without warping or losing structural integrity. Plastic lumber can be made into required sizes, and is available in a range of colours, making it suitable for diverse applications such as decking, fencing, park benches, and outdoor furniture. Even though the initial cost of plastic lumber making is higher than that of traditional wood, the long lifespan and low maintenance requirements of plastic lumber make it a cost-effective choice over time.

7.3.9.4 Textile Production: Fabric from Plastic Bottles

The process of creating textiles and fabrics from plastic waste primarily involves recycling PET, the plastic used in bottles and containers. Recycled PET fibres are now broadly used in the production of clothing, upholstery, and industrial materials, offering an eco-friendly alternative to virgin polyester and other conventional fabrics. The production of textiles and fabric from waste plastics involves many steps. A few of them are as follows. Plastic waste is collected from recycling centres, oceans, and landfills. PET is separated from other types of plastic for further processing. The collected PET is cleaned to remove labels, caps, and contaminants, then shredded into small flakes. The PET flakes are melted and extruded into fine threads or fibres. The recycled fibres are spun into yarn, which is woven or knitted into fabric. The resulting fabric is dyed and finished to meet aesthetic and functional requirements.

The major applications of textiles from plastic waste include recycled fabrics in activewear, jackets, and footwear. The resulting fabrics are lightweight, durable, and versatile. Recycled PET fibres are used in carpets, curtains, rugs, and cushion covers. rPET fabrics are ideal for tents, backpacks, and sleeping bags due to their strength and water resistance. These fabrics are increasingly utilized in automotive upholstery, insulation materials, and geotextiles. While the production of textiles from plastic waste is an important step towards sustainability, using waste plastics for textile production has many challenges. Washing synthetic fabrics releases microplastics, which can enter waterways and harm marine life. Although more sustainable than virgin polyester, the recycling process still requires energy and infrastructure. Recycled fibres may degrade over multiple recycling cycles, limiting their reuse potential. Many countries lack the infrastructure to effectively collect, sort, and process plastic waste for textile production.

The future of textiles from plastic waste lies in the advancement of technology development and expanding its applications. Innovations such as depolymerization of plastic waste, which breaks down plastics into their molecular components, can improve the quality and recyclability of fibres. Additionally, blending recycled PET fibres with natural fibres such as cotton or wool can create hybrid fabrics that combine sustainability with comfort and functionality. Continued research and development in biodegradable and recyclable materials will further enhance the environmental benefits of this approach.

7.4 CONCLUSIONS AND PERSPECTIVES

The sustainable management and treatment of plastic waste demand a comprehensive, integrated approach that combines prevention, innovation, recycling, and robust policy frameworks. As the global plastic waste crisis intensifies, it becomes increasingly clear that no single solution can address the complexity of the issue. Instead, a multifaceted strategy is essential, one that begins with designing products for circularity, continues through advanced recycling technologies, and culminates in systemic changes in production and consumption patterns. Design for sustainability plays a pivotal role in reducing plastic waste at the source. By considering the entire life cycle of plastic products, from raw material extraction to final disposal, manufacturers can minimize environmental impact and facilitate easier recycling or reuse. This proactive approach reduces dependency on virgin plastics and supports the transition to a circular economy, where materials are kept in use for as long as possible.

Closed-loop (plastic-to-plastic) recycling emerges as one of the most promising pathways to achieving this vision. It enables the continuous reuse of plastic materials without significant degradation in quality. However, realizing its full potential requires overcoming several challenges, including contamination of waste streams, insufficient recycling infrastructure, and limited market demand for recycled plastics. Addressing these barriers will require coordinated efforts among policymakers, industry leaders, researchers, and consumers. In parallel, upcycling offers a creative and impactful solution by transforming waste plastics into high-value products with enhanced functionality or aesthetics. This not only diverts plastics from landfills and oceans but also adds economic value to waste materials. While scalability and efficient sorting remain hurdles, the growing interest in upcycling, fuelled by innovation and consumer awareness, positions it as a vital component of sustainable plastic management. Eventually, the path forward hinges on collaboration and innovation. Governments must enact and enforce supportive policies, industries must invest in sustainable technologies, communities must embrace responsible consumption, and individuals must adopt eco-conscious behaviours. Although this is a collective effort, we can foster a future where plastic waste is not a burden but a resource managed wisely, reused creatively, and minimized effectively.

REFERENCES

Al-Salem, S. M., Lettieri, P., & Baeyens, J. (2009). Recycling and recovery routes of plastic solid waste (PSW): A review. *Waste Management*, *29*(10), 2625–2643. https://doi.org/10.1016/j.wasman.2009.06.004

Ángeles-Hurtado, L., Rodríguez-Reséndiz, J., Salazar-Colores, S., Torres-Salinas, H., & Sevilla-Camacho, P. Y. (2021). Viable disposal of Post-Consumer polymers in Mexico: A review. *Frontiers in Environmental Science*, *9*. https://doi.org/10.3389/fenvs.2021.749775

Asgher, M., Qamar, S. A., Bilal, M., & Iqbal, H. M. (2020). Bio-based active food packaging materials: Sustainable alternative to conventional petrochemical-based packaging materials. *Food Research International*, *137*, 109625. https://doi.org/10.1016/j.foodres.2020.109625

Awasthi, A. K., Shivashankar, M., & Majumder, S. (2017). Plastic solid waste utilization technologies: A review. *IOP Conference Series Materials Science and Engineering*, *263*, 022024. https://doi.org/10.1088/1757-899x/263/2/022024

Breslin, V. T., Senturk, U., & Berndt, C. C. (1998). Long-term engineering properties of recycled plastic lumber used in pier construction. *Resources Conservation and Recycling*, *23*(4), 243–258. https://doi.org/10.1016/s0921-3449(98)00024-x

Courtene-Jones, W., Rodríguez, A. M., & Handy, R. D. (2022). From microbes to ecosystems: A review of the ecological effects of biodegradable plastics. *Emerging Topics in Life Sciences*, *6*(4), 423–433. https://doi.org/10.1042/etls20220015

Evode, N., Qamar, S. A., Bilal, M., Barceló, D., & Iqbal, H. M. (2021). Plastic waste and its management strategies for environmental sustainability. *Case Studies in Chemical and Environmental Engineering*, *4*, 100142. https://doi.org/10.1016/j.cscee.2021.100142

George, N., & Kurian, T. (2014). Recent developments in the chemical recycling of postconsumer poly(ethylene terephthalate) waste. *Industrial & Engineering Chemistry Research*, *53*(37), 14185–14198. https://doi.org/10.1021/ie501995m

Geyer, R., Jambeck, J. R., & Law, K. L. (2017). Production, use, and fate of all plastics ever made. *Science Advances*, *3*(7). https://doi.org/10.1126/sciadv.1700782

Gu, L., & Ozbakkaloglu, T. (2016). Use of recycled plastics in concrete: A critical review. *Waste Management*, *51*, 19–42. https://doi.org/10.1016/j.wasman.2016.03.005

Hamad, K., Kaseem, M., & Deri, F. (2013). Recycling of waste from polymer materials: An overview of the recent works. *Polymer Degradation and Stability*, *98*(12), 2801–2812. https://doi.org/10.1016/j.polymdegradstab.2013.09.025

Herrera, J. P., Bedoya-Ruiz, D., & Hurtado, J. E. (2023). Recycled Plastic Lumber walls for one and two-story housing: An assessment of their seismic performance. *Journal of Building Engineering*, *65*, 105822. https://doi.org/10.1016/j.jobe.2023.105822

Hopewell, J., Dvorak, R., & Kosior, E. (2009). Plastics recycling: Challenges and opportunities. *Philosophical Transactions of the Royal Society B Biological Sciences*, *364*(1526), 2115–2126. https://doi.org/10.1098/rstb.2008.0311

Hosni, A. S. A., Pittman, J. K., & Robson, G. D. (2019). Microbial degradation of four biodegradable polymers in soil and compost demonstrating polycaprolactone as an ideal compostable plastic. *Waste Management*, *97*, 105–114. https://doi.org/10.1016/j.wasman.2019.07.042

Huang, J., Veksha, A., Chan, W. P., Giannis, A., & Lisak, G. (2022). Chemical recycling of plastic waste for sustainable material management: A prospective review on catalysts and processes. *Renewable and Sustainable Energy Reviews*, *154*, 111866. https://doi.org/10.1016/j.rser.2021.111866

Jeswani, H., Krüger, C., Russ, M., Horlacher, M., Antony, F., Hann, S., & Azapagic, A. (2021). Life cycle environmental impacts of chemical recycling via pyrolysis of mixed plastic waste in comparison with mechanical recycling and energy recovery. *Science of the Total Environment*, *769*, 144483. https://doi.org/10.1016/j.scitotenv.2020.144483

Jiang, J., Shi, K., Zhang, X., Yu, K., Zhang, H., He, J., Ju, Y., & Liu, J. (2022). From plastic waste to wealth using chemical recycling: A review. *Journal of Environmental Chemical Engineering*, *10*(1), 106867. https://doi.org/10.1016/j.jece.2021.106867

Kumar, S., Panda, A. K., & Singh, R. (2011). A review on tertiary recycling of high-density polyethylene to fuel. *Resources Conservation and Recycling*, *55*(11), 893–910. https://doi.org/10.1016/j.resconrec.2011.05.005

Lee, G. H., Kim, D., Jin, Y. H., Kim, S. M., Lim, E. S., Cha, M. J., Ko, J. K., Gong, G., Lee, S., Um, Y., Han, S. O., & Ahn, J. H. (2023). Biotechnological plastic degradation and valorization using systems metabolic engineering. *International Journal of Molecular Sciences*, *24*(20), 15181. https://doi.org/10.3390/ijms242015181

Locaspi, A., Pelucchi, M., & Faravelli, T. (2023). Towards a lumped approach for solid plastic waste gasification: Polystyrene pyrolysis. *Journal of Analytical and Applied Pyrolysis*, *171*, 105960. https://doi.org/10.1016/j.jaap.2023.105960

Lu, D., Wu, J., Jin, S., Wu, Q., Wang, F., Deng, L., & Nie, K. (2024). Dual-Enzyme-Cascade catalysis for PET biodegradation based on a Variable-Temperature program. *Catalysts*, *14*(8), 543. https://doi.org/10.3390/catal14080543

Maqsood, T., Dai, J., Zhang, Y., Guang, M., & Li, B. (2021). Pyrolysis of plastic species: A review of resources and products. *Journal of Analytical and Applied Pyrolysis*, *159*, 105295. https://doi.org/10.1016/j.jaap.2021.105295

Moshood, T. D., Nawanir, G., Mahmud, F., Mohamad, F., Ahmad, M. H., & AbdulGhani, A. (2022). Biodegradable plastic applications towards sustainability: A recent innovations in the green product. *Cleaner Engineering and Technology*, *6*, 100404. https://doi.org/10.1016/j.clet.2022.100404

Mosio-Mosiewski, J., Warzala, M., Morawski, I., & Dobrzanski, T. (2007). High-pressure catalytic and thermal cracking of polyethylene. *Fuel Processing Technology*, *88*(4), 359–364. https://doi.org/10.1016/j.fuproc.2006.10.009

Munir, D., Irfan, M. F., & Usman, M. R. (2018). Hydrocracking of virgin and waste plastics: A detailed review. *Renewable and Sustainable Energy Reviews*, *90*, 490–515. https://doi.org/10.1016/j.rser.2018.03.034

Narancic, T., Cerrone, F., Beagan, N., & O'Connor, K. E. (2020). Recent advances in bioplastics: Application and biodegradation. *Polymers*, *12*(4), 920. https://doi.org/10.3390/polym12040920

Payne, J., McKeown, P., & Jones, M. D. (2019). A circular economy approach to plastic waste. *Polymer Degradation and Stability*, *165*, 170–181. https://doi.org/10.1016/j.polymdegradstab.2019.05.014

Sahin, O., & Kirim, Y. (2018). 2.31 material recycling. In *Elsevier eBooks* (pp. 1018–1042). https://doi.org/10.1016/b978-0-12-809597-3.00260-1

Schyns, Z. O. G., & Shaver, M. P. (2020). Mechanical Recycling of Packaging Plastics: A review. *Macromolecular Rapid Communications*, *42*(3). https://doi.org/10.1002/marc.202000415

Sharuddin, S. D. A., Abnisa, F., Daud, W. M. a. W., & Aroua, M. K. (2016). A review on pyrolysis of plastic wastes. *Energy Conversion and Management*, *115*, 308–326. https://doi.org/10.1016/j.enconman.2016.02.037

Singh, D., & Chen, S. (2008). The white-rot fungus Phanerochaete chrysosporium: Conditions for the production of lignin-degrading enzymes. *Applied Microbiology and Biotechnology*, *81*(3), 399–417. https://doi.org/10.1007/s00253-008-1706-9

Sonawane, Y., Shindikar, & Khaladkar, M. (2014). Onsite conversion of thermoplastic waste into fuel by catalytic pyrolysis. *International Journal of Innovative Research in Science Engineering and Technology*, *03*(09), 15903–15908. https://doi.org/10.15680/ijirset.2014.0309016

Sun, J., Pang, Y., Lei, Z., OuYang, B., Lai, W., Wang, Y., & Lan, D. (2024). Enzymatic depolymerization of plastic materials by a highly efficient two-enzyme system. *Biochemical Engineering Journal*, *204*, 109222. https://doi.org/10.1016/j.bej.2024.109222

Thew, C. X. E., Lee, Z. S., Srinophakun, P., & Ooi, C. W. (2023). Recent advances and challenges in sustainable management of plastic waste using biodegradation approach. *Bioresource Technology*, *374*, 128772. https://doi.org/10.1016/j.biortech.2023.128772

Tournier, V., Topham, C. M., Gilles, A., David, B., Folgoas, C., Moya-Leclair, E., Kamionka, E., Desrousseaux, M., Texier, H., Gavalda, S., Cot, M., Guémard, E., Dalibey, M., Nomme, J., Cioci, G., Barbe, S., Chateau, M., André, I., Duquesne, S., & Marty, A. (2020). An engineered PET depolymerase to break down and recycle plastic bottles. *Nature*, *580*(7802), 216–219. https://doi.org/10.1038/s41586-020-2149-4

Venkatesh, K R, Hu, J, Tierney, J W, & Wender, I (1995). Hydrocracking of polyolefins to liquid fuels over strong solid acid catalysts. *Preprints of Papers, American Chemical Society, Division of Fuel Chemistry*, *40*(4).

Verma, R., Vinoda, K., Papireddy, M., & Gowda, A. (2016). Toxic pollutants from plastic waste- A review. *Procedia Environmental Sciences*, *35*, 701–708. https://doi.org/10.1016/j.proenv.2016.07.069

Vilaplana, F., & Karlsson, S. (2008). Quality concepts for the improved use of recycled polymeric materials: A review. *Macromolecular Materials and Engineering*, *293*(4), 274–297. https://doi.org/10.1002/mame.200700393

Webb, H., Arnott, J., Crawford, R., & Ivanova, E. (2013). Plastic degradation and its environmental implications with special reference to poly (ethylene terephthalate). *Polymers*, *5*(1), 1–18. https://doi.org/10.3390/polym5010001

8 Sustainable Treatment and Management of E-Plastic Waste

Vishnu Jee[1], Chinmay Prakash Chordiya[1], Mohammad Alikarami[2], and Bikashbindu Das[1]*

[1]Department of Chemical Engineering, Malaviya National Institute of Technology, Jaipur, India

[2]Department of Chemical & Petroleum Engineering, Schulich School of Engineering, University of Calgary, Alberta, Canada

8.1 INTRODUCTION

The rapid growth of the electrical and electronic equipment sector has resulted in a notable rise in electronic waste (e-waste), which has become a major global issue. In 2019, the world generated around 53.6 million metric tonnes of e-waste, which is estimated to reach 74.7 million metric tonnes by 2030. However, of the 53.6 million metric tonnes generated in 2019, only about 9.32 million metric tonnes, or 17.4%, were recorded as recycled. As of 2022, global electronic waste (e-waste) generation reached approximately 62 million metric tons, averaging 7.8 kilograms per person. This indicates an annual increase in the e-waste generation rate of 3–5%. China is the world's largest e-waste producer, with 10.13 million metric tonnes in 2019, followed by the United States and India (Jadhao et al. 2020). E-waste could lead to serious health and environmental risks due to the presence of toxic chemicals and harmful substances. E-plastic waste, in particular, originates from various sources (Figure 8.1), including household electronics, computers, printers, mobile phones, small electrical appliances, kitchen equipment, and personal care devices (Alassali et al. 2020). The disposal of electronic waste has severely impacted ecosystems in developing countries, contaminating the atmosphere, groundwater, air, and soil.

E-waste has a varied composition (Figure 8.2), with metals and plastics as major components. Metals make up around 60%, while plastics account for 15.2% of e-waste. Additionally, cathode ray tubes (CRT) and liquid crystal displays (LCD) contribute 11.87%, and other components comprise 12.72 % (Zhang and Xu 2016). Together, metals and plastics constitute approximately 75.41% of e-waste. This high proportion highlights the importance of recycling e-waste to recover valuable metals and repurpose plastics into new products, promoting sustainable resource use.

The economic unsustainability of current processes for efficient recycling is a key reason why only a minimal amount of e-waste (around 17%) is recycled. Mechanical recycling is one widely used method; however, due to the difficulty in effectively removing hazardous pollutants, this process incurs high sorting costs (Mtibe et al. 2023) Another approach is thermal and chemical recycling, specifically for e-plastic waste, where chemical and thermal techniques are applied. These treatments yield various useful products that can be repurposed, promoting the reuse of valuable resources.

One of the primary benefits is the recovery of valuable materials such as gold, silver, copper, platinum, and rare earth elements from e-waste. It can be reused in the manufacturing of new electronic devices. Also, recycling e-waste consumes considerably less energy compared to traditional

* Corresponding author's email: bbdas.chem@mnit.ac.in

DOI: 10.1201/9781003581000-11

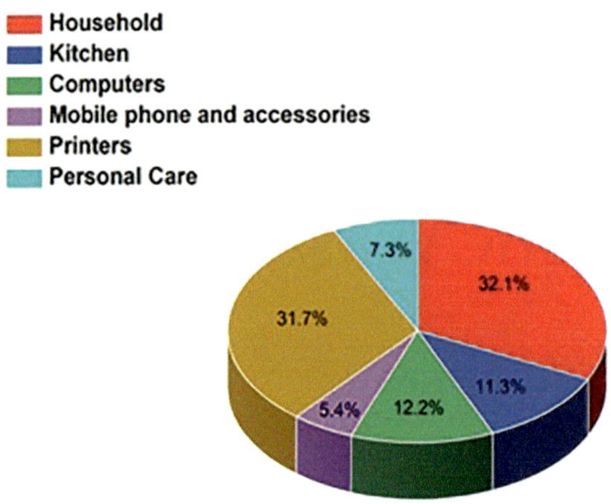

FIGURE 8.1 E-waste generation from various sources.

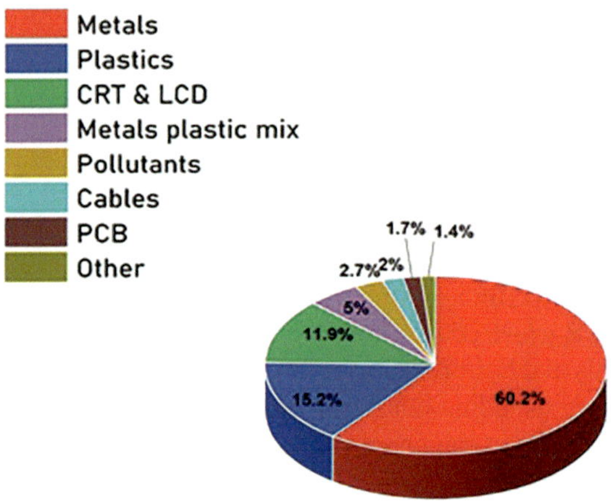

FIGURE 8.2 Composition of e-waste.

mining and refining processes, contributing to energy conservation. The formal e-waste recycling sector also promotes job creation in areas such as collection, dismantling, sorting, and material recovery. Effective e-waste management supports the circular economy by encouraging practices like reuse, repair, and recycling of electronic products.

Despite its potential benefits, e-waste poses several serious drawbacks, particularly when improperly managed. One of the most significant concerns is environmental pollution. Toxic substances such as lead, mercury, cadmium, and brominated flame retardants can leach into soil and water or be released into the air through open burning and informal recycling practices. These pollutants not only harm ecosystems but also pose severe health risks to humans, including respiratory problems, neurological damage, and developmental disorders, especially in children. Additionally, the high cost of establishing and maintaining advanced recycling infrastructure also presents a barrier, particularly in low-income regions, limiting the ability to safely and efficiently manage the growing volume of electronic waste.

The disposal of e-waste in landfills, inadequate recycling practices, and the unchecked build-up of electronic waste contribute to the release of harmful substances into the soil and water, causing environmental contamination. These toxic substances can be absorbed by plants from the soil, while harmful metals leach into water sources, causing water pollution and necessitating wastewater treatment to remove these contaminants (Islam et al. 2020). Furthermore, toxic substances can enter the human body through the food chain, potentially causing serious health issues, including damage to the kidneys, brain, lungs, and various respiratory disorders (Islam et al. 2020). Recycling e-plastic waste is an effective approach to mitigate environmental strain and address associated hazards, making sustainability in e-waste recycling essential. Therefore, this review centres on sustainable methods for treating e-plastic waste, alongside a discussion of management techniques for efficient e-plastic waste handling.

8.2 TYPES OF PLASTICS IN E-PLASTIC

Recycling e-waste plastics is techno-economically challenging, primarily due to the mixing of materials and their highly variable composition. In e-plastic waste, contaminant composition differs based on the number and type of plastics, which vary by appliance type. For instance, large cooling appliances and CRT monitors/televisions contain as many as 10 types of plastics, whereas CPUs and smaller devices have over 13 types (Martinho et al. 2012). E-plastics also contain hazardous components, like brominated flame retardants (BFRs), which can cause food chain contamination, affecting human and animal health.

Within e-waste plastics, more than 15 distinct polymers are present, including acrylonitrile butadiene styrene (ABS), polyurethane (PU), polystyrene (PS), polypropylene (PP), high-impact polystyrene (HIPS), polyphenylene oxide (PPO), polycarbonate (PC), polyamide (PA), polyethylene terephthalate (PET), polyphenylene ether (PPE) (Figure 8.3), polymethylmethacrylate (PMMA), polyoxymethylene (POM), styrene-butadiene (SB), high-density polyethylene (HDPE), and polyvinyl chloride (PVC). For example, analysing 150 kg of waste electrical and electronic components reveals approximately 55 kg of ABS, 23 kg of PP, 20 kg of HIPS, 16 kg of PA, 14 kg of PS, 7 kg of (PC + ABS), 3.5 kg of PA, 3 kg of (PPE + PS), 2.5 kg of HDPE, 1.5 kg of PC, and 4.5 kg of other materials (Alassali et al. 2020), as shown in Figure 8.3.

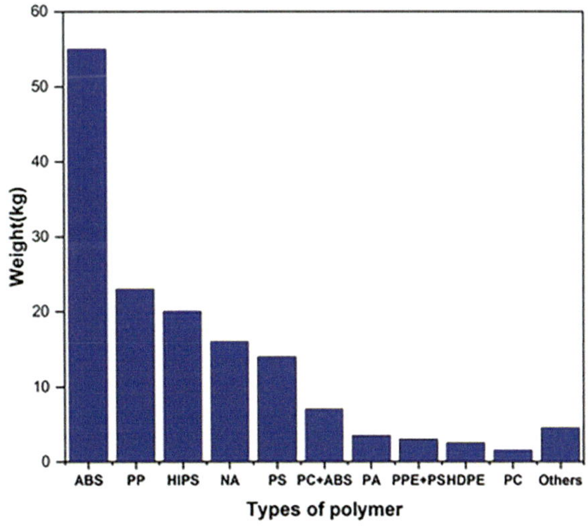

FIGURE 8.3 Different plastic types in e-waste.

8.3 TYPES OF CONTAMINANTS IN E-PLASTIC WASTE AND THEIR ADVERSE EFFECTS

E-plastics are derived from discarded electronic waste which contain a variety of hazardous contaminants (Table 8.1) and have serious environmental and health issues. The essential contaminants consist of heavy metals, BFRs, phthalates, polychlorinated biphenyls (PCBs), chlorinated compounds, and different additives (Anderson et al. 2022). Lead, mercury, cadmium, chromium, and beryllium are examples of heavy metals (Alassali et al. 2020). Due to their characteristics, these metals are used in various electrical components. Lead (Pb) is commonly found in lead-acid batteries, CRTs, and solders. Its presence in the environment can cause neurological harm and cognitive impairments (Wang and Xu 2014). Mercury (Hg) is found in switches, relays, and certain types of fluorescent lamps. Due to its high toxicity, even a small quantity can harm the nervous and immune systems (Wang and Xu 2014). Cadmium (Cd) is present in rechargeable batteries, switches, and housings. It causes kidney damage and bone demineralization. Chromium [Cr (VI)] is mostly used for corrosion resistance in metal coatings. It is very harmful, causing respiratory issues and skin irritation. Thermal interface materials and connectors contain beryllium (Be). Exposure to beryllium leads to chronic beryllium disease and lung cancer.

Flame retardants (FRs) are added to plastics for the safety of consumers (Alassali et al. 2020) FRs are particularly necessary when plastic components are in close proximity to heating elements. Among all flame retardants, BFRs are the most effective. However, they are also the most hazardous additives used in e-plastics. The primary types of BFRs include polybrominated diphenyl ethers, decabromodiphenylethane, and tetrabromobisphenol A (TBBPA) (Hennebert and Filella 2018). In e-plastic waste, BFR concentration varies between 3% and 25% w/w (Hennebert and Filella 2018) PBDEs are added to plastics to make them inflammable. PBDEs are also known as persistent organic pollutants, which can bioaccumulate and cause endocrine disruption and various health issues (Butturi et al. 2020). TBBPA is used to manufacture printed circuit boards and casings. It can affect the immune system.

Phthalates are mostly used as plasticizers in electronic devices to improve the flexibility of plastics such as cables, connectors, and different plastic components. Di(2-ethylhexyl) phthalate, dibutyl phthalate, and butyl benzyl phthalate are common phthalates found in e-plastic waste (Jadhao et al. 2022). These substances are referred to as endocrine disruptors, which can interfere with hormone work and have been connected to reproductive issues. It especially affects the health of youngsters.

PCBs are found in older electronic equipment because they are now banned in many regions due to their toxicity. These substances were used as coolants, insulating fluids, and in capacitors and

TABLE 8.1
Different Types of Contaminants Found in E-Waste

Components	Sources
Lead	CRT in PCs and televisions, lead-acid batteries, PVC cables
Arsenic	Integrated circuit boards
Beryllium	Connectors including finger clips and Mother boards
Cadmium	Switches, PVC cables, solder joints, etc.
Mercury	Batteries, relays, switches, flat screen electronic displays
Chromium	Steel housing
Barium	CRT
Polybrominated diphenyl ethers	Casings
Polychlorinated biphenyls	Electrical transformers, capacitors, PVC
Phthalates (e.g., dibutyl phthalate)	PVC

Source: Wang and Xu, 2014.

transformers. PCBs are highly persistent in the environment and tend to bioaccumulate through the food chain, resulting in prolonged exposure risks.

Some additives are also used to enhance the performance and durability of electronic devices. These added substances include stabilizers, colorants, and fillers. Stabilizers are generally lead-based compounds that protect plastics from degradation caused by heat and light. But it can leach out and cause soil and water contamination. Colorants enhance appearance but often contain toxic metals like cadmium and chromium, which can cause damage once discharged (Jadhao et al. 2022). Fillers are used to improve material properties and reduce costs. It can include hazardous substances that contribute to e-waste generation.

Despite the ban on many toxic substances in production and use, some BFRs, particularly persistent organic pollutants, have been found in significant quantities in recycled materials and end-of-life products. These items often contain traces of these banned substances, with some even exceeding the permitted concentration limits. (Jadhao et al. 2022). A study examined plastics obtained from various types of waste electrical and electronic equipment (WEEE) to assess their recyclability by measuring the levels of potentially toxic elements (PTEs), with a particular focus on cadmium, lead, mercury, and chromium. The study also analysed the presence of BFRs by detecting indicators such as bromine and antimony. To evaluate the material quality, the study adhered to the Restriction of Hazardous Substances (RoHS) directive 2002/95/EC, which sets limits on the use of certain hazardous substances in electrical and electronic equipment. RoHS compliance, required for such equipment in the European Union, specifies maximum allowable concentrations for ten restricted substances.

8.4 E-WASTE LEGISLATION

For the management of e-waste, different countries have passed legislation in their countries (Table 8.2). These legislations manage e-waste effectively by focusing on reducing environmental

TABLE 8.2
E-Waste Legislation in Different Countries

Country	Regulations
India	i) Environmental Protection Act, 1986 ii) E-waste Management Rules, 2016 iii) E-waste Management Amendment Rules, 2018
China	i) Ban on WEEE for Managing Waste Import, 2000 ii) Cleaner Production Law, 2002 iii) Extended Producer Responsibility System, 2012
Australia	i) National Waste Policy, 2009 ii) Product Stewardship Act, 2011
Japan	i) Law for Promotion of Effective Utilization of Resources (LPUR) ii) Law for the Recycling of Specified Kinds of Home Appliances (LRHA), 2001 iii) Small Electrical and Electronic Equipment Recycling Act, 2013
Korea	i) Introduction of Waste Deposit—Refund System, 1992 ii) Extended Producer Responsibility System, 2003 iii) Resource Recycling of WEEE and Vehicles Act, 2007
European Union	i) Waste Shipment Regulation (WSR), 1993 ii) Battery Directive, 2006 iii) WEEE Directive (2012/19/EU), 2012
United Kingdom	i) Packaging Regulations, 2003 ii) Producer Responsibility Obligations Regulations, 2007
New Zealand	i) Resource Management Act, 1991 ii) Local Government Act, 2002

Source: Islam et al., 2020.

impacts and promoting recycling among citizens. However, it totally depends upon the implementation and frameworks of these legislations (Mtibe et al. 2023). Regulatory bodies and policymakers in various countries have established different laws to address e-waste concerns. Key international regulations include the Basel Convention, which ensures that the transboundary movement of e-waste, regardless of whether it is hazardous, requires the consent of both the importing and the transit countries. Additionally, the EU's WEEE directive mandates the separation and sorting of plastics containing BFRs from other e-waste plastics. In many Asian countries, illegal e-waste import remains a significant problem. In India, despite having relatively progressive e-waste legislation, challenges such as poor implementation, lack of public awareness, and inadequate infrastructure hinder effective e-waste recycling (Mtibe et al. 2023). In many Asian countries, the illegal import of e-waste is a significant problem. In India, despite having more advanced e-waste regulations, challenges such as poor enforcement, limited public awareness, and inadequate infrastructure hinder effective e-waste recycling.

Many developing countries do not yet have any proper legislations made for the proper tackling of e-waste, which leads to poor management of e-waste. These e-wastes are not taken into accountability, leading to illegal recycling centres, where they are poorly managed. Several factors contribute to the inadequate recycling of e-waste in various countries, including inconsistent e-waste categorization, insufficient collection centres, low public awareness, and a shortage of proper recycling facilities. These challenges could be addressed by revising current regulations and improving enforcement strategies. The absence of clear guidelines for law enforcement agencies is a key factor in the ineffective implementation of such laws. For monitoring e-waste properly, skilled authorities are required who can coordinate with the other agencies for smooth management (Kirchherr et al. 2017) Also, there are a few important points that every developing country should focus on. The first one is promoting the principles of extended producer responsibility (EPR). EPR will make the producer responsible for their product after the end-of-the-service cycle. The second is promoting the 9R's, that is, refuse, rethink, reduce, reuse, refurbish, recycle, remanufacture, repurpose, recover (Mtibe et al. 2023). It will help in controlling the generation and consumption of more e-waste. The third one is developing collection banks near the household and offices so that people can easily dispose of their e-waste there. And the last one is monitoring the system and rewarding the different stages of e-waste management systems. It will encourage management authorities to do work more properly. Overall, effective management of e-waste requires robust legislation combined with strict enforcement of these laws.

8.5 E-PLASTIC TREATMENT AND MANAGEMENT

E-plastic treatment and management are necessary for minimizing the impact on environment and health. Effective treatment methods include mechanical, thermal, and chemical recycling, which aim to separate valuable materials and neutralize harmful components (Naik and Eswari 2022). Management strategies focus on minimizing environmental and health risks by ensuring safe disposal, reducing the release of toxic substances like brominated flame retardants, and promoting circular economy practices where recovered plastics are reused in new products (Figure 8.4).

Sustainable e-plastic treatment and management can be achieved by following a series of steps such as (i) collection, (ii) e-plastic pretreatment and processing, (iii) physical treatment (separation/size reduction), and (iv) metal recovery. These steps are as follows.

8.5.1 E-Waste Collection

The initial stage of e-plastic waste management involves the collection of e-waste, which is gathered from households and businesses. This collection process can be carried out through several methods (Shahabuddin et al. 2023), including the following: (a) official take-back systems, (b) mixed residual waste streams, (c) collection outside official systems in countries with advanced waste management

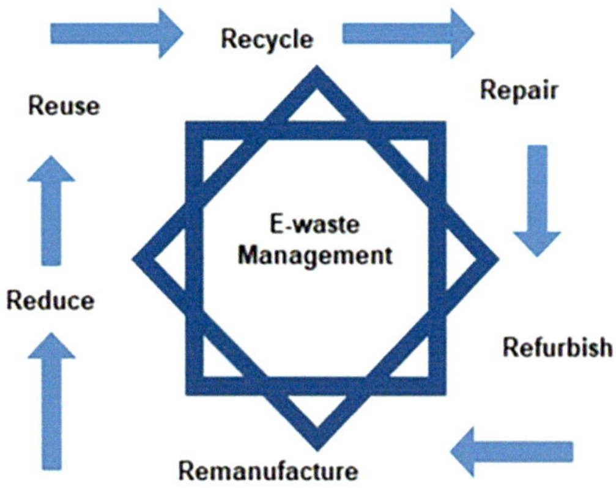

FIGURE 8.4 E-waste management circular economy (adapted from Naik and Satya Eswari 2022).

infrastructure, and (d) collection outside official systems in countries with less-developed waste management systems.

The first stage in managing e-waste is its collection, which can occur through various methods depending on the source. E-waste from households and businesses can be gathered under an official take-back system, where designated organizations such as government agencies, councils, or producers following national e-waste regulations are responsible for the process (Baldé et al. 2017). This method is considered optimal, as the collected e-waste is typically subjected to pretreatment before being transported to treatment facilities for the recovery of valuable materials. On the other hand, e-waste may also be collected alongside other waste in designated bins, as part of a mixed residual waste collection system. Unfortunately, this often leads to e-waste being incinerated or disposed of improperly rather than being recycled or separated. In regions with well-established waste management systems, e-waste collection can take place outside the formal take-back framework (Shahabuddin et al. 2023). In these areas, authorized waste dealers or organizations collect e-waste and sell it to recycling companies, offering a more informal yet organized approach. Conversely, in countries without e-waste legislation, informal collection occurs, with unauthorized individuals or groups gathering e-waste without adequate oversight or planning (Baldé et al. 2017).

The choice of the e-waste collection method in a region is influenced by factors such as legislation, waste management infrastructure, public awareness, and economic conditions (Shahabuddin et al. 2023). For instance, many developing countries in Northern Africa and Southeast Asia lack robust e-waste laws. Consequently, these governments often rely on informal sectors where impoverished individuals collect e-waste door-to-door. These collectors manually separate e-waste and sell it to recycling businesses. However, this informal method poses significant health risks to the collectors and is inefficient for proper recycling and resource recovery. Strengthening legislation, improving infrastructure, and raising public awareness are essential to transitioning towards safer and more effective e-waste management practices.

8.5.2 Physical Treatment: Separation/Size Reduction

Once e-waste is collected, it undergoes a pretreatment phase before being directed to final processing facilities. The primary aim of pretreatment is to extract valuable and reusable materials from the overall e-waste mixture, enabling their efficient transfer to specialized recovery plants. There are three main types of pretreatment methods: (i) manual dismantling, (ii) mechanical dismantling and separation, and (iii) a combination of both manual and mechanical approaches.

The manual dismantling method involves carefully removing hazardous and valuable components, including printed circuit boards, casings, monitors, and batteries. These components require labour-intensive processes to extract due to their value and complexity. Under the mechanical dismantling and separation approach, e-waste is subjected to various mechanical processes, including shredding, crushing, size reduction, and separation of metals from non-metals. Techniques like magnetic separation are used for extracting ferrous metals, while tools based on electrical conductivity are employed for separating non-ferrous metals (Shahabuddin et al. 2023). A combination of manual and mechanical methods is followed when dealing with complex mixtures of e-waste; a hybrid method combining manual and mechanical processes is often employed. This approach is cost effective and time efficient, particularly for intricate assemblies and diverse materials.

Manual dismantling is primarily practiced in developing countries. To facilitate separation, magnets are employed to extract ferrous metals from e-waste, while tools that rely on electrical conductivity are used to separate non-ferrous metals (Shahabuddin et al. 2023). Gravity separation method by means of airflow and water flow tables followed by sifting is used quite often. After the metals are separated from the other materials, they are moved to the final recovery stage, where pure metals are extracted using methods like pyrometallurgy (heat-based techniques) and hydrometallurgy (chemical-based techniques).

8.5.3 Metal Recovery

E-waste mainly contains plastics and metals, with plastics being separated and recycled. The metal content of e-waste includes both base metals (such as Ni, Cu, Al, Pb, Sn) and precious metals (Pd, Au, Pt, etc.) (Akcil et al. 2015), making it a valuable source for metal recovery. Currently, approximately 50% of indium, antimony, and ruthenium, along with over 30% of silver, copper, and tin are consumed in the making of electronic equipment (Buekens and Yang 2014). Furthermore, the metal concentration in e-waste is often greater than in natural ores. For instance, Cu and Au ores contain up to 1 wt% Cu and 1–10 g/ton of Au, while e-waste offers a significantly richer metal source (Tuncuk et al. 2012).

PCBs in computers contain 20% Cu by weight and 250 g/ton of Au, concentrations much higher than those found in ore (Tuncuk et al. 2012). This highlights the importance of recovering metals from e-waste to avoid environmental pollution. In addition to their value, many metals found in e-waste, such as Pb, Cd, Hg, Cr, Ni, Zn, and Be, are toxic and contribute to environmental pollution through leaching (Islam et al. 2020). Therefore, metal recovery from e-waste offers several advantages over extracting metals from ores, including cost and energy efficiency, as well as reducing environmental pollution caused by the presence of these harmful metals in water and soil.

Metal recovery can be achieved through two primary technologies (Figure 8.5): pyrometallurgy and hydrometallurgy (Pant et al. 2012). Pyrometallurgy typically involves processes like smelting, roasting, and calcination. In contrast, hydrometallurgy involves leaching, a process that utilizes various agents like cyanide, thiosulfate, thiourea, halides, and acids to extract metals from e-waste (Pant et al. 2012).

Pyrometallurgy is a high-temperature technology (ranging from 800°C to 1200°C) that includes processes such as roasting, smelting, and calcination. Only a limited number of pyrometallurgical plants are specifically designed to recover metals from e-waste. In Sweden, metals like gold, silver, copper, nickel, palladium, selenium, and zinc are recovered through pyrometallurgical processes in Rönnskär smelter. E-waste is first introduced into the Kaldo furnace, where it is combusted along with lead concentrates in an oxygen-rich environment, with oil added for the reaction. Materials with high copper content are then sent for further processing in the converting stage, while the copper alloy produced is transferred to a Cu converter for additional refinement. After refining, valuable metals (Au, Ag, Ni, Pd, Cu, Se, and Zn) are extracted. The gases generated in the process are treated to produce sulfuric acid and SO_2, while volatile metals like Pb, Sb, In, and Cd are captured

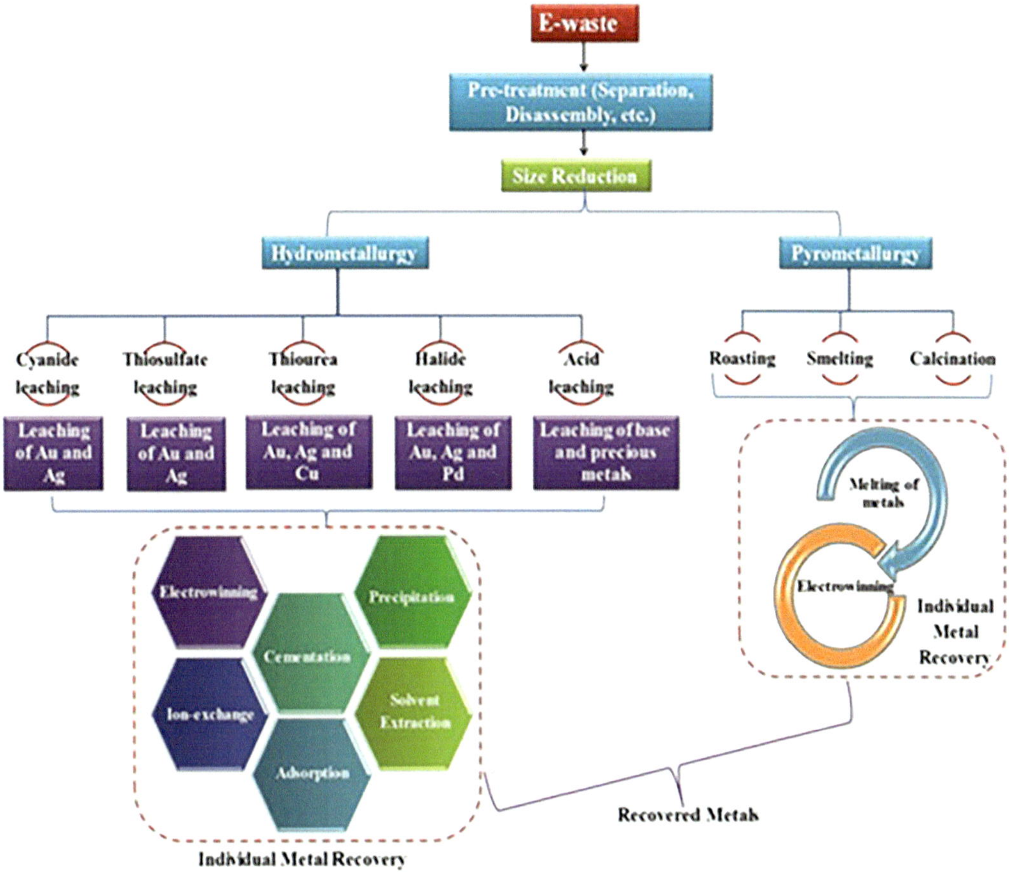

FIGURE 8.5 E-waste metal recovery by chemical recycling. (Adapted from Jadhao et al. 2022.)

through a separate recovery process. The overall recovery efficiency of this system is greater than 95% (Jadhao et al. 2022).

Although pyrometallurgical processes are effective for recovering metals, they come with certain limitations. A major issue is the production of dioxins and furans, which occur when halogenated flame retardants, such as chlorinated and brominated biphenyls found in e-waste, are processed (Hao et al. 2020). The requirement of additional facilities to prevent the toxic gas release into the environment could also increase the overall cost of the process. Another limitation is metals such as iron and aluminium recovery, which remain in the slag phase and are difficult to extract (Hong and Valix 2014). Additionally, the high temperature required for pyrometallurgical processes makes them energy intensive. As a result, hydrometallurgy is often preferred, as it offers a more environmentally friendly and cost-effective alternative for metal recovery.

Hydrometallurgy is a metallurgical process that utilizes various leaching agents to extract metals from materials. Common leaching agents include cyanide, thiosulfate, thiourea, halides, and acids. Precious metals are dissolved in aqueous solutions of these agents and subsequently recovered by solvent extraction, ion exchange, precipitation, adsorption, and electrowinning (Jadhao et al. 2022). A variety of leaching agents are employed in hydrometallurgical processes. Cyanide is a frequently used reagent for dissolving Au in aqueous solutions and has been utilized since 1887 (Egan et al. 2016). It remains the primary leaching agent for gold recovery, accounting for approximately 75%

of global gold production. Despite its efficiency, cyanide poses significant health risks due to its high toxicity, leading to restrictions on its industrial use. Thiosulfate is another leaching agent that can effectively extract gold (Tripathi et al. 2012). Studies have shown that increasing the concentration of thiosulfate enhances gold leaching. However, beyond a certain threshold, leaching efficiency decreases due to the formation of undesirable by-products, such as sulfate, trithionate, and tetrathionate (Tripathi et al. 2012). These by-products can interfere with the process, reducing its overall effectiveness. Thiourea is also used as a leaching agent, offering better recovery rates for gold and silver (Gurung et al. 2013). Ferric ions are often added as oxidizing agents to promote the dissolution of metals during thiourea leaching. The addition of ferric ions has been shown to increase gold recovery by up to fourfold (Gurung et al. 2013). However, excessive ferric ion concentrations lead to the formation of undesired products like sulfur and cyanamide. A thin sulfur layer may form on the metal surface, hindering mass transfer and reducing the recovery rates of gold and silver. Each leaching agent has advantages and limitations, requiring careful optimization to balance efficiency, environmental safety, and economic feasibility.

Halides were also used as leaching agents for the metal recovery process. Halide leaching employs halides such as chloride and iodide to extract metals, including gold, by forming metal–halide complexes. For gold recovery, halide leaching has shown promising results, particularly when paired with pretreatment methods like ball milling and supercritical water. Gold recovery rates were observed to be 89% with ball milling and as high as 99% with supercritical water, highlighting the effectiveness of these pretreatment techniques (Jadhao et al. 2022).

Acid leaching is a widely utilized method for extracting metals from e-waste. This process can be classified into three categories: (i) leaching with acid alone, (ii) leaching with acid combined with an oxidizing agent, and (iii) multistage acid leaching. A study investigated the efficiency of five acids for metal recovery: hydrochloric acid, sulfuric acid, acetic acid, nitric acid, and citric acid (Jadhav and Hocheng 2015). Among these, only hydrochloric acid and sulfuric acid achieved 100% metal recovery, requiring approximately 22 hours and 96 hours, respectively. In contrast, acetic acid, citric acid, and nitric acid showed much lower recovery efficiencies, with rates of 8.8%, 9.89%, and 19.57%, respectively. The use of acid with the oxidant method combines acids with oxidizing agents to enhance leaching efficiency. For instance, using sulfuric acid with hydrogen peroxide as an oxidant resulted in 100% copper recovery (Birloaga et al. 2014). The addition of oxidants accelerates the dissolution process and improves recovery rates. Multistage acid leaching is particularly effective for recovering precious metals like gold. This gold recovery using this method involves a three-stage process (Sheng and Etsell 2007). In the first stage, nitric acid is used to dissolve base metals. It is followed by Ag recovery and base metals from the residue using nitric acid in the second stage. In the third stage, aqua regia is used to recover gold from the remaining material.

8.5.4 Uses in Construction

The plastic derived from e-waste is being utilized in construction for various applications, aiming to protect natural resources and decrease the environmental burden of plastic waste. E-plastics can serve as fine or coarse aggregates in concrete production, offering an eco-friendly alternative to traditional materials (De la Colina Martínez et al. 2019). One significant advantage of using e-waste plastic in concrete is its cost effectiveness, as it reduces reliance on conventional resources. Additionally, concrete made with e-plastic is lighter, making transportation more convenient and fuel-efficient. It was found that concrete incorporating e-waste plastic exhibits improved compressive, flexural, and tensile strength compared to traditional concrete (Mtibe et al. 2023).

However, excessive e-waste plastic incorporation can negatively impact the mechanical properties (tensile strength and hardness) of the concrete. To address this, e-waste plastic is often combined with other composites with superior mechanical properties. Furthermore, nanofillers are added to the mix to reduce porosity, resulting in concrete with enhanced durability compared to conventional options (Mtibe et al. 2023). Despite its benefits, concerns remain about the environmental effects of producing concrete from e-waste plastic. The manufacturing process may release harmful and toxic substances into the environment, necessitating further research and innovation to mitigate these effects.

8.6 E-PLASTIC CONVERSION TECHNOLOGIES

E-waste is mainly composed of plastic and metals. E-plastic makes up a substantial portion of the e-waste. It results in a significant recycling challenge for different types of plastics. Also, the increasing volume of e-waste plastic demands innovative solutions for reuse and recycling. Hence, for complete recycling of e-plastic waste, it is necessary to recover and then convert it into valuable products. E-plastics can be converted through two main approaches: physical methods and chemical methods. E-plastic is recovered using physical methods, while chemical methods are used to transform e-plastics into fuels. Chemical methods include conversion technologies like pyrolysis, catalytic pyrolysis, and gasification. These chemical processes break down e-plastics into their monomers, valuable products, or energy.

8.6.1 Pyrolysis of E-Plastic Waste

Pyrolysis is a process to decompose or break down organic materials in the absence of oxygen under heat. In pyrolysis, polymers are thermally decomposed into fuel oils, gases, and chars in an oxygen-free environment. This method is useful for those e-waste plastics that are mixtures of different types of polymers. Depending upon the holding time, temperature, and heating rate, pyrolysis is categorized into various types (conventional, slow, and fast pyrolysis) (Wu and Williams 2013). Conventional pyrolysis operates at a medium heating condition, reaching a highest temperature around 550°C to 600°C within 5 to 30 minutes. Slow pyrolysis, on the other hand, has a much slower heating rate and typically operates at temperatures below 400°C, while fast pyrolysis occurs at temperatures above 600°C and involves a rapid heating rate. Slow pyrolysis requires a residence time of 24 hours, whereas fast pyrolysis has a much shorter residence time, ranging from 0.5 to 5 seconds (Demirbas 2005). E-waste plastics, which consist of long polymer chains, are broken down into shorter chains and low molecular weight hydrocarbons during these processes.

The factors affecting the pyrolysis process are temperature, heating rate and residence time, feed composition, and use of a catalyst. Higher temperatures favour gas production, whereas moderate temperatures favour liquid yield. Increasing heating rate could increase oil yield and longer residence times promote secondary cracking. Composition differences could affect the final product characteristics. For instance, PE and PP produce more oil, whereas PVC releases HCl, complicating the process. Moreover, catalyst types, such as zeolites (e.g., ZSM-5), improve selectivity towards aromatics and reduce char. Similarly, during gasification, feedstock composition such as mixed or halogenated plastics can lead to variation in syngas quality and pollutant load. In addition, use of air, oxygen, or steam can alter the H_2/CO ratio in syngas.

Pyrolysis of e-waste plastics produces three main products: pyrolysis oil, pyrolysis gas, and solid residue (char). The pyrolysis oil is composed of various aliphatic and aromatic hydrocarbons such as benzene, toluene, xylene, styrene, and long-chain paraffins. The pyrolysis gas mainly includes hydrogen (H_2), methane (CH_2), carbon monoxide (CO), carbon dioxide (CO_2), and light hydrocarbons like ethane and propane.

Pyrolysis oil can be used as an alternative fuel or chemical feedstock for producing new plastics and petrochemicals. Pyrolysis gas can be utilized as a heating source for the pyrolysis process or as fuel for power generation. Char can be applied in metal recovery, in solid fuel production, or as a raw material for activated carbon.

8.6.1.1 Temperature Effect on Pyrolysis

The impact of temperature on PCB pyrolysis varies within the range of 300°C to 700°C. Research showed that the solid residue yield decreased as the temperature increased from 300°C to 500°C

(Jie et al. 2008). However, beyond 500°C, there was a little change in the solid residue yield, and the carbon and hydrogen content remain constant. In contrast, the yield of liquid products steadily increased up to 500°C and levelled off beyond this point. Additionally, increasing the temperature was also found to raise the gross calorific value of the liquid product from (Jie et al. 2008). Similar to the solid and liquid products, the gas stream composition was stabilized above a certain temperature (around 500°C). Furthermore, while the heating rate does not influence the final pyrolysis yield, it does accelerate the pyrolysis process.

8.6.1.2 Feed Composition Effect on Pyrolysis

The composition of the feedstock significantly influences the yield of pyrolysis. To explore how feed composition affects pyrolysis yield, three types of feed materials were investigated: PCBs from discarded computers, mobile phones, and televisions (Hall and Williams 2007). The pyrolysis process was carried out at a temperature of 800°C for approximately 135 minutes. It was found that the metal content in feedstock could influence the metals in the solid residue (Hall and Williams 2007). Additionally, the liquid products obtained from the pyrolysis of various feedstocks could produce considerable variation in their composition. It was noted that the PCB composition is affected by the type of resin used during manufacturing and the curing extent used, which in turn impacts the composition of the resulting liquid products after pyrolysis. The e-waste metals also play a role in shaping the pyrolysis outcomes. Specifically, the bromophenols formed during the pyrolysis process were found to vary in amount due to the presence of these metals.

The liquid products derived from the three main types of PCBs—computers, mobile phones, and televisions—include phenols, brominated compounds, and triphenyl phosphate (Blazso and Cze 2002). Phenols were identified as the predominant components, originating from bisphenol epoxy resin or epoxy novolac resin decomposition. The presence of brominated complexes is attributed to the epoxy resin in the PCBs, while triphenyl phosphate, primarily sourced from televisions, was also detected. Additionally, the liquid products contained 2,4-dibromophenol, 2,6-dibromophenol, and tetrabromobisphenol A. The pyrolysis of the three PCB types yielded gases primarily consisting of up to 88 wt% CO, CO_2, and CH_4. Other gaseous by-products included Hethene, ethane, propane, butane, propene, and trace amounts of chlorine and bromine (Jadhao et al. 2022). The formation of CO and CO_2 was linked to the breakdown of the epoxy groups and $CaCO_3$, while CH_4 was generated from the methyl groups.

During pyrolytic cleavage of the brominated resin, several brominated compounds were formed, including 1-bromo-2-(3-cyclopropylpent-4-yn-2-yl) benzene and 5-bromo-2–(2-(3-ethoxy phenol)-2-oxoethyl) benzoic acid (Jadhao et al. 2022) Pyrogenic cleavage also produced a range of benzenes (benzene, ethylbenzene, toluene, xylene, styrene, cumene, etc.) and phenolic compounds (phenol, cresol, propylphenol, ethylphenol, etc.). Biphenyl was formed by the dehydration of two benzene molecules. The Diels-Alder reaction between biphenyl and acetylene leads to the formation of phenanthrene (Kislov et al. 2005). Naphthalene was the dehydrogenation product of butylbenzene and dibenzofuran was generated through the dehydrogenation of biphenyl-2-ol.

8.6.1.3 Use of Catalysts in E-Waste Pyrolysis

The use of catalysts in pyrolysis has been observed to enhance the desired product yield through minimizing unwanted char and other by-products formation. Under the presence of catalysts, catalytic cracking of plastic, and thermal decomposition as well as the formation of carbocations, occurs. Degradation of polymers takes place by beta-scission and disproportionation reactions (Areeprasert and Khaobang 2018). Along with cracking, a few other reactions such as aromatisation, oligomerization and cyclization take place. With cracking reactions, skeletal isomerization also occurs enhancing the branched products. Cracking and reforming reactions produce various products with the help of different catalysts such as Y-zeolite, ZSM-5, HUSY, and HZSM-5 (Bozi et al. 2008). Catalytic pyrolysis has advantages including low operating temperature, less reaction time, higher conversion, selectivity towards desired products, and better fuel quality than only thermal degradation.

A study was conducted to investigate catalytic pyrolysis using zeolite-based catalysts with CRT plastics, refrigerators, HIPS, and ABS as feedstocks (Muhammad et al. 2015). The results indicated that Y-zeolite produced higher liquid fuel yields compared to ZSM-5 under experiments performed at 500°C. For CRT plastics, the oil yield was 84 wt% without a catalyst, which decreased to 80 wt% with Y-zeolite and further to 77.5 wt% with ZSM-5. The use of catalysts in this case reduced the oil yield but increased the gas yield. In the case of refrigerator plastics, the oil yield was 80 wt% without a catalyst, which dropped to 70 wt% with Y-zeolite and 63 wt% with ZSM-5 (Muhammad et al. 2015). Catalytic pyrolysis of both refrigerator and CRT plastics resulted in a notable decrease in the oil styrene concentration, while the benzene and toluene levels increased.

In another study conducted to explore the catalytic pyrolysis of plastics using zeolites HUSY and HZSM-5, the research focused on a plastic mixture composed of 4 wt% ABS, 33 wt% HIPS, and 3 wt% polybutylene terephthalate, subjected to pyrolysis at 400°C in the presence of these catalysts. The results indicated that HZSM-5 produced a higher yield of liquid products than HUSY due to wider pore in HUSY, which allowed for further degradation of liquid products into gaseous compounds and coke. CH_4, CO_2, C_2H_4, C_2H_6, H_2, and CO were found in the gaseous product. Similarly, the gaseous composition from HZSM-5 pyrolysis closely resembled the thermally cracked gaseous products (Santella et al. 2016). Char yields were 13.2 wt% and 7.1 wt% for HUSY and HZSM-5, respectively. The use of HZSM-5 reduced the char yield while increasing the oil yield. However, changes in the operating conditions for HZSM-5 could lead to an increase in both char and gaseous product yields. A higher Si:Al ratio in the catalyst could enhance the silica content and result in greater hydrophobicity and a stronger affinity for hydrocarbons (Santella et al. 2016). The liquid products from catalytic pyrolysis contained fewer compounds (toluene, ethylbenzene, and styrene monomers) than those from thermal cracking. This indicates that zeolite catalysts promote the formation of lighter compounds in the liquid products.

8.6.2 E-Waste Plastic Gasification

The gasification process involves heating feedstock under oxygen, producing CO, and H_2 as the primary components of the gas stream. Other components, such as CO_2, CH_4, and N_2, are also present in smaller amounts (Devi et al. 2003). One of the key advantages of gasification is the generation of electricity and heat from the resulting syngas. Additionally, syngas can be used to produce chemicals like dimethyl ether and methanol (Guan et al. 2016). In gasification, plastics with diverse compositions can be effectively processed than pyrolysis. However, tar formation is a significant drawback, which is more prevalent in gasification of plastics than biomass. As a result, cleaning the feed gas before using it for energy production is necessary. During the gasification of waste plastics, challenges arise due to characteristics such as low thermal conductivity, sticky behaviour, high volatile content, and tar formation, which complicate the use of conventional gasifiers. Some traditional gasification reactors for biomass include plasma reactors, updraft, downdraft, fixed bed, and fluidized bed reactors (He et al. 2009). However, these conventional gasifiers are not ideal for e-waste plastics due to the aforementioned characteristics. Rotary kilns and rotary ovens struggle with the sticky nature of plastic waste, while updraft and downdraft gasifiers are unsuitable for plastics with low fixed carbon content. Fluidized bed gasifiers, however, are considered the most suitable and widely used for gasification of plastics due to their ability to handle these challenges effectively.

A study was conducted on e-waste plastics containing brominated flames retardants (Yamawaki 2003) at 1150°C, followed by shock cooling of the gases to 50°C. The decomposition of polybrominated dibenzodioxins and polybrominated dibenzofurans was found to be 99.99% or higher. E-plastic waste and automobile shredder dust were also gasified under heating and shock-cooling temperature of 1150°C and 70°C. The results showed a significant reduction in the generation of brominated dioxins. The resulting syngas was then used as a feedstock for various chemical industry applications.

A study explored the impact of nickel with carbonates for the steam gasification of PCBs (Salbidegoitia et al. 2015). The research indicated that hydrogen production during steam gasification occurs through the following specific reactions:

$$C + H_2O \rightarrow CO + H_2 \quad eq^n(1) \quad \Delta H298 = 131\,KJ/Mol$$

$$CO + H_2O \rightarrow CO_2 + H_2 \quad eq^n(2) \quad \Delta H298 = 41.2\,KJ/Mol$$

It was observed that the use of LNK-carbonates could lower the temperature required for steam gasification. When nickel powder was added to the LNK-carbonate, hydrogen production increased by a factor of 3.2 (Salbidegoitia et al. 2015). Additionally, the presence of nickel reduced char formation from 49 wt% to 23 wt%. Notably, no tar was formed in the presence of nickel powder. It was concluded that the catalytic effect of Ni increased the H_2 formation as well as reduced the char yield.

8.6.3 Depolymerization of E-Waste Plastic

Depolymerization is a promising method for recycling e-waste plastic, as it breaks down polymer chains into smaller molecules or monomers that can be reused. Supercritical fluids are recognized as highly effective for converting waste materials into valuable resources. A supercritical fluid is one that exists at a temperature and pressure above its critical point, exhibiting unique properties such as low viscosity, high mass transfer coefficient, low dielectric constant, and high diffusivity (Wang and Xu 2014). Several supercritical fluids, including acetone, water, carbon dioxide, and methanol, have been successfully applied in the depolymerization of waste plastics. A study was conducted to explore the use of sub/supercritical water to recycle waste PCBs from computers (Xing and Zhang 2013). The researchers observed a maximum debromination concentration of 97.8% at 400°C. As the temperature increased, a decrease in the dielectric constant was noted, which weakened hydrogen bonding in the water. This facilitated the oil formation through brominated epoxy resin decomposition (Jin et al. 2011). Additionally, higher temperatures caused the breaking of polymer bonds into free radicals, a key mechanism of polymer degradation in supercritical fluids.

8.7 LIFE-CYCLE ASSESSMENT ANALYSIS OF E-PLASTIC WASTE

A life-cycle assessment (LCA) of e-plastic waste is essential to evaluate its environmental impact and associated human risks. Given the relatively low environmental burden in categories such as human toxicity, terrestrial ecotoxicity, marine ecotoxicity, and freshwater ecotoxicity, reusing e-waste and managing its disposal at the end of life can significantly benefit the environment. As previously reported (Naik and Satya Eswari 2022), the goal of the LCA study takes consideration of different stages such as (i) functional unit and system boundaries, (ii) inventory analysis, (iii) informal treatment, (iv) end-of-life alternate scenarios, (v) impact assessment, and (vi) interpretation of results.

Functional unit is basically the plastics derived from electronic waste after collection. The system boundaries encompass informal treatment processes and end-of-life scenarios, such as incineration, landfilling, and the reuse of e-plastics in construction materials (Naik and Satya Eswari 2022). An illustration of the system boundaries for informal treatment processes and end-of-life scenarios can be found in Figure 8.6. Inventory data for this analysis can be obtained from scientific studies, reports, and the Eco-invent 3.0 LCA database (Butturi et al. 2020), which includes information on e-waste quantities, composition, and recycling rates from WEEE management research in various countries. The informal treatment begins with manual dismantling to recover valuable materials,

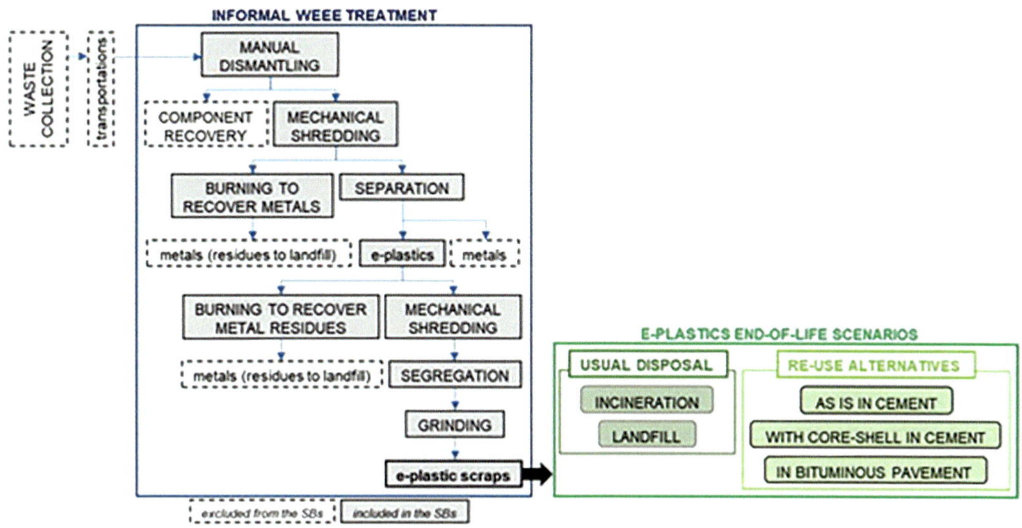

FIGURE 8.6 System boundaries for the informal treatment and end-of-life scenarios for e-plastic. (Adapted with permission from Naik and Satya Eswari, 2022.)

followed by mechanical shredding to separate metals. Some fractions are burned to recover metal scraps, with incineration residues disposed of in landfills. The remaining materials undergo various end-of-life treatments, including incineration, landfill, or reuse in construction materials (cement, bituminous pavement).

The impact assessment examines the environmental profile of informal e-waste treatment and e-plastic end-of-life scenarios, with a focus on human toxicological and ecotoxicological effects. The impact categories include freshwater ecotoxicity and human toxicity (both cancer and non-cancer), along with additional considerations of human health and ecosystem damage (Butturi et al. 2020). Improper handling of hazardous electronic waste and the use of inadequate methods to recover components can release harmful substances into the environment, leading to detrimental effects on both human health and ecosystems. The overall findings highlight that environmental impact values are highly dependent on the specific end-of-life scenarios. Reusing e-plastic scrap proves to be particularly beneficial in reducing environmental issues, especially when compared to landfill disposal or incineration.

In a Life-Cycle Impact Assessment (LCIA) study conducted by (Butturi et al. 2020), it was found that burning e-waste to recover metals significantly contributes to all impact categories. For "human toxicity, cancer," burning accounted for 59% of the impact, primarily due to emissions of nickel, dioxins, and mercury. Other significant contributions came from shredding (14%), manual dismantling (10%), and landfill residues (9%). In the "human toxicity, non-cancer" category, emissions of mercury, antimony, cadmium, and arsenic to air, soil, and groundwater were key contributors. "Freshwater ecotoxicity" was largely driven by copper emissions to soil and groundwater. According to the study, the LCIA results for alternative end-of-life scenarios revealed that, except for "human toxicity, non-cancer," the landfill scenario had the most significant impact at the mid-point level, contributing nearly 40% to "human toxicity, cancer" and 50% to "freshwater ecotoxicity," mainly due to nickel and copper releases into groundwater. Treatment processes contributed about 25%. For "human toxicity, non-cancer," incineration was the largest contributor, accounting for nearly 50% of the impact, mainly from mercury and antimony emissions. Reusing e-plastic in cement contributed 15% to "human toxicity, cancer," 20% to "human toxicity, non-cancer," and

10% to "freshwater ecotoxicity" due to nickel, mercury, and copper releases. The core-shell strategy for reusing e-plastic in cement helped reduce environmental impacts by almost 6% across all categories. Reuse in bituminous pavement accounted for about 9% of the "human toxicity, cancer" impact and less than 5% in other categories. The treatment phase findings confirmed that improper e-waste handling and recovery methods release harmful substances, impacting both human health and ecosystems. Specifically, burning e-plastics for metal recovery and landfilling residues in open dumps had the most significant impact across all categories due to mercury and copper emissions.

A comparison of alternative end-of-life scenarios indicates that recycling and reusing e-plastics in construction materials are more environmentally beneficial than incineration and disposal, with their impacts being less than 50% of those from the other methods. Sensitivity analyses confirm that changes in parameters do not alter this overall conclusion. Among recycling alternatives, reusing e-plastics in bituminous pavement presents the lowest environmental impact when compared to reuse in cement (Butturi et al. 2020). However, when potential leaching and percolate emissions are included in the sensitivity analysis, the environmental impact of the bituminous pavement scenario is reduced, narrowing the gap with cement reuse by approximately 10%.

8.8 CHALLENGES DURING E-PLASTIC PROCESSING

E-plastic waste processing presents several challenges, largely due to the complex nature of electronic waste. One of the key difficulties is identifying and segregating different plastic types, as e-plastic waste consists of various polymers, each with distinct chemical properties that require separate recycling processes. Additionally, the presence of additives, flame retardants, and other chemicals complicates processing, necessitating costly pre-processing steps that increase both time and expenses (Hopewell et al. 2009).

Technological limitations also hinder e-plastic processing. Mechanical recycling methods often degrade plastic quality, limiting the potential for recycled materials to be used in higher-value applications (Hopewell et al. 2009). While advanced chemical recycling techniques can handle a wider range of plastics and contaminants, they are not yet widely adopted in industry due to their high costs and technical complexities. Economically, the costs of collection, sorting, and processing e-plastics can exceed the value of the recovered materials, making it unprofitable for many industries. Additionally, the infrastructure needed for efficient e-plastic recycling is still inadequate.

However, ongoing research and innovation offer promising solutions. Improvements in sorting and identification technologies are essential, along with a greater focus on designing products with recyclability in mind. Implementing EPR policies can incentivize manufacturers to assume more accountability for the management of their products at the end of their life cycle. With these advancements, the future of e-plastic processing can become more efficient, economically viable, and environmentally sustainable.

8.9 CONCLUSIONS AND PERSPECTIVES

As the population grows and the use of electronic devices increases, the generation of e-waste is rising rapidly. Improper disposal of e-waste presents significant environmental and health risks. However, it also offers an opportunity to recover valuable resources. E-waste contains a variety of valuable metals, and improper disposal, such as landfilling, results in the loss of these recoverable materials. Given that e-waste is rich in both metals and plastics, effective recycling could provide considerable economic benefits. Various methods, such as pyrolysis, gasification, and supercritical fluid technology, were explored as effective means for converting e-waste plastics into useful products. Pyrometallurgy, a common technique for metal recovery, is energy-intensive and produces toxic emissions. Hydrometallurgy, another method, involves the use of harmful leaching agents. However, using eco-friendly alternatives, such as chelating agents and ionic liquids, could improve metal recovery process while minimizing environmental harm. Ultimately, an integrated recycling approach that combines the conversion of e-waste plastics with environmentally friendly metal recovery techniques can greatly enhance the economic and environmental sustainability of e-waste management. This dual approach promotes resource conservation while reducing ecological impact, making large-scale sustainable e-waste processing a more achievable goal.

REFERENCES

Akcil, A., Erust, C., Gahan, C.S., Ozgun, M., Sahin, M., Tuncuk, A., 2015. Precious metal recovery from waste printed circuit boards using cyanide and non-cyanide lixiviants—A review. *Waste Manag*. 45, 258–271. https://doi.org/10.1016/j.wasman.2015.01.017

Alassali, A., Barouta, D., Tirion, H., Moldt, Y., Kuchta, K., 2020. Towards a high quality recycling of plastics from waste electrical and electronic equipment through separation of contaminated fractions. *J. Hazard. Mater*. 387, 121741. https://doi.org/10.1016/j.jhazmat.2019.121741

Anderson, L., Yu, E., Chen, W.T., 2022. Chemical recycling of mixed plastics in electronic waste using solvent-based processing. *Processes* 10. https://doi.org/10.3390/pr10010066

Areeprasert, C., Khaobang, C., 2018. Pyrolysis and catalytic reforming of ABS/PC and PCB using biochar and e-waste char as alternative green catalysts for oil and metal recovery. *Fuel Process. Technol*. 182, 26–36. https://doi.org/10.1016/j.fuproc.2018.10.006

Baldé, C.P., Forti, V., Gray, V., Kuehr, R., Stegmann, P., The Global E-waste Monitor., 2017. *United Nations University (UNU), International Telecommunication Union (ITU) & International Solid Waste Association (ISWA)*. Bonn/Geneva/Vienna.

Birloaga, I., Coman, V., Kopacek, B., Vegliò, F., 2014. An advanced study on the hydrometallurgical processing of waste computer printed circuit boards to extract their valuable content of metals. *Waste Manag*. 34, 2581–2586. https://doi.org/10.1016/j.wasman.2014.08.028

Blazso, M., Czégény, Z., Csoma, C., 2002. Pyrolysis and debromination of flame retarded polymers of electronic scrap studied by analytical pyrolysis. *J. Anal. Appl. Pyrolysis* 64, 249–261.

Bozi, J., Czégény, Z., Blazsó, M., 2008. Conversion of the volatile thermal decomposition products of polyamide-6,6 and ABS over Y zeolites. *Thermochim. Acta* 472, 84–94. https://doi.org/10.1016/j.tca.2008.03.018

Buekens, A., Yang, J., 2014. Recycling of WEEE plastics: A review. *J. Mater. Cycles Waste Manag*. 16, 415–434. https://doi.org/10.1007/s10163-014-0241-2

Butturi, M.A., Marinelli, S., Gamberini, R., Rimini, B., 2020. Ecotoxicity of plastics from informal waste electric and electronic treatment and recycling. *Toxics* 8, 1–19. https://doi.org/10.3390/toxics8040099

De la Colina Martínez, A.L., Martínez Barrera, G., Barrera Díaz, C.E., Ávila Córdoba, L.I., Ureña Núñez, F., Delgado Hernández, D.J., 2019. Recycled polycarbonate from electronic waste and its use in concrete: Effect of irradiation. *Constr. Build. Mater*. 201, 778–785. https://doi.org/10.1016/j.conbuildmat.2018.12.147

Demirbas, A., 2005. Pyrolysis of ground beech wood in irregular heating rate conditions. *J. Anal. Appl. Pyrolysis* 73, 39–43. https://doi.org/10.1016/j.jaap.2004.04.002

Devi, L., Ptasinski, K.J., Janssen, F.J.J.G., 2003. A review of the primary measures for tar elimination in biomass gasification processes. *Biomass Bioenergy* 24, 125–140. https://doi.org/10.1016/S0961-9534(02)00102–2

Egan, J., Bazin, C., Hodouin, D., 2016. Effect of particle size and grinding time on gold dissolution in cyanide solution. *Minerals* 6. https://doi.org/10.3390/min6030068

Guan, G., Kaewpanha, M., Hao, X., Abudula, A., 2016. Catalytic steam reforming of biomass tar: Prospects and challenges. *Renew. Sustain. Energy Rev*. 58, 450–461. https://doi.org/10.1016/j.rser.2015.12.316

Gurung, M., Adhikari, B.B., Kawakita, H., Ohto, K., Inoue, K., Alam, S., 2013. Recovery of gold and silver from spent mobile phones by means of acidothiourea leaching followed by adsorption using biosorbent prepared from persimmon tannin. *Hydrometallurgy* 133, 84–93. https://doi.org/10.1016/j.hydromet.2012.12.003

Hall, W.J., Williams, P.T., 2007. Separation and recovery of materials from scrap printed circuit boards. *Resour. Conserv. Recycl*. 51, 691–709. https://doi.org/10.1016/j.resconrec.2006.11.010

Hao, J., Wang, Y., Wu, Y., Guo, F., 2020. Metal recovery from waste printed circuit boards: A review for current status and perspectives. *Resour. Conserv. Recycl*. 157, 104787. https://doi.org/10.1016/j.resconrec.2020.104787

He, M., Xiao, B., Hu, Z., Liu, S., Guo, X., Luo, S., 2009. Syngas production from catalytic gasification of waste polyethylene: Influence of temperature on gas yield and composition. *Int. J. Hydrogen Energy* 34, 1342–1348. https://doi.org/10.1016/j.ijhydene.2008.12.023

Hennebert, P., Filella, M., 2018. WEEE plastic sorting for bromine essential to enforce EU regulation. *Waste Manag*. 71, 390–399. https://doi.org/10.1016/j.wasman.2017.09.031

Hong, Y., Valix, M., 2014. Bioleaching of electronic waste using acidophilic sulfur oxidising bacteria. *J. Clean. Prod*. 65, 465–472. https://doi.org/10.1016/j.jclepro.2013.08.043

Hopewell, J., Dvorak, R., Kosior, E., 2009. Plastics recycling: Challenges and opportunities. *Philos. Trans. R. Soc. B Biol. Sci.* 364, 2115–2126. https://doi.org/10.1098/rstb.2008.0311

Islam, A., Ahmed, T., Awual, M.R., Rahman, A., Sultana, M., Aziz, A.A., Monir, M.U., Teo, S.H., Hasan, M., 2020. Advances in sustainable approaches to recover metals from e-waste-A review. *J. Clean. Prod.* 244, 118815. https://doi.org/10.1016/j.jclepro.2019.118815

Jadhao, P.R., Ahmad, E., Pant, K.K., Nigam, K.D.P., 2020. Environmentally friendly approach for the recovery of metallic fraction from waste printed circuit boards using pyrolysis and ultrasonication. *Waste Manag.* 118, 150–160. https://doi.org/10.1016/j.wasman.2020.08.028

Jadhao, P.R., Ahmad, E., Pant, K.K., Nigam, K.D.P., 2022. Advancements in the field of electronic waste Recycling: Critical assessment of chemical route for generation of energy and valuable products coupled with metal recovery. *Sep. Purif. Technol.* 289, 120773. https://doi.org/10.1016/j.seppur.2022.120773

Jadhav, U., Hocheng, H., 2015. Hydrometallurgical recovery of metals from large printed circuit board pieces. *Sci. Rep.* 5, 1–10. https://doi.org/10.1038/srep14574

Jie, G., Ying-Shun, L., Mai-Xi, L., 2008. Product characterization of waste printed circuit board by pyrolysis. *J. Anal. Appl. Pyrolysis* 83, 185–189. https://doi.org/10.1016/j.jaap.2008.08.007

Jin, Y.-qi, Tao, L., Chi, Y., Yan, J.-hua, 2011. Conversion of bromine during thermal decomposition of printed circuit boards at high temperature. *J. Hazard. Mater.* 186, 707–712. https://doi.org/10.1016/j.jhazmat.2010.11.050

Kirchherr, J., Reike, D., Hekkert, M., 2017. Conceptualizing the circular economy: An analysis of 114 definitions. *Resour. Conserv. Recycl.* 127, 221–232. https://doi.org/10.1016/j.resconrec.2017.09.005

Kislov, V.V., Islamova, N.I., Kolker, A.M., Lin, S.H., Mebel, A.M., 2005. Hydrogen abstraction acetylene addition and Diels-Alder mechanisms of PAH formation: A detailed study using first principles calculations. *J. Chem. Theory Comput.* 1, 908–924. https://doi.org/10.1021/ct0500491

Martinho, G., Pires, A., Saraiva, L., Ribeiro, R., 2012. Composition of plastics from waste electrical and electronic equipment (WEEE) by direct sampling. *Waste Manag.* 32, 1213–1217. https://doi.org/10.1016/j.wasman.2012.02.010

Mtibe, A., Mokhena, T.C., John, M.J., 2023. Sustainable valorization and conversion of e-waste plastics into value-added products. *Curr. Opin. Green Sustain. Chem.* 40, 100762. https://doi.org/10.1016/j.cogsc.2023.100762

Muhammad, C., Onwudili, J.A., Williams, P.T., 2015. Catalytic pyrolysis of waste plastic from electrical and electronic equipment. *J. Anal. Appl. Pyrolysis* 113, 332–339. https://doi.org/10.1016/j.jaap.2015.02.016

Naik, S., Satya Eswari, J., 2022. Electrical waste management: Recent advances challenges and future outlook. *Total Environ. Res. Themes* 1–2, 100002. https://doi.org/10.1016/j.totert.2022.100002

Pant, D., Joshi, D., Upreti, M.K., Kotnala, R.K., 2012. Chemical and biological extraction of metals present in E waste: A hybrid technology. *Waste Manag.* 32, 979–990. https://doi.org/10.1016/j.wasman.2011.12.002

Salbidegoitia, J.A., Fuentes-Ordóñez, E.G., González-Marcos, M.P., González-Velasco, J.R., Bhaskar, T., Kamo, T., 2015. Steam gasification of printed circuit board from e-waste: Effect of coexisting nickel to hydrogen production. *Fuel Process. Technol.* 133, 69–74. https://doi.org/10.1016/j.fuproc.2015.01.006

Santella, C., Cafiero, L., De Angelis, D., La Marca, F., Tuffi, R., Vecchio Ciprioti, S., 2016. Thermal and catalytic pyrolysis of a mixture of plastics from small waste electrical and electronic equipment (WEEE). *Waste Manag.* 54, 143–152. https://doi.org/10.1016/j.wasman.2016.05.005

Shahabuddin, M., Uddin, M. Nur, Chowdhury, J.I., Ahmed, S.F., Uddin, M.N., Mofijur, M., Uddin, M.A., 2023. A review of the recent development, challenges, and opportunities of electronic waste (e-waste). *Int. J. Environ. Sci. Technol.* 20, 4513–4520. https://doi.org/10.1007/s13762-022-04274-w

Sheng, P.P., Etsell, T.H., 2007. Recovery of gold from computer circuit board scrap using aqua regia. *Waste Manag. Res.* 25, 380–383. https://doi.org/10.1177/0734242X07076946

Tripathi, A., Kumar, M., C. Sau, D., Agrawal, A., Chakravarty, S., Mankhand, T.R., 2012. Leaching of gold from the waste mobile phone printed circuit boards (PCBs) with ammonium thiosulphate. *Int. J. Metall. Eng.* 1, 17–21. https://doi.org/10.5923/j.ijmee.20120102.02

Tuncuk, A., Stazi, V., Akcil, A., Yazici, E.Y., Deveci, H., 2012. Aqueous metal recovery techniques from e-scrap: Hydrometallurgy in recycling. *Miner. Eng.* 25, 28–37. https://doi.org/10.1016/j.mineng.2011.09.019

Wang, R., Xu, Z., 2014. Recycling of non-metallic fractions from waste electrical and electronic equipment (WEEE): A review. *Waste Manag.* 34, 1455–1469. https://doi.org/10.1016/j.wasman.2014.03.004

Wu, C., Williams, P.T., 2013. Advanced thermal treatment of wastes for fuels, chemicals and materials recovery. In Hester, R.E., Harrison, R.M., Harrison, R., Hester, R. (Eds.), *Waste as a Resource*. https://doi.org/10.1039/9781849737883-00001

Xing, M., Zhang, F.S., 2013. Degradation of brominated epoxy resin and metal recovery from waste printed circuit boards through batch sub/supercritical water treatments. *Chem. Eng. J.* 219, 131–136. https://doi.org/10.1016/j.cej.2012.12.066
Yamawaki, T., 2003. The gasification recycling technology of plastics WEEE containing brominated flame retardants. *Fire Mater.* 27, 315–319. https://doi.org/10.1002/fam.833
Zhang, L., Xu, Z., 2016. A review of current progress of recycling technologies for metals from waste electrical and electronic equipment. *J. Clean. Prod.* 127, 19–36. https://doi.org/10.1016/j.jclepro.2016.04.004

9 Sustainable Treatment of Microplastic and Nanoplastic Waste

Surajit Ghosh[1*] *and Urmimala Naha*[2]

[1]Department of Chemical Engineering, Malaviya National Institute of Technology, Jaipur, India

[2]Department of Biomedical Engineering, Jaipur National University, India

9.1 INTRODUCTION

Micro- and nanoplastics (MNPs) are an increasing concern as they are environmental contaminants that are found everywhere and pose risks. Microplastics are plastic particles less than 5 mm, and nanoplastics are those with sizes from 1 to 100 nm. They result from the breakdown of larger plastics or are manufactured for various purposes, such as cosmetics, specialty coatings, and targeted drug delivery. MNPs are pertinent in that their presence can last for long periods in the biosphere and can unleash both toxic chemicals and additives that are detrimental to living beings.

Microplastics (MPs) and nanoplastics (NPs) are present in all spheres of the environment: the atmosphere, hydrosphere, lithosphere, and biosphere. Their presence is mainly due to the overuse and mishandling of plastic waste, which begins with the fragmentation of larger plastic products into smaller components by mechanical wearing and natural destruction over time. The resources can be further divided into primary sources and secondary sources. An illustration of micro- and nanoplastic pollution resources in the terrestrial environment is shown in Figure 9.1.

Primary sources are manufactured mainly for specific applications such as cosmetics, personal care products, cleaning agents, and industrial processes such as fibers for textiles. They are often used for their unique properties. Microbeads of polyethylene (PE) are widely used in facial scrubs, toothpaste, and shower gels as exfoliants and binding agents. Polypropylene (PP) is used in creams and shampoos as an exfoliant, like PE. Polyethylene terephthalate (PET) is used in products for its film-forming properties. Poly (methyl methacrylate) (PMMA) is employed as an opacifying agent in various cosmetics (Rahim et al., 2022). Commonly identified MPs in domestic cleaning agents are PP, PE, polyamide (PA), and polyurethane (PU).

Secondary sources include the generation of MPs and NPs due to the disintegration of larger plastic materials due to use or natural degradation. As such, microfibers are shed from larger plastic items, clothing, and other textiles. All of this degradation would be due to environmental aspects such as UV, heat, and wear and tear. Clothing, such as polyester and nylon, sheds microfiber while being washed. A fraction of these fibers could be disposed of in wastewater systems and ultimately end up in water bodies. The degradation of tires chipped within high-contact surfaces on vehicles releases microfibers into the atmosphere. Such particles could be poured into river bodies while it rains or settle down to the base and into the soil. Other sources would be some industrial activities such as the production and processing of plastics, which also discharge microfibers into the air space due to dumping, leakage, and unstructured disposal. Microfibers could also find

* Corresponding author's email: surajit.chem@mnit.ac.in

 DOI: 10.1201/9781003581000-12

FIGURE 9.1 Various resources of microplastics and nanoplastics in the terrestrial environment.

their way out of landfills as waste biodegrades, which contributes to soil and water table pollution. Micro- and nanoplastics could also escape into the atmosphere through the disintegration of larger plastics, where they can be transported by wind over long distances until they fall back on the ground (Shen et al., 2022).

MNPs have made their way into nearly every corner of our environment—land, water, and even the air we breathe. They spread easily through natural processes, moving between these ecosystems with surprising efficiency. In water, MNPs pose a serious threat because they can be ingested by marine and freshwater creatures, potentially building up in their systems and causing harmful effects. MNPs can penetrate plant and human health through the food chain since these MNPs accumulate in soil (Benson et al., 2022). MNPs can travel long distances through air, leading to greater concern for our indoor and outdoor environments.

The physicochemical and biological properties of the environment can be altered by MNPs, posing a great risk to ecosystems. MNPs infiltrate ecosystems, wildlife, and human bodies through food chains, ingestion, and inhalation, leading to severe health issues such as blood–brain barrier blockages, reduced fertility, and behavioral abnormalities (Chen et al., 2024; Liu et al., 2024; Luo et al., 2024; Prattichizzo et al., 2024). Other risks include a reduction in the growth of crop and microbial diversity and an increase in greenhouse gas emissions in soil ecosystems (Chen et al., 2024). MNPs can alter the oxidative stress, energy metabolism, and immune responses of organisms (Eliso et al., 2024). Exposure to airborne MNPs poses potential health risks to humans. These particles accumulate in the human body through inhalation and ingestion. Studies have detected microplastics in human stool samples, indicating exposure through food and water (Jiang et al., 2020). MNPs can release toxic additives and adsorb environmental chemicals, which may exhibit endocrine-disrupting and other toxic effects. Researchers are discovering more health impacts of MNPs, and the interactions of such particles with human health suggest significant risks (Amobonye et al., 2021; Jiang et al., 2020).

Traditional methods, such as light scattering and mass detection techniques for detecting particulate matter (PM) in the air, are well established and cost-effective. However, modern concerns are associated with the small size and diverse chemical nature of MNPs that are undetectable by traditional methods. Moreover, distinguishing between plastic and non-plastic particles or identifying specific types of plastics is challenging. Modern techniques utilize a variety of approaches in determining the quality of the MNPs, such as fractionating them into smaller identifiable compounds and studying the kinetics of these decompositions. Analytical methods such as pyrolysis-gas chromatography-mass spectrometry (Py-GC/MS) (Thomas et al., 2024), microscopic and spectroscopic techniques (Rai et al., 2021), pyrolysis gas chromatography-ion mobility mass spectrometry (pyr-GCxcIMS) (Hashemihedeshi et al., 2024), matrix-assisted laser desorption/ionization-time of flight mass spectrometry (MALDI-TOF MS) (Wu et al., 2020), and surface-enhanced Raman spectroscopy (Mogha & Shin, 2023) are effective in detecting the true nature of MNPs. However, standardization of measurement protocol for sampling and size fractionation of MNPs is needed to redevelop the aggregation and comparison of data across studies. Moreover, analytical machinery is quite expensive and is unable to provide point-of-care testing, which is a modern-day requirement. The development of portable and low-cost detection methods, such as handheld microscopes for microscopic identification, is a promising direction for making MNP detection more accessible and efficient (Thomas et al., 2024).

Effective waste management and remediation approaches are essential to mitigate the environmental and health impacts of microplastics and nanoplastics. Commercial interventions and policy framing are necessary in deriving waste management strategies. Long-term sustainability and the protection of ecosystems and human health through those strategies will be crucial in mitigating the plastic pollution problem.

This chapter highlights the current research and technological advancements toward the remediation of MNPs from air, water, and soil. In addition, it discusses challenges in developing sustainable solutions to tackle MNP contamination. Finally, the chapter proposes innovative solutions and offers insights into future research directions.

9.2 EXISTING WASTE MANAGEMENT AND REMEDIATION APPROACHES

To solve the menace of microplastic and nanoplastic pollution, several strategies were adopted and each solution has its pros and cons. Conventional approaches to solving the plastic management problem include landfilling, which is itself the biggest resource for microplastic and nanoplastic pollution. Targeting to separate the microplastics from water bodies, conventional sand filtration methods were adopted, which can separate micrometer-sized particles but are incapable of separating smaller particles. This challenge gradually leads to the development of a membrane-based filtration system. Membrane-based filtration methods are found to be impressive in separating smaller microplastics and nanoplastics. However, challenges like fouling of membranes by microplastics and nanoplastics on the membrane surface and pores created huge challenges for long-term operation. To mitigate the fouling problem, other physical separation methods, such as magnetic separation and centrifugal separation, and chemical methods, such as coagulation and flocculation, were developed. Such methods are quite effective in separating MPs from water; however, they are not suitable for separating MPs from air medium. Electrostatic methods were developed to separate MPs from the air and showed promising results. Such advanced methods are quite successful in removing MPs, but the separation of nanoplastics is quite challenging for them. To separate NPs, chemical-based methods, that is, adsorption methods, are quite effective. Adsorption methods are highly successful because of the selective separation of targeted plastics. The cost of adsorbents is a crucial commercial challenge; hence, low-cost adsorbents like biochar or modified biochar turn into a center of attraction for all the adsorption-based separation research. However, chemical-based methods require harsh chemicals that pose another challenge to disposal after use. To protect the environment from these after-treatment residues, biological techniques were adopted for sustainability. Biological methods such as microbial degradation, enzymatic breakdown, and bioremediation are widely researched and found to be effective in degrading microplastics and nanoplastics into less harmful products. However, the slow kinetics of these biological methods is a great challenge for commercial adoptability. Researchers have identified that the integration of such methods based on the optimization of variable MNPs can be removed through commercial and sustainable approaches. Various treatment methods for the separation of MNPs are discussed in the following sections. This chapter focuses on various physical separation techniques for MNPs and gives an overview of other separation techniques since the number of research studies in those directions is limited, and the insight into such methods is still evolving. The advantages and limitations of these methods are reported in each section, and the relevant optimization factors for these methods are identified.

9.3 PHYSICAL TREATMENT TECHNOLOGIES

Physical treatment technologies are generally efficient and easy to adapt to separate particulate matters from different media, that is, air and water. Filtration, centrifugation, magnetic separation, and ultrasonication are some common physical treatment methods for wastewater treatment, and electrostatic precipitation is generally employed for air purification. Filtration processes, including membrane technologies, can remove MNPs from wastewater, and removal efficiencies of over 90% are easily achieved. Filtration methods require a filtering media, and membranes are typically the most popular ones. Centrifugation and electrostatic precipitation are also employed to separate and capture plastic particles based on their physical properties, while magnetic separation utilizes magnetic fields to extract plastics that have been treated with magnetic particles (Kundu et al., 2021). Ultrasonication, on the other hand, uses high-frequency sound waves to agitate and break down plastic particles, facilitating their removal from contaminated environments. These physical treatment technologies are discussed in the following sections.

9.3.1 Filtration Techniques

Filtering is a process based on particle size, where the filter medium is able to hold back particles that are larger than the pores of the filter medium. Some filtration techniques include gravity filtration and membrane filtration, which have very high removal efficiencies, with research indicating that microplastics larger than 100 μm can be removed completely during water treatment. For the separation of smaller-sized particles, the filter medium must contain pores that are smaller than a micron. Other filtration techniques being discussed include ultrafiltration and nanofiltration, as well as their ability to remove microplastics from freshwater bodies efficiently, even though there are challenges concerning the costs and implementation. The integration of nanotechnology into filtration systems, for instance, the use of electrospun nanofiber membranes, has the potential to improve the filtration efficiency and multifunctionality of the systems, including air filtration and water treatment. Hence, as the research advances, the discovery of new filtration techniques, including granular media-based technologies and modified media filters, is also enhanced with the objective of seeking better ways of dealing with micro- and nanoplastics (Zhang et al., 2020; Juraij et al., 2023; Oza & Pathak, 2024).

9.3.1.1 Gravity Filtration

Sand filtration technologies, particularly rapid sand filtration and granular filtration, are effective methods for removing microplastics from water. These technologies offer high removal efficiencies, especially for larger microplastics, and can be optimized through careful selection of filter media size and integration with pretreatment processes.

9.3.1.1.1 Rapid Sand Filtration

Rapid sand filtration (RSF) is a promising technology for microplastic removal from water, including agricultural wastewater. In this procedure, water is led through a bed of sand, which adsorbs microplastic particles. Research reports that RSF can reach a microplastic removal efficiency of up to 97% and that this approach is a means to mitigate plastic pollution in the water supplies. The performance of RSF may depend on such parameters as the size of the filter media and loading rate. For example, smaller filter media dimensions can be expected to raise removal efficiency due to a larger surface area available to bind particles. Furthermore, RSF is especially efficient at removing larger microplastic particles >200 μm in size. Whereas RSF is less effective than certain advanced technologies, such as membrane bioreactors, it is a cost-effective and broadly applicable microplastic removal technique that is applicable in many water treatment applications. RSF can be implemented in many locations, such as urban wastewater treatment works and plastic recycling industry effluent, which illustrates RSF's versatility and efficacy in a wide range of water treatment (Bayo et al., 2020b; Bodzek & Pohl, 2022; Sembiring et al., 2021; Talvitie et al., 2017).

Advantages

1. RSF is efficient in particle removal, with removal efficiencies as high as 97% for municipal wastewater (Wolff et al., 2021).
2. Specifically, it is able to remove particulate microplastic forms as efficiently as fibers. RSF shows greater removal yield of micrometaplastics (from 95.53 to 98.83) than fibers (from 53.83 to 57.65) (Bayo et al., 2020a, b, 2021).
3. RSF can also be incorporated into the routine of water treatment processes, and thus, it is a very flexible option for a set of applications.

Disadvantages

1. RSF exhibits lower removal efficiency for microplastics <20 μm, where removal efficiency could be as low as around 86.9% (Bayo et al., 2020b; Fajar et al., 2022).
2. The process may involve pretreatment stages (e.g., coagulation, sedimentation) to improve efficiency and avoid fouling.

Optimization Factors

1. The effective size of the filter media is crucial. Evidence shows that smaller filter media, for example, 0.39 mm, yield higher removal efficiencies for microplastics. For example, removal efficiencies of 97.7% and 94.3% were achieved at 0.39 mm and 0.68 mm medium for plastics above (or less) 200 μm in diameter, respectively (Fajar et al., 2022; Sembiring et al., 2021). In the same way, a 0.4-mm media size has a microplastic efficiency of 91.30% for microplastics <400 μm and 95.80% for larger sizes (Wulandari et al., 2024). Shorter media sizes cause increased head loss, which could affect the filtration cycle. As an example, a 0.39-mm media diameter produced significantly more head loss than 0.68-mm media, altering the duration of the filtration cycle (Fajar et al., 2022).
2. Shifting the filtration rate and adding pretreatment steps such as coagulation can increase the reduction of smaller microplastics (Lapointe et al., 2020; Nakazawa et al., 2021; Zhang et al., 2020).

9.3.1.1.2 Granular Filtration

Granular filtration is a relatively universal method for MNP removal in water treatment plants. This technique involves propagating water flow through a bed of granular material [e.g., granular activated carbon (GAC)], which has inherent characteristics to filter particulates. Granular filtration has also been confirmed to be extremely effective in microplastic removal, with removal efficiencies between 86.9% and almost 99.9% for particles >100 μm. The granular filtration removal rate is associated with the morphology and size of the microplastics, and the removal rate of the microplastic with the smallest size (10–20 μm) is relatively low. At an advanced stage of large-scale water treatment plants, treatment using GAC has been shown to be effective in the removal of microplastics by ~56.8–60.9% depending on the size of the microplastics but with a substantially reduced effect for the smaller-sized microplastics. The adsorption capacity of GAC is also enhanced due to electrostatic interactions that are enhanced by the dissolved organic matter and divalent species and, in turn, improved removal efficiencies. In addition, as a cheap material, biochar has also been explored for inclusion in sand filter systems with a high capacity for microplastic sphere removal and encapsulation. With these findings, it is shown that granular filtration is a potent technique that holds the potential to control microplastic pollution in water treatment units (Zhang et al., 2020; Wang et al., 2020b).

Advantages

1. Effective, nearly complete removal of microplastics >100 μm can be achieved with granular filtration, with efficiency as high as 99.9%. This high efficiency is due to the physical confinement of microplastics in the granular media that efficiently collects and store them.
2. Optimum MNP removal efficiency can be obtained with an optimized reactor design or longer residence time. GAC can remove microplastic to the efficiency of up to 95.5% in synthetic wastewater and improve its effectiveness by increasing the bed length of GAC from 7.5 to 17.5 cm (Napi et al., 2023).
3. It is efficient when combined with other treatment strategies, that is, activated carbon filtration, to improve the removal efficiency as a whole. The presence of GAC filtration decreased microplastic levels, indicating its contribution to making the overall treatment process better.

Disadvantages

1. The biofilm on microplastic formation can complicate the removal process, but it may also increase removal efficiency in certain situations.
2. However, filtration by itself may not be effective enough to remove the smallest microplastics, and other treatment steps need to be taken.
3. Granular filtration may also break microplastics into smaller pieces, contributing to a rise in smaller particles in the effluent.

Optimization Factors

1. *Coagulating aids*: The removal of microplastics, particularly smaller particles, can be significantly enhanced by the addition of coagulant aids during the coagulation/flocculation stage. The addition of coagulant aids such as PolyDADMAC has been shown to improve the removal efficiency of microplastics during the coagulation/flocculation stage. However, even with these aids, the removal efficiency for smaller microplastics (e.g., 45–53 μm) can be limited, achieving a maximum of 13.6% removal in some cases. Different coagulant aids have varying levels of effectiveness. For instance, anionic polyacrylamide (PAM) and activated silicic acid (ASA) have been tested, with ASA achieving a removal efficiency of 54.70% under conventional dosages, while PAM reached up to 91.45% at higher dosages (Zhang et al., 2020, 2021).
2. *Regular maintenance*: Optimization of this process also includes a number of parameters, such as routine and periodic maintenance and optimization of filtration media to prevent fouling and maintain high removal efficiency. Maintenance operations are based on washing and replacing the filtering material to prevent clogging and fouling, which strongly reduces the system's performance, as mentioned in the characterization of the filter. For instance, granular activated carbon (GAC) filters have been shown to be effective

in removing microplastics but may become less effective as a function of time because of fouling. Fouling can be reduced by tuning the operating conditions and monitoring the system performance on a regular basis. Due to the precipitation and deposition of dissolved organic matter (DOM) and particulate matter, fouling takes place, and the adsorption capability of the filtration medium is subsequently reduced. Therefore, understanding the interaction between microplastics, DOM, and filtration materials is of great importance for developing strategies to prevent fouling and overcome low removal efficiencies (Napi et al., 2023; Zhang et al., 2020; Ramirez Arenas et al., 2021).

9.3.1.2 Membrane Filtration

Due to the failure of the sand filtration process to effectively remove media with <100 μm MPs, new advanced membrane filtration methods have been developed. They work based on fractionation of the contaminants according to their size, charge, or other characteristics, so they are very effective in removing plastic particles; that is, membrane filtration is a class of technologies that include microfiltration (MF), ultrafiltration (UF), and nanofiltration (NF) methods, each of them adapted to remove different particle sizes. Such pore sizes of membranes are depicted in Table 9.1. For example, microfiltration and ultrafiltration are highly efficient at removing microplastics, with removal efficiencies greater than 90% in many. Apart from size-based separation, membrane technologies can remove MNPs by charging the particles, hydrophobicity, and adsorption. Advanced filtration methods, including electrospun nanofiber membranes, utilize surface charge modifications to enhance the removal of nanoplastics, even those smaller than the membrane pores, through electrostatic attraction. Hydrophobicity can also be induced into the membrane to trap the hydrophobic plastic particles using hydrophobic polymer. Modified media filters, such as those incorporating carbon nanoparticles, offer innovative approaches to simultaneously address water disinfection and microplastic removal. However, all the membrane technologies are susceptible to fouling due to pore clogging and pore plugging, which leads to a reduction in the flow rate of treated water. Longer operation creates cake layers that are hard to remove. Figure 9.2 shows a cake formation mechanism on the membranes. These technologies collectively provide a comprehensive framework for mitigating the pervasive issue of plastic pollution and will be discussed in detail to explore the mechanisms and applications (Pramanik et al., 2021; Acarer, 2023).

9.3.1.2.1 Membrane Microfiltration

Membrane filtration is a process by which particles are removed via size-selective exclusion, namely, microfiltration. It does it through the phenomenon of allowing water and other smaller molecules to permeate through a porous membrane while keeping larger molecules, such as microplastics, on the membrane surface. The efficiency of microfiltration for microplastic removal is mainly controlled by the pore of the membrane. Typically, microfiltration membranes have pore sizes ranging from 0.1 to 10 μm, which are suitable for capturing microplastics and even some nanoplastics. For example,

TABLE 9.1
Various Membrane-Based Filtration Methods Used Based on the Size of Micro- and Nanoplastics

Serial Number	Filtration Technique	Particle Size Range
1	Microfiltration (MF)	0.1 μm to 10 μm
2	Ultrafiltration (MF)	1 nm to 0.1 μm
3	Nanofiltration (NF)	1 nm to 10 nm
4	Electrospun nanofiber membrane	10 nm to 1 μm

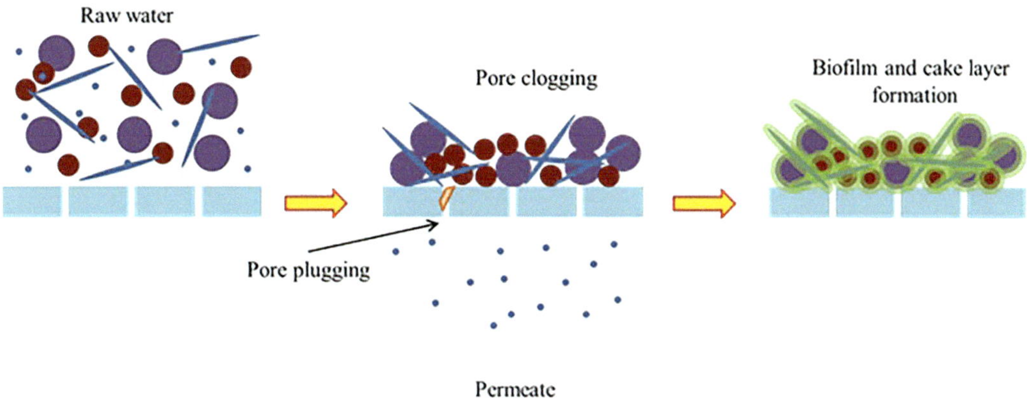

FIGURE 9.2 Fouling of the membranes with the microplastics and formation of cake layer on the membrane.

membranes with 0.1 μm pore size have been reported to furnish up to 98% removal efficiency of microplastics from wastewater discharges (Yahyanezhad et al., 2021). Microfiltration is generally carried out at relatively low pressure, a property that gives it an advantage over other membrane process techniques, that is, ultrafiltration or reverse osmosis. Pressure requirements for microfiltration often vary between 0.5 and 2 bar (7 and 30 psi) depending on the application and membrane configuration (Wang et al., 2017, 2020b).

Advantages

1. *High efficiency in removing microplastics*: Compared to other filtration methods, such as ultrafiltration, nanofiltration, and reverse osmosis, microfiltration requires less energy. This is due to its higher permeability, which allows for efficient water flow through the membrane with lower pressure requirements. This energy efficiency makes microfiltration a cost-effective option for large-scale applications.
2. *Pretreatment and process integration*: Microfiltration is often used as a pretreatment step before other processes (UF, NF, RO), significantly reducing fouling and extending the life span of these membranes. This leads to lower cleaning and maintenance costs and improves the overall efficiency of the system.

Disadvantages

1. *Inability to remove nanoplastics*: Although microfiltration is highly selective for microplastics, it is ineffective for nanoplastics because of their smallness. Conventional microfiltration membranes do not have pores that are large enough to trap particles smaller than the pores made in the membrane; therefore, their maximum efficiency in removing all kinds of plastic pollutants is limited.
2. *Fouling*: Fouling is a significant challenge for microfiltration membranes. Accumulation of particles and microorganisms on the membrane surface impairs both its efficiency and lifetime. This requires constant cleaning and intervention to maintain a good level of performance, which adds to operational costs.
3. *Limited lifespan of membranes*: Physical and chemical loads accumulated by microfiltration membranes during operation may result in deterioration over time. This restricts their life span and necessitates replacement at regular intervals, which increases the total cost and complexity of its implementation for long-term environmental cleanup.

4. *Maintenance requirements*: Regular maintenance is critical for controlling fouling and maintaining microfiltration membrane effectiveness. This encompasses cleaning methodologies and possible chemical interventions, which may be labor intensive and need special expertise to be carried out efficiently.

Optimization Factors

1. *Membrane material*: The choice of membrane material affects the interaction between the membrane and microplastics. Materials like cellulose acetate and polycarbonate have been used effectively, with cellulose acetate showing high performance in long-term applications. They are also known for their durability and resistance to fouling, making them suitable for long-term applications. Membranes must resist degradation from chemicals present in the water or used in cleaning processes. Cellulose acetate generally performs well, although it can be affected by certain aggressive chemicals.
2. *Pore size distribution*: The pore size of the membrane is critical for determining the size range of particles that can be removed. Narrower pore size generally offers better removal capacities for smaller microplastics and nanoplastics. This is consistent with models for the prediction of pore blocking and permeability loss, which emphasizes that size exclusion is the main mechanism of particle retention in microfiltration operations. The capability for modulating and controlling the pore size distribution is an important requirement for the separation performance of membranes. Small pore sizes, on the other hand, have the trade-off of enhancing the removal of finer particles, as well as increasing fouling and higher pressures. Membranes with smaller pores are more susceptible to plugging and can result in decreased permeability and higher operational costs. There is a critical trade-off between pore size and fouling resistance, as seen from studies that demonstrate the benefits of narrowing the pore size distribution to gain both antifouling performance and long-term stability. Transmembrane pressure (TMP), flow rate, and membrane fouling must be controlled to keep the performance at its maximum level.
3. *Transmembrane pressure*: TMP is an important system parameter for microfiltration devices that affects not only the filtration performance but also the life of the membrane. High TMP can increase the driving force for filtration, which can lead to improved water microplastic removal efficiency. However, the elevated TMP can further increase the effect of membrane fouling and reduce the performance of the membrane owing to the higher compaction and pore blocking caused by the higher TMP. Low-pressure operation (up to 0.6 psi, ~4 kPa) has been demonstrated as effective for the removal of polystyrene nanoplastics, particularly in the case of membranes with improved surface characteristics. (Wang et al., 2020a).
4. *Flow rate*: The flux is also one of the critical performance parameters of microfiltration systems. When balanced flow rates are used, sufficient contact time between the water and the membrane is provided to maximize the removal efficiency of microplastics and nanoplastics. Experiments have demonstrated that a water flow rate of 2 L/min plus gas scouring is able to limit the fall in water flux as well as reduce fouling (Enfrin et al., 2021a). This equilibrium plays a key role in ensuring high removal efficiency without suffering from premature fouling, which can occur at higher flow rates.
5. *Fouling removal strategies*: Hydrophilic surface coatings (e.g., acrylic acid) have been reported to decrease nanoparticle/microplastic adsorption by more than 60%, thus increasing the water flux and decreasing fouling (Enfrin et al., 2021b). Furthermore, physical cleaning approaches, such as periodic gas scouring, can greatly reduce fouling by generating shear forces that interfere with the adsorption of microplastics onto the membrane surface.

9.3.1.2.2 Membrane Ultrafiltration

Ultrafiltration (UF) is a membrane filtration process that separates particles based on size through a semi-permeable membrane. The process operates under low pressure, typically ranging from 0.1 to 1 bar, which forces water and small solutes through the membrane while retaining larger particles, such as microplastics and certain nanoplastics, on the surface. The pore size of UF membranes is typically between 1 and 100 nm, and thus UF membranes are suited for removing very large particles (microplastics whose diameter is larger than 100 nm) and some nanoplastics (as small as a few nm). Several ultrafiltration membranes, including those based on poly (sulfone), PVDF, and poly(ethersulfone), have been fabricated to discriminate the size and kind of micro and nano plastics with the aid of size exclusion and adsorption to improve separation efficiency. Enfrin et al. (2020) found that poly (sulfone) membranes could exclude MNPs with sizes between 13 and 690 nm; however, fouling of the membrane due to adsorption of the MNPs can substantially decrease the flux over time. The same team also stated that by plasma treatment and sour gas bubbling of the acrylic acid, cyclopropylamine, and hexamethyldisiloxane membranes, it is possible to mitigate the decrease in flux produced by fouling (Enfrin et al., 2021b).

Advantages

1. *High removal efficiency*: Ultrafiltration membranes are very efficient in microplastic and nanoplastic removal, thanks to the filtering action on particles of small size. Ultrafiltration can provide removal efficiencies in excess of 90 for microplastic and nanoplastic particles, therefore making it a viable technology for water treatment purposes.
2. *Moderate pressure operation*: Ultrafiltration is done under medium pressure levels as opposed to other membrane engineering technologies, such as nanofiltration, where higher pressure is applied. This property allows ultrafiltration to be energy efficient and financially more efficient for microplastics and nanoplastics removal from water.
3. *Versatility across applications*: Ultrafiltration membranes are versatile and can be applied in various water treatment settings, including drinking water treatment plants and wastewater treatment facilities. They work well in various water matrices and provide a versatile solution for a wide range of environmental conditions.

Disadvantages

1. *Fouling*: Like microfiltration membranes, one of the main limitations of ultrafiltration membranes is their susceptibility to fouling, particularly when dealing with complex water matrices containing microplastics. Fouling can lead to decreased membrane performance and increased maintenance requirements. The presence of microplastics can exacerbate fouling by forming a cake layer on the membrane surface, which can increase TMP and reduce flux.
2. *Higher energy and maintenance requirements*: In contrast to microfiltration, ultrafiltration demands much higher energy input and maintenance, much greater pressure, regularity of regular cleaning, plus continuous fouling suppression efforts. This may lead to an increase in the operational costs and management of the water-treating systems.
3. *Complexities in handling complex water matrices*: The presence of multiple pollutants, such as humic acid and bisphenol A, can complicate the filtration process. These can result in increased fouling and a decline in membrane performance, and therefore, the use of highly advanced filtration systems or precipitation steps is desirable.

Optimization Factors

1. *Transmembrane pressure*: Just as microfiltration membranes, TMP has an important effect on the performance of the ultrafiltration membrane in that it directly affects the speed

of filtration as well as the degree of fouling. It has been demonstrated that TMP is positively related to microplastic loading with the potential to aggravate membrane fouling. In order to counteract this, strategies such as aluminum-based flocculants have been used to develop loose cake layers on the membrane surface and thus greatly reduce TMP and fouling. Moreover, it has been shown that low-pressure run conditions can achieve excellent removal efficiency while consuming little energy (Liao et al., 2020).

2. *Pore size selection*: Removal is significantly improved when the pore size is larger than the particle size by electrostatic interactions, as shown for membranes with controlled surface charges. This emphasizes the need to trade off pore size against other membrane characteristics in order to achieve the desired performance optimization.
3. *Flow or pressure conditions*: Complex flow and pressure environments, for example, the dynamic rotation of membrane modules can be utilized to control fouling by controlling the thickness of cake deposits that are located on the membrane surface (Li et al., 2021). Computational fluid dynamics (CFD) modeling has been employed to parameterize these conditions, optimizing them from laminar to turbulent flow, where the shear stress is maximized and fouling is minimized.
4. *Membrane modification with treatments or additives*: Additional treatments and additives play a significant role in enhancing the filtration performance of ultrafiltration membranes. Surface modifications, such as the deposition of zwitterionic polymers and poly(ethylene glycol), have been shown to improve antifouling properties without compromising membrane permeability. The incorporation of nanomaterials, such as multi-walled carbon nanotubes, can also enhance antifouling capabilities and improve the removal of specific contaminants. Furthermore, the use of silver nanoparticles has been explored for their antimicrobial properties, although care must be taken to prevent rapid depletion of the active material (Zodrow et al., 2009). The combination of blending and surface modification strategies can further enhance both permeability and fouling resistance, leveraging the strengths of each approach. Plasma treatments have a positive impact on ultrafiltration membranes by altering their hydrophilicity. Plasma-treated ultrafiltration membranes of acrylic acid and cyclopropylamine show low flux declination, as low than 8% flux reduction after 6 hours of crossflow filtration, while reducing the adsorption of nano/microplastics by more than 60% compared to untreated membranes (Enfrin et al., 2020).

9.3.1.2.3 Membrane Nanofiltration

Nanofiltration is a process that sits between ultrafiltration and reverse osmosis for the purpose of selectively removing particles according to their molecular size and charge. In particular, it is highly effective for 1–10-nm particles and, therefore, viable for the removal of nanoplastics in an aqueous environment. NF membranes have a pore size of 1 nm and a molecular weight cut-off (MWCO) between 300 and 500 Da, which can essentially exclude ions and most small organic molecules by cooperative action of steric hindrance, electric exclusion, dielectric effects, and hydration processes (Suhalim et al., 2022). Both size exclusion and charge repulsion mechanisms act as separation mechanisms in NF, and the pore size and surface charge of the membrane have a considerable impact on the selectivity and performance of the filtration process. Nanofiltration membranes are extremely useful for the removal of nanoplastics that are typically smaller than 100 nm. For instance, polystyrene nanoplastics down to 50 nm have been effectively removed by positively charged membranes, independent of the particle size being less than the pore size of the membrane. This is based on electrostatic interactions, which result in selective adsorptions of both the nanoplastic and the membrane.

The constitutions of nanofiltration membranes are critical to their performance and lifetime. Polymers and polyacrylonitrile (PAN), poly(sulfone), and covalent organic frameworks (COFs) are a few examples of materials used. These materials are selected with regard to mechanical resistance, chemical stability, and their ability to be tailored to achieve the best compromise in terms of surface charge and hydrophilicity. For instance, a study implemented a highly porous PAN electrospun membrane followed by layer-by-layer buildup to adjust the surface charge, which led to significant improvement in nanoplasticle removal capacity (Wang et al., 2020a). In addition, COF membranes have also been described for their high porosity and completely controlled pore structure, which are capable of providing high nanofiltration behavior.

Advantages

1. *High removal efficiency*: Nanofiltration membranes, with their subnanometer pore sizes and the capability of using multiple separation mechanisms, including steric hindrance and electrostatic interactions, are capable of high removal efficiencies. It has been demonstrated that nanofiltration can remove >90% of nanoplastic particles, thus representing a promising technology for the treatment of water.
2. *Selective separation*: Nanofiltration membranes have the benefit of sieving separation, and the nanoplastics are retained, and water and some dissolved salts can permeate. This selectivity is advantageous for achieving water quality protection and for pollutant removal. The capability to fine-tune membrane characteristics (e.g., surface charge), in turn, augments the selectivity and efficiency in the removal of nanoplastic particles (Chen et al., 2021; Wang et al., 2020a).

Disadvantages

1. *Fouling*: Similarly to other membrane technologies, fouling may arise from the deposition of nanoplastics and other organic/inorganic materials on the membrane surface, thus inducing decreased permeability with increased operation cost. This is a task that requires routine cleaning and maintenance on a regular basis, which is time consuming and expensive.
2. *Pressure requirements and energy consumption*: Processes based on nanofiltration generally require higher pressure conditions (0.4–1 bar) than other techniques, which can result in higher energy consumption. This requirement can be a serious limitation, particularly when large applications are considered where energy efficiency plays an important role (Suhalim et al., 2022; Wang et al., 2020a).
3. *Higher initial costs and maintenance challenges*: The fabrication of specialized membranes with, for example, controlled surface charges to improve selectivity and permeability requires advanced manufacturing processes with increased cost. As an example, the fabrication of electrospun polyacrylonitrile (PAN) membrane with a layer-by-layer assembly strategy to develop targeted surface charges is a laborious and expensive task (Wang et al., 2020a). Interaction of nanoplastics with membrane surfaces can result in fouling and decreased performance with time, requiring cleaning and replacement more often. Since membranes with enhanced surface characteristics have fewer irreversible fouling episodes, maintenance is still the key problem (Mohana et al., 2021).
4. *Complexities in handling complex water matrices*: The interaction between nanoplastics and other water constituents (e.g., biofilms) adds a challenge to the removal process. Biofilms formed on nanoplastics may increase the removal efficiency of certain treatment processes like coagulation/flocculation, but they also introduce variability and unpredictability in treatment outcomes (Castillo-Ruiz et al., 2024).

Optimization Factors

1. *Surface charge and membrane material*: Surface charge is of great importance in improving the selection and capture of nanoplastics. For example, positively charged membranes have shown high performance in nanoplastic removal of smaller nanoplastics, for example,

50 nm polystyrene particles, using electrostatic attraction that promotes high selectivity while maintaining high performance and avoiding irreversible fouling (Wang et al., 2020a). Alternatively, membrane composition (e.g., the presence of hydrophilic coatings) can also limit the amount of nano/microplastics adsorbed to the membrane, which in turn can be prevented from accumulating on the membrane (membrane fouling) and so from impeding high water fluxes. These changes not only increase the removal efficiency of nanoplastics but also increase the mechanical strength and hydrophilicity of membranes, which is effective in complex water matrices.

2. *Membrane structure and composition*: Nanofiltration membranes are now increasingly acknowledged as having the ability to effectively remove nanoplastics from water, with membrane process and configuration as important parameters of optimization. The efficacy of nanofiltration is primarily dependent on the pore size distribution and the surface charge of the membrane, which can be modified to improve the selectivity and permeability. Polyamide nanofiltration membranes that can be finely engineered to have a specific pore structure and high surface charge density have shown excellent >98% rejection performance to small molecules and ions, a key role in the efficient removal of nanoplastic particles (Cheng et al., 2021).
3. *Fouling resistance and maintenance*: In general, using thermoresponsive interfaces, for example, grafting poly(N-isopropyl acrylamide) (PNIPAM) chains onto membranes to improve antifouling ability, has been shown to decrease membrane fouling due to the formation of a buffer layer and the consequent decrease of membrane-fouling interaction (Daliang Xu et al. (2022). Mixed-matrix membranes, like membranes containing MXene and cellulose acetate, can improve antifouling properties and bacterial proliferation inhibition and are thus suitable for water purification applications (Pandey et al., 2020).

9.3.1.2.4 Electrospun Nanofiber Membranes

Electrospun nanofiber membranes (ENMs) are very efficient for nanoplastics removal due to their ultrafine fiber morphology and high surface area-to-volume ratio. These characteristics allow ENMs to capture nanoscale particles efficiently. The high porosity and interconnected structures of the membranes favor filtration due to the presence of numerous channels for water movement, which trap the nanoplastics. The ultrafine fibers provide a high level of robustness that forms a densely packed network capable of capturing particles smaller than the apparent pore size, thereby conferring the membrane with an increased potential to capture nanoparticles. Figure 9.3 presents types of electrospun nanofibers applied to water treatment.

Advantages

1. *High surface area-to-volume ratio*: The high surface area-to-volume ratio of nanofiber membranes can greatly improve the filtration or adsorption performance of the membrane as it offers a more active surface for contaminant interactions. This property enables adsorptive and filtration properties with an increased efficiency because, with the rise of the surface area, the contact area between pollutants and material, is improved.
2. *Tunable pore size*: Tuning pore size capability is of great importance to improve the filtration performance of these membranes, especially in capturing nanoplastics below the membrane pore size. By the layer-by-layer (LbL) assembly method, the pore size can be modulated. This method comprises the serial deposition of complementary charged polymers on the surface of the membrane, which can affect the surface charge and the pore size. A study demonstrated the use of polyethylenimine (PEI) and poly(acrylic acid) (PAA) to create membranes with varying charges, which effectively removed polystyrene nanoplastics as small as 50 nm with a removal efficiency of 89.9% under low-pressure conditions (R. Wang et al., 2020a). Polymer chains grafting onto the nanofiber surface can also

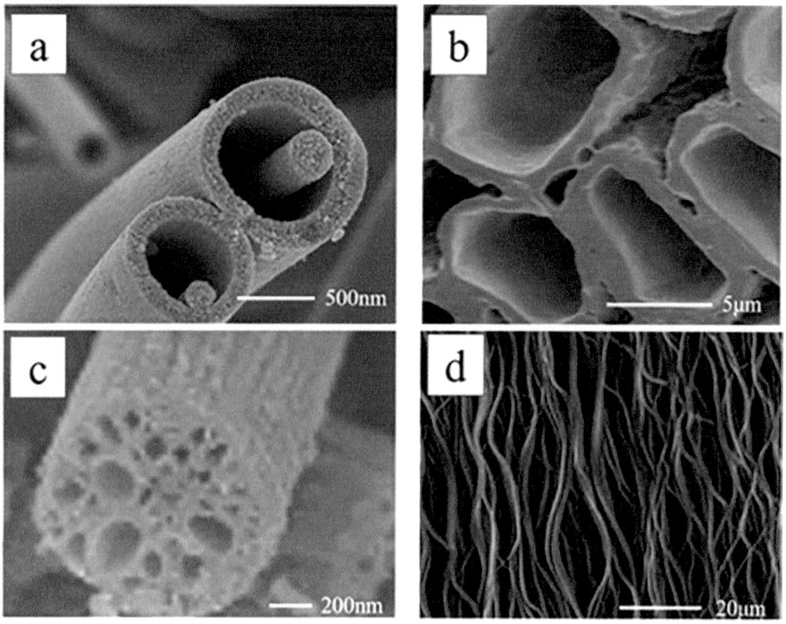

FIGURE 9.3 SEM images of electrospun fibers with different morphologies. (a) Core-shell structure; (b) hollow structure; (c) porous structure; (d) aligned structure. (The figure is adapted from Cui et al., 2020, with permission.)

decrease the pore size through an increase in the diameter of the fibers or by sealing the pores with gel-like structures. Grafting water-insoluble poly(HEMA) to regenerated cellulose nanofibers reduced pore sizes and facilitated the exclusion of nanoparticles >40 nm (Wang et al., 2017).

Disadvantages

1. *Mechanical strength and durability*: Mechanical strength and durability of ENMs are some of their challenges. Although these properties provide a high porosity and surface area, they may, at times, lead to a lack of the structural robustness of the membranes and decreased durability under certain operational conditions.
2. *Fouling*: Like other membranes, ENMs are also susceptible to fouling. Despite being addressed to a certain degree through surface modifications, fouling continues to be a major problem that can be strongly detrimental to the long-term performance and serviceability of membranes.
3. *High production costs*: The synthesis of electrospun nanofiber membranes (ENMs) comes with a number of cost inputs that make the process a commodity. The substrates employed in the synthesis of ENMs—polyacrylonitrile (PAN), polyethyleneimine (PEI), and poly(acrylic acid) (PAA)—are expensive. The LbL polymerization process needs to be precisely controlled and requires several attempts, including alternative assembly of different polymer layers that require more time and manpower. The nature of these operations renders production costs high due to the need for special-purpose equipment and skilled personnel. The size of the process is usually limited to maintain the homogeneity and uniformity in the membrane structure and properties, thereby escalating the costs (Li et al., 2023; Pervez et al., 2022; Wang et al., 2020a).

9.3.2 Centrifugation

Centrifugation is a technique that uses centrifugal force to separate particles based on particle size and particle density. If a sample containing nanoplastics is centrifuged, centrifugal acceleration of the drug solution forces the heavier nanoplastics to the bottom of the centrifuge tube and the lighter nanoplastics to the top. This partitioning is suitable for nanoplastics due to their specific density difference with respect to the surrounding medium. For instance, ultracentrifugation has been applied to efficiently remove polystyrene nanoplastics from river water with a recovery rate of 87.1% (Cai et al., 2021). It is applicable to natural or artificial water systems in order to reduce the pollution of nanoplastics. Given the ability of the study to concentrate and enrich nanoplastics, this is a valuable tool for environmental monitoring and research. More precisely, continuous flow centrifugation, especially, offers the possibility to sample and enrich nanoplastics in outdoor water and could thus solve the problem of contamination and discriminate by particle size (Hildebrandt et al., 2019).

Advantages

1. *Density difference*: The efficiency of centrifugation is mainly determined by the density difference between nanoplastics and the matrix. A greater density difference facilitates more efficient separation.
2. *Centrifuge speed and time*: The speed and duration of centrifugation are critical parameters. Increased speeds and increased centrifugation time are generally more efficient for the separation of a given constituent. As an illustration, centrifugation at 10,000 rpm for 10 min has been shown to remove up to 99% of nanoplastics (Murray & Örmeci, 2020).
3. *Operational scale*: The scalability of centrifugation methods is a consideration. Although lab-scale experiments provide encouraging results, for large-scale applications, investigations are needed to ensure an acceptable level of efficiency and economic viability.

Disadvantages

1. *Energy requirements*: Centrifugation is energy intensive (especially at high speeds dictated by the separation requirement of particles below nanoplastics).
2. *Scalability*: Scaling up from lab-scale experiments to mass production is challenging due to the energy consumption and capital cost of the equipment.
3. *Pretreatment and post-treatment*: Nevertheless, additional strategies still need to be developed in order to improve the separation efficiency to a greater extent, including pretreatment to facilitate the aggregation of nanoplastics and post-treatment for the adsorption of the separated fractions.

9.3.3 Electrostatic Separation Method

Electrostatic precipitation and triboelectric separation are two techniques based on electric fields for separating charged particles from a mixture. The electrostatic precipitation method is a process that charges the nanoplastic particles, and an electric field is applied to the nanoplastic particles to be collected at a collector electrode. The performance of this method strongly depends on field strength and homogeneity acquired in the electric field, and to a much greater extent in the charge field, on the particles. For example, the Korona-Walzen-Scheider (KWS) system has been tried with the gauze to separate microplastics with mineral content samples, with good results for the detection of large particles but poorer results for the detection of microparticles, due to the hydrocarbon fraction of particulates and charge distribution difficulties. Triboelectric separation is the process of charging particles via contact friction (the triboelectric effect) and subsequently applying an electric field, whereby the particles are separated. This methodology is very efficient in processing plastic waste particles, as shown by the vibrating table-type triboelectrostatic separator (VTE), in which particles are tribocharged before separation by a rotary-cylinder device (Achouri et al., 2024). The performance of such a process can be optimized by changing parameters such as vibration amplitude, area of the electrode, and the force of the applied voltage. Triboelectrically generated static electric charges play a key role in efficient electrostatic separation.

Optimization Factors

1. *Electric field intensity and uniformity*: The stability and reproducibility of the electric field are very important to good separation. Larger voltages and homogeneous fields have an advantageous effect on the deflection and charging of the particles.
2. *Material compatibility*: The choice of materials for the tribocharging process (i.e., the type of belts and rollers) impacts the charge generation and separation efficiency. Materials such as wool, polyester, and nylon have been subjected to triboelectric assessment (Dizdar et al., 2018).
3. *Particle size and charge distribution*: The smaller particles are more problematic, with a lower charge-to-mass ratio, which leads to a lowered separation efficiency. Shifts in the tribocharging procedure can contribute to an improvement in charge distribution (Achouri et al., 2022; Enders et al., 2020).
4. *Environmental conditions*: Some effects (e.g., humidity and temperature) can affect triboelectric charging and electric field behavior and, thus, the separation efficiency.

Operational Challenges and Advancements

- Also, some operational difficulties are related to energy demands for high-voltage electric field support and to consistent triboelectric charging with varying particle size and composition. Recent developments are directed at furthering the design of tribocharging devices and the optimization of the control variables to increase the separation efficiency. Further, the emergence of designs and materials for triboelectric separators is leading to increased scalability and utility of these methods.

9.3.4 Magnetic Separation Method

Magnetic separation techniques, including high-gradient magnetic separation (HGMS) and magnetic filtration, have potential use in nanoplasticle removal from multiple environments. Critical parameters, including the strength of the magnetic field, methods for the labeling of particles, and environmental factors, determine the effectiveness of these techniques. Despite the difficulties in efficiency and scalability, improvements in materials and procedures further improve the performance of magnetic separation techniques for nanoplastics removal.

9.3.4.1 High-Gradient Magnetic Separation

High-gradient magnetic separation (HGMS) is a technique that utilizes strong magnetic fields to capture particles that either are inherently magnetic or have been tagged with magnetic materials. Magnetic field magnitude and gradient play an important role in their efficient separation. With a higher experience level, a stronger magnetic field is favorable to improving the efficiency of tagged nanoplastics. Treatment with hydrophobic iron oxide nanoparticle coatings has been shown to effectively magnetize nanoplastics in aqueous solutions, thereby rendering them tractable for water extraction. The velocity of medium flow through the magnetic separator may affect the efficiency of separation as a slower velocity generally leads to a more efficient capture factor for particles (Ramage et al., 2022; Soylak et al., 2021).

The size and magnetic nature of nanoplastics, either natural or induced, are also important for optimal separation. Smaller particles could be suited to stronger magnetic fields or better tagging strategies. The chemical makeup of nanoplastics might influence the interaction between magnetic nanoparticles and the system in which they are separated. Factors including pH, ionic strength, and the presence of interfering species can affect magnetic separation efficiency (Rhein et al., 2019).

9.3.4.2 Magnetic Filtration Techniques

The magnetic filtration process, based on a magnetic field, is used to remove particles from a suspension through magnetic filtration. This method can be applied to a fine polymer particle (e.g., microplastics) separation from a dilute suspension. The process is based on the magnetic nature of the particles or the magnetic seeds employed to immobilize them. The surface charge of the particles and the ionic strength of the solution are crucial parameters. When customized, it is possible to attenuate the electrostatic repulsion and to increase separation effectiveness. Magnetic seed concentration during the filtration stage can substantially influence separation efficiency, and higher concentrations are commonly associated with more favorable results (Rhein et al., 2019).

Although HGMS and magnetic filtration work well to retain nanoplastics, scaling up these techniques to large-scale applications presents challenges. The amount of energy used to maintain high magnetic field strength can be high, and it is not straightforward to optimize the process for various environmental conditions and plastic materials. Recent progress has been made in the areas of magnetic nanoparticle (including nanoparticle coating) development with more effective magnetic nanoparticle properties, which are aimed at improving the performance of the capture of nanoplastics. As an example, applications of magnetic carbon nanotubes have been demonstrated as acting in the extraction of microplastics from liquid media, and it is environmentally friendly in terms of recyclability (Martin et al., 2022; Ramage et al., 2022; Tang et al., 2021).

9.3.5 Ultrasonication Method

Ultrasonication is an attractive method for nanoplastics removal, which uses sonic standing waves to physically and chemically disrupt the medium. In batch ultrasonication, the sample is subjected to an ultrasonic wave in a stationary vessel, and it is possible to obtain uniform exposure of the sample to the ultrasonic wave. Continuous flow ultrasonication, however, refers to the flow of the sample through continuous exposure of the sample to ultrasonic waves and is hence seen in large-scale applications. The major process by which ultrasonication removes nanoplastics is cavitation. Ultrasonic waves induce the transient pressure variations within the liquid medium, which result in transient generation and retraction of microbubbles. This process induces high local energy, which may disaggregate or lyse the nanoplastics. The high temperature and pressure local to the ultrasonication field can induce polymer degradation through energy transfer mechanisms, which dissociate the polymer chain (predominantly C—C bonds and C—F bonds). Moreover, microstreaming, the fast fluidic motion around moving bubbles, increases the mixing and dilution of particles and, therefore, contributes to the separation of nanoplastics. High frequency degrades perfluorinated chemicals like PTFE more efficiently (Shende et al., 2021).

Advantages

- Size, chemical speciation, and aggregation state of nanoplastics determine the susceptibility to ultrasonication. Smaller particles may need more frequencies for efficient removal, and chemical composition can influence their response to ultrasonic waves. The nanoplastic's dispersion state (aggregated state or well-dispersion state) also plays a role in process efficiency. The characteristics of the medium, such as viscosity and density, are also of substantial importance in ultrasonication efficiency. Increased viscosity will help suppress the cavitating effect, thereby decreasing the efficacy of the removal of nanoplastics. In the same way, the medium has an influence on the way ultrasonic waves are propagating and how the cavitation bubbles are being formed.

Disadvantages

- An important issue in ultrasonication is energy economy. Despite their effectiveness, the process is energy hungry, especially at high power levels. Parameter optimization (frequency, power, and duration) is crucial so as not to consume too much energy. Scalability is another concern, as batch systems are limited in capacity, whereas continuous flow systems require careful design to maintain efficiency at larger scales. The treatment time used in ultrasonication influences the degree of nanoplastics degradation. Higher exposure times usually lead to better removal efficiency; however, it has to be offset against energy consumption and is subject to degradation of the medium. In continuous flow devices, flow characteristics, including flow rate and turbulence, have a significant effect on achieving homogeneous exposure to ultrasound waves and avoiding blockage or nonuniform treatment.

9.4 CHEMICAL TREATMENT TECHNOLOGIES

Physical methods concentrate on the cleanup of nanoplastics from their matrices, whereas chemical methods consist of the conversion/degrading pollutants into less toxic species. Techniques including coagulation and flocculation, photocatalytic degradation, and biochar adsorption have emerged to positively address the challenge of smaller-scale plastics and complicated environmental matrices. Coagulation and flocculation, for example, have been successful in removing microplastics from urban and wastewater plant effluents, removing microplastics with efficiencies of up to 99.4% using their effective coagulants and conditions. (Rajala et al., 2020; Xu et al., 2021). Photocatalytic degradation then promotes this degradation process by decomposing microplastics into safer by-products. Furthermore, biochar, especially after the introduction of metals such as Mg and Zn and with the adsorbed nanoplastics, showed high adsorption capacities and nanoplastics degrading capabilities, which is a low-cost and environment-friendly solution. Not only did these chemical approaches overcome the disadvantages associated with physical methodology, but they also outlined a possible route toward full nanoplastic pollution cleanup, to be discussed in the following sections.

9.4.1 Coagulation and Flocculation Method

Coagulants (aluminum-based compounds and ferric chloride, for instance) are generally used to neutralize the surface charges of nanoplastic particles to aggregate particles. Coagulants (aluminum-based compounds and ferric chloride, for instance) are generally used. These coagulants work on the principle of compressing the electrical double layer of the particles, resulting in less repulsion and bringing the particles to adhere to one another and clump together. The ability of coagulation to be good may be affected by the surface features of the nanoplastics, including their size and charge, as well as by surfactants that can change the zeta potential and aggregation properties (Lapointe et al., 2020). Flocculation follows coagulation and consists of the careful agitation of the water so to induce the formation of macroscopic floccules by the aggregates formed by the particles. Flocculants (e.g., polyacrylamide or natural bio-flocculants, such as lysozyme amyloid fibrils) are introduced for improvement by bridging between particles and creating larger, more easily dissociable flocs. The flocculation step plays an important role in the efficient removal of nanoplastics because by doing so, it is possible for the collected flocs to sediment or be filtered (Zhang et al., 2020).

9.4.1.1 Inorganic Coagulants

The selection of the coagulant and the coagulant dosage is the most important factor in the successful elimination of nanoplastic. For example, ferric chloride and polyaluminum chloride have been reported to have high removal efficiencies of up to 99.4 in some reports; optimal dosages are variable depending upon the water chemistry and the nature of the nanoplastic (Keawchouy et al. 2022; Rajala et al., 2020; Yujian Zhang et al., 2021). Coagulant aids (such as anionic polyacrylamide) can afford drastic improvement in removal efficiency (Yu et al., 2021; Zhang et al., 2021). Ali et al. (2023) and Girish et al. (2023) reported on a plethora of inorganic coagulants for microplastic/nanoplastics removal, including alum, polyaluminum chloride, ferric chloride, aluminum chlorohydrate, cobalt ferrites, magnetic magnesium hydroxide, and layered double hydroxide. The main mechanism for microplastic removal is adsorption bridging, in which coagulants and flocculants produce huge, dense agglomerates that sequester MPs and so can be removed. Sweep flocculation is another mechanism that is dominant in systems using alum and other similar coagulants, where MPs are removed by being swept into the floc matrix (Skaf et al., 2020; Sun et al., 2023). Scaling up the method for MNP removal faces challenges such as process optimization and variability of the different MNP resources in the targeted stream. Efficiency depends on multiple factors, such as coagulant type, dosage, pH, and water chemistry. The presence of natural organic matter (e.g., humic acid) and other competing substances can inhibit removal efficiency, requiring site-specific adjustments. Excessive usage of flocculants to treat smaller-size nanoplastics can also increase the viscosity and settling time, which causes operation constraints.

9.4.1.2 Organic Coagulants

Polyamine is a cationic coagulating agent and is widely used in wastewater treatment. Compared with ferric chloride and PAC, it is relatively less effective (65% for microplastic removal) (Rajala et al., 2020). Shahi et al. (2020) used polyamine-coated sand as a coagulant aid to remove polyethylene (PE) of 500 μm and humic acid (HA), which showed up to 98% removal efficiency. According to the study, PE and HA carry a negative surface charge, which is neutralized by cationic polyamine through charge neutralization. However, the high concentrations of these flocculants decrease the removal efficiencies through self-aggregation. Lysozyme amyloid fibrils showed better performance than conventional coagulants, with removal efficiencies of 98.2% for MPs. Peydayesh et al. (2021) reported that lysozyme amyloid fibrils could effectively flocculate MPs across a range of pH from 2 to 10 with optimum flocculation at pH 2 as a result of the positive charges of the amyloid fibrils contributing to electrostatic interaction. Scaling up requires precise control of coagulant and flocculant dosages. Hybrid systems (organic coagulant + polyelectrolyte) can reduce the amount of flocculant needed but require careful balancing to maintain efficiency and cost effectiveness. Hybrid systems can produce larger, denser, and more shear-resistant flocs, which is beneficial for separation. However, maintaining these properties consistently at scale can be difficult, especially with variable water matrices.

9.4.1.3 Composite Flocculants

Composite flocculants polyferric titanium sulfate–polydimethyldiallylammonium chloride (PFTS–PDMDAAC) has been synthesized, and the highest loading capacity (85.60% of polystyrene microplastics removal) has been demonstrated (Sun et al., 2023). PAC, together with anionic polyacrylamide, can also be effectively used for separation (91.45 of PET microplastics in the size range of 100–400 μm) (Yujian Zhang et al., 2021). Sludge management is a critical scalability challenge for such treatments since dewatering and abundant disposal of the sludge that includes the MNPs can create the same problems as persistent MNPs.

9.4.2 Photocatalytic Degradation

This method relies on the use of light, as UV or visible spectrum, in the presence of a photocatalyst to degrade plastic particles into smaller molecules or even minerals at one or the other extreme. Photocatalytic degradation is based on the application of photocatalysts such as titanium dioxide (TiO_2) and zinc oxide (ZnO) for the absorption of light energy and the formation of reactive oxygen species (ROS) that degrade plastic polymers. When these photocatalysts are illuminated, they receive photons, excite electrons from the valence band to the conduction band, and generate electron–hole pairs. These pairs react with water and oxygen to produce ROSs (e.g., hydroxyl radicals and superoxide anions), which are very reactive and can break down the polymer structure of microplastics and nanoplastics. Electrode structures of anodized TiO_2 have been used in the degradation of polystyrene nanoparticles (PS-NPs). Various TiO_2 structures (e.g., nanotubes and nanograss) have demonstrated effectiveness in PS-NPs removal from aqueous solutions, and particularly mixed structures have yielded a greater effect compared with photolysis. Zinc oxide nanorods have been studied for low-density polyethylene (LDPE) microplastic residue removal. The photocatalytic reaction using visible light leads to further oxidation and brittleness of the microplastics and surface modifications like wrinkles and cracks.

Other than the generally implemented TiO_2 and ZnO photocatalysts, a range of other materials with improved electronic properties are used to exploit light to degrade microplastics and nanoplastics. MXene/$Zn_XCd_{1-X}S$ photocatalysts for the degradation of PET microplastics are synthesized. These photocatalysts show high activity in photocatalytic H_2 evolution and in PET degradation, producing useful organic photoproducts like glycolate, acetate, and ethanol from PET. Enhanced performance is ascribed to the photocarrier separation improvement, and the optimized band structure (Cao et al., 2022). BiOCl photocatalysts [spherical (BOC-S) and nanosheet (BOC-N) forms] have been employed for PET microplastic degradation. These photocatalysts, especially with hydrothermal pretreatments, notably improve the electrophilic degradation efficiency of PET microplastics by promoting the hydrophobicity and specific surface area, which enable optimal contact between the catalyst and the microplastics (Jiang et al., 2021; Lin et al., 2024). MoO_3 nanoflakes and nanobelts, as well as MoO_3/SWCNT nanocomposites, have been used for the photooxidative degradation of polystyrene nanoplastics. These photocatalysts exhibit high degradation efficiency under UV irradiation and are able to remodel the morphology and decrease the size of PS nanoplastics (Singh et al., 2024). CuMgAlTi-layered double hydroxide (LDH) composites have been used to degrade PS and PE microplastics under visible light. These photocatalysts also reduce the PS particle size to a large extent, meaning that they exhibit good degradation efficiency (Jiang et al., 2023).

Advantages

1. *High efficiency*: Designing high-performance photocatalysts, such as BiOCl, MXene/$Zn_XCd_{1-X}S$, can greatly enhance the MNPs degradation rate.
2. *Degradation into harmless by-products*: Photocatalytic degradation offers to detoxify plastic into less toxic degradation products. For example, the degradation of polyethylene microplastics using zinc oxide nanorods or polystyrene nanoparticles using TiO_2 resulted in increased oxidation, leading to the formation of carbonyl groups and a more brittle structure, which suggests a breakdown into smaller, less harmful molecules.
3. *Use of renewable light energy sources*: In many photocatalytic processes, visible/ultraviolet light sources is used as they are a renewable energy resource. This makes the process environmentally friendly and sustainable.

Disadvantages

1. *Need for specific light conditions*: One of the most important drawbacks of photocatalytic degradation is the need for certain light conditions in order to trigger the photocatalysts. This may restrict the versatility of the technique for conditions where such a situation is not available or difficult to create.
2. *Potential generation of secondary pollutants*: Photocatalysis is designed to degrade plastics to environmentally benign degradation products, while it is also possible to form secondary pollutants. For instance, the degradation process may lead to the formation of intermediate products which are still environmentally toxic. Ecological applications of the intermediates must be critically evaluated to ensure the global safety of the process.
3. *Challenges in catalyst recovery and reusability*: The recovery and reusability of photocatalysts pose practical challenges. Process efficiency may be hampered by the difficulty encountered with recovering the catalyst after the degradation process, which also results in higher costs and higher resource consumption. Also, the cycling activity and stability of catalysts multiple times are problems that should be solved.

9.4.3 Adsorption

Adsorption for the removal of nanoplastics is used for the physical or chemical adsorption of nanoparticles onto the surface of the adsorbents. These interactions can involve electrostatic force, hydrogen bonds, Van der Waals forces, π-π interaction, and so forth. In particular, electrostatic interactions are most important when the surface charges on the surface of the adsorbent and of nanoplastics are equal, and nanoplastics are thus attracted and adsorbed to the adsorbent surface. Chemical interactions such as complexation and π-π conjugation are also critical to the adsorption process, especially when surface functional groups of the adsorbent are grafted onto the nanoplastics (Zhao et al., 2023).

Types of Adsorbents

1. *Biochar*: Biochar is a highly porous and nanostructured carbon-based material that shows good potential as an adsorbent for nanoplastics. Modified biochars, for example, Mg/Zn-modified magnetic biochar, have extremely high polystyrene removal efficiencies due to further increased electrostatic interactions and chemical bonding.
2. *Activated carbon*: Granular activated carbon (GAC) is popular because of its high surface area and porous nature. Essentially, it is the electrostatic interaction responsible for the adsorption of nanoplastic, and its effectiveness is dictated by the associated presence of dissolved organic matter and ionic components in water.
3. *Clay minerals*: The adsorptive ability of clay minerals is manipulated by tailoring them. Magnetic zeolites use iron oxide functionalities to adsorb nanoplastics using electrostatic charging and complexation.
4. *Functionalized nanomaterials*: Materials such as graphene-based composites and metal–organic frameworks (MOFs) with relatively high adsorption capacities provided by their structural and surface functionalities turn out to be excellent adsorbents. Most of them are especially useful as adsorbents for organic pollutants and have been designed for targeted interactions with nanoplastics.

Factors Influencing Adsorption

1. *Surface area and porosity*: Greater surface area and porosity allow for greater adsorption capacity due to increased active sites for nanoplastic binding.
2. *Surface chemistry*: The functional groups on the surface of the adsorbent can clearly influence adsorption via chemical interaction.
3. *pH and temperature*: These parameters influence the charge and solubility of nanoplastics and adsorbents, thus directing the adsorption process. Increased temperature not only improves adsorption capacity but also results in stronger molecular interaction.
4. *Chemical composition of nanoplastics*: The polymer type and surface properties may influence the level of interactions with the adsorbent.

Adsorption Kinetics and Capacity

- The adsorption kinetics of micro- and nanoplastics are very important for knowing how fast these particles can be adsorbed from the aqueous solutions. A variety of studies have revealed the adsorption kinetics of nanoplastics; for example, they tend to be described by pseudo-second-order kinetics, which suggests that the adsorption of nanoplastics is governed by chemisorption involving valency forces through electron sharing or exchange between adsorbent and adsorbate. For example, the adsorption of polystyrene nanoplastics onto GAC and biochar obtained from date palm fibers are well described by the pseudo-second-order kinetic model, implying that the adsorption rate is limited by both the concentration of nanoplastics and available adsorption sites (Ramirez Arenas et al., 2021; Rezaei Kahkha et al., 2024).

- The adsorption property of materials employed for the removal of micro- and nanoplastics is significantly dependent on the adsorption capability of the adsorbent and the environmental factors. GAC has been shown to have a maximum adsorption capacity of 2.20 mg/g for polystyrene nanoplastics in ultrapure water, which increases to 6.33 mg/g in the presence of dissolved organic matter and divalent ions (Ramirez Arenas et al., 2021). In the same way, biochar from polluted corncobs has been theoretically shown to have a maximum adsorption capacity of 56.02 mg/g for polystyrene nanoparticles, making biochar an effective adsorbent (Zhu et al., 2022). Magnetic carbon nanotubes have also been shown to be good candidates, with a maximum adsorption capacity of 1650 mg/g for a type of polyethylene microplastics (Tang et al., 2021).

9.5 BIOLOGICAL TECHNIQUES

Contamination of the environment by even pervasive micro- and nanoplastics is increasingly addressed by biological methods, which offer environmentally acceptable and sustainable interventions. These techniques take advantage of the natural processes such as microbial degradation, enzymatic hydrolysis, and bioremediation to restrain from the traditional physical and chemical methods. Microbial remediation, for example, is presented as a more sustainable technology that employs different microbes and their enzymatic potential for plastic degradation in different materials—wastewater, municipal sludge, and composter facilities. The use of enzymatic processes, based on enzymes like PETase and MHETase, for the depolymerization of tough polymers, which ultimately leave behind less toxic by-products, has been shown to be an effective strategy for managing plastic pollution (Bhandari et al., 2021; Choi et al., 2024). The use of living organisms, including bacteria, algae, and fungi, which carries the name "bioremediation," in particular has ability in aquatic environments and is able to present itself as a tool to clean up or degrade microplastic that is widely dispersed. Such biological techniques play a key role in tackling the ubiquitous problem of micro- and nanoplastics and will be the basis for comprehensively describing and implementing certain methods in different ecosystems.

9.5.1 Microbial Degradation

Microbial degradation is the application of microorganisms, including bacteria, fungi, and algae, to degrade microplastics and nanoplastics. These microbes have the ability to use plastic polymers as a carbon substrate and subsequently degrade them. Bacteria and fungi are most efficient; some strains are able to degrade broad ranges of plastics, including PE, PET, and PS. Bacterial consortia and biofilms also exhibited promising MP degradation capabilities through modification of the surface microstructure and decrease in the molecular weight. The performance of microbial degradation is governed by environmental conditions, including temperature, pH, and co-substrate availability (Zhou et al., 2022).

Advantages and Challenges

- Microbial degradation is considered an eco-friendly and sustainable approach. Nevertheless, there are challenges, such as the low degradation rate and the requirement for optimal microbial growth conditions under specific environmental conditions (Masiá et al., 2020; Zhou et al., 2022).

9.5.2 Enzymatic Breakdown

Enzymatic degradation is a biodegradation process of plastic polymers using enzymes secreted by microbes. Enzymes (PETase and MHETase) known to lyse PET and lipases and hydrolases known to lyse various other polymer types have been identified. These digestive enzymes act by executing a cleavage reaction on polymer chains, adding them to smaller, more innocuous pieces. The activity of these enzymes is highly dependent on the optimum temperature, pH, and substrate concentration (Bhandari et al., 2021; Sharma et al., 2018).

Advantages and Challenges

- Enzymatic depolymerization is a fast and highly selective reaction, providing a selective treatment method to degrade plastics. However, the production and stability of the enzymes can be a bottleneck; genetic engineering is being investigated to improve enzyme activity (Saravanan et al., 2021; Sharma et al., 2018).

9.5.3 Bioremediation

The process of bioremediation uses living organisms (bacteria, algae, and higher aquatic eukaryotes) to degrade MPs and NPs in the environment. This approach has strong promise in wastewater treatment plants, where microorganisms can be accurately contained and controlled. Bioremediation is a process that may be based on direct microbial degradation or on the bioaccumulation and containment of plastics for straightforward removal.

Advantages and Challenges

- Bioremediation is a customizable and ecologically sustainable solution applicable to different kinds of ecosystems. Yet confining and regulating microorganisms in a free state is still an ungraded issue, and some improvement in this area is required to improve these practices. Recent work has shown the ability of certain bacterial strains, for example, *Pseudomonas aeruginosa*, to break down nanoplastics to considerably reduce the size of particles and modify their chemical composition (El-Kurdi et al., 2024). These breakthroughs emphasize the possibility of microbial and enzymatic solutions to be used in the sea and on land, providing solutions for plastic waste management.

9.6 CONCLUSIONS AND PERSPECTIVES

In this chapter, the irrefutable importance of prevention of pollution by microplastics and nanoplastics by means of environmentally friendly and highly efficient treatment technologies is addressed. Physical, chemical, and biological remediation technologies have their own advantages and disadvantages, and all demonstrate the requirement for a solution that integrates all three approaches to this ubiquitous issue of plastic pollution. Sophisticated physical approaches, such as membrane filtration and electrostatic, as well as chemical strategies, such as adsorption and photocatalytic degradation, are promising for MNP isolation and remediation of plastic into less hazardous residual materials, respectively. Biological approaches, such as microbial and enzymatic degradation, are created as environmentally friendly examples, but scalability and efficiency are still open problems.

There is no uniform treatment modality that can be applied everywhere, and the use of ensembles of technologies in hybrid frameworks could finally lead to better solutions. Scaling up degradation technologies is partially hindered by the need for cost effectiveness. The need for low-cost, process-, and material-friendly materials with practical side-to-side implementation capability at a large scale is critical for true public adoption. It is essential to avoid the introduction of environmental pollutants from the degradation processes. This comprises reducing the leaching of toxic intermediate and guaranteeing that the degradation products are non-toxic and environmentally friendly. The development of precise regulatory frameworks is required to facilitate the design and use of degradation technologies. This includes the specification of degradation efficiency, environmental safety, and renewable resource utilization.

With increasing worries over microplastic and nanoplastic pollution in terrestrial ecosystems, the requirement for efficient, commercially viable, and environmentally friendly solutions is unprecedented. Research in novel materials and their underlying new and hybrid technologies is crucial. Advancements in the use of nanomaterials and sophisticated photocatalytic methods for increasing microplastic degradation efficiency have emerged as a tempting prospect.

Photocatalytic nanomaterial grafted in membranes can thus multifunction as an MNP trap and as a degradation unit to break down the MNPs into smaller, non-polluting end products that can also overcome the fouling issue. Filtration systems that incorporate filter pore sizes first and then methods of further oxidation can be commercially viable for water treatment. Increasing the rate of enzymatic reactions by using enzymes such as PETase and MHETase illustrates the potential of the biological treatment modalities. The development of reusable multifunctional materials with functions that can participate in degradation processes is of paramount importance for sustainable applications. Energy efficiency is one of the key parameters in the viability of the degradation techniques. Combining renewable energy sources like solar energy with photocatalytic processes holds great potential to decrease the energy consumption of these processes. Physical, chemical, and biological processes can be combined in hybrid techniques to also increase overall efficiency and effectiveness.

REFERENCES

Acarer, S. (2023). A review of microplastic removal from water and wastewater by membrane technologies. *Water Science and Technology*, *88*(1), 199–219. https://doi.org/10.2166/wst.2023.186

Achouri, I.-E., Boukhoulda, M.-F., Medles, K., Richard, G., Zeghloul, T., & Dascalescu, L. (2022). Electrostatic separation of tribocharged granular mixtures of two or more plastics originating from WEEE. *IEEE Transactions on Industry Applications*, *58*(6), 7701–7708. https://doi.org/10.1109/TIA.2022.3197544

Achouri, I. E., Richard, G., Zeghloul, T., Medles, K., & Dascalescu, L. (2024). New vibrating-table-type tribo-electrostatic separator for selective sorting of granular plastic wastes. *IEEE Transactions on Industry Applications*, *60*(2), 3537–3542. https://doi.org/10.1109/TIA.2023.3340633

Ali, I., Tan, X., Xie, Y., Peng, C., Li, J., Naz, I., Duan, Z., Wan, P., Huang, J., Liang, J., Rui, Z., & Ruan, Y. (2023). Recent innovations in microplastics and nanoplastics removal by coagulation technique: Implementations, knowledge gaps and prospects. *Water Research*, *245*, 120617. https://doi.org/10.1016/j.watres.2023.120617

Amobonye, A., Bhagwat, P., Raveendran, S., Singh, S., & Pillai, S. (2021). Environmental impacts of microplastics and nanoplastics: A current overview. *Frontiers in Microbiology*, *12*(December). https://doi.org/10.3389/fmicb.2021.768297

Bayo, J., López-Castellanos, J., & Olmos, S. (2020a). Abatement of microplastics from municipal effluents by two different wastewater treatment technologies. *WIT Transactions on Ecology and the Environment*, *242*, 15–26. https://doi.org/10.2495/WP200021

Bayo, J., López-Castellanos, J., & Olmos, S. (2020b). Membrane bioreactor and rapid sand filtration for the removal of microplastics in an urban wastewater treatment plant. *Marine Pollution Bulletin*, *156*, 111211. https://doi.org/10.1016/j.marpolbul.2020.111211

Bayo, J., Olmos, S., & López-Castellanos, J. (2021). Assessment of microplastics in a municipal wastewater treatment plant with tertiary treatment: Removal efficiencies and loading per day into the environment. *Water (Switzerland)*, *13*(10), 1–22. https://doi.org/10.3390/w13101339

Benson, N. U., Agboola, O. D., Fred-Ahmadu, O. H., De-la-Torre, G. E., Oluwalana, A., & Williams, A. B. (2022). Micro(nano)plastics prevalence, food web interactions, and toxicity assessment in aquatic organisms: A review. *Frontiers in Marine Science*, *9*(March), 1–19. https://doi.org/10.3389/fmars.2022.851281

Bhandari, S., Poudel, D. K., Marahatha, R., Dawadi, S., Khadayat, K., Phuyal, S., Shrestha, S., Gaire, S., Basnet, K., Khadka, U., & Parajuli, N. (2021). Microbial enzymes used in bioremediation. *Journal of Chemistry*, *2021*. https://doi.org/10.1155/2021/8849512

Bodzek, M., & Pohl, A. (2022). Removal of microplastics in unit processes used in water and wastewater treatment: A review. *Archives of Environmental Protection*, *48*(4), 102–128. https://doi.org/10.24425/aep.2022.143713

Cai, H., Chen, M., Du, F., Matthews, S., & Shi, H. (2021). Separation and enrichment of nanoplastics in environmental water samples via ultracentrifugation. *Water Research*, *203*, 117509. https://doi.org/10.1016/j.watres.2021.117509

Cao, B., Wan, S., Wang, Y., Guo, H., Ou, M., & Zhong, Q. (2022). Highly-efficient visible-light-driven photocatalytic H2 evolution integrated with microplastic degradation over MXene/ZnxCd1-xS photocatalyst. *Journal of Colloid and Interface Science*, *605*, 311–319. https://doi.org/10.1016/j.jcis.2021.07.113

Castillo-Ruiz, M., Negrete, C., Espinoza, J. P., Martínez, I., Daille, L. K., González, C., & Rodríguez, B. (2024). Antibiofilm effects of modifying polyvinylidene fluoride membranes with polyethylenimine, poly(acrylic acid) and graphene oxide. *Polymers*, *16*(23). https://doi.org/10.3390/polym16233418

Chen, L., Qiu, T., Huang, F., Zeng, Y., Cui, Y., Chen, J., White, J. C., & Fang, L. (2024). Micro/nanoplastics pollution poses a potential threat to soil health. *Global Change Biology*, *30*(8), e17470. https://doi.org/10.1111/gcb.17470

Chen, T., Li, B., Huang, W., Lin, C., Li, G., Ren, H., Wu, Y., Chen, S., Zhang, W., & Ma, H. (2021). Highly crystalline ionic covalent organic framework membrane for nanofiltration and charge-controlled organic pollutants removal. *Separation and Purification Technology*, *256*, 117787. https://doi.org/10.1016/j.seppur.2020.117787

Cheng, X., Qin, Y., Ye, Y., Chen, X., Wang, K., Zhang, Y., Figoli, A., & Drioli, E. (2021). Finely tailored pore structure of polyamide nanofiltration membranes for highly-efficient application in water treatment. *Chemical Engineering Journal*, *417*, 127976. https://doi.org/10.1016/j.cej.2020.127976

Choi, J., Kim, H., Ahn, Y. R., Kim, M., Yu, S., Kim, N., Lim, S. Y., Park, J. A., Ha, S. J., Lim, K. S., & Kim, H. O. (2024). Recent advances in microbial and enzymatic engineering for the biodegradation of micro- and nanoplastics. *RSC Advances*, *14*, 9943–9966. https://doi.org/10.1039/d4ra00844h

Cui, J., Li, F., Wang, Y., Zhang, Q., Ma, W., & Huang, C. (2020). Electrospun nanofiber membranes for wastewater treatment applications. *Separation and Purification Technology*, *250*, 117116. https://doi.org/10.1016/j.seppur.2020.117116

Dizdar, T. O., Kocausta, G., Gülcan, E., & Gülsoy, Ö. Y. (2018). A new method to produce high voltage static electric load for electrostatic separation—Triboelectric charging. *Powder Technology*, *327*, 89–95. https://doi.org/10.1016/j.powtec.2017.12.065

El-Kurdi, N., Elbaghdady, K., El-Shatoury, S., Hammad, S., & Ghazy, M. (2024). Sustainable removal of nanoplastics: Exploiting the lipolytic activity of Pseudomonas aeruginosa O6 isolated from Mariout Wetland, Egypt. *Egyptian Journal of Aquatic Biology and Fisheries*, *28*(2), 247–272. https://doi.org/10.21608/EJABF.2024.347346

Eliso, M. C., Billè, B., Cappello, T., & Maisano, M. (2024). Polystyrene micro- and nanoplastics (PS MNPs): A review of recent advances in the use of -omics in PS MNP toxicity studies on aquatic organisms. *Fishes*, *9*(3), 1–25. https://doi.org/10.3390/fishes9030098

Enders, K., Tagg, A. S., & Labrenz, M. (2020). Evaluation of electrostatic separation of microplastics from mineral-rich environmental samples. *Frontiers in Environmental Science*, *8*(July), 1–9. https://doi.org/10.3389/fenvs.2020.00112

Enfrin, M., Lee, J., Fane, A. G., & Dumée, L. F. (2021a). Mitigation of membrane particulate fouling by nano/microplastics via physical cleaning strategies. *Science of the Total Environment*, *788*, 147689. https://doi.org/10.1016/j.scitotenv.2021.147689

Enfrin, M., Lee, J., Le-Clech, P., & Dumée, L. F. (2020). Kinetic and mechanistic aspects of ultrafiltration membrane fouling by nano- and microplastics. *Journal of Membrane Science*, *601*, 117890. https://doi.org/10.1016/j.memsci.2020.117890

Enfrin, M., Wang, J., Merenda, A., Dumée, L. F., & Lee, J. (2021b). Mitigation of membrane fouling by nano/microplastics via surface chemistry control. *Journal of Membrane Science*, *633*, 119379. https://doi.org/10.1016/j.memsci.2021.119379

Fajar, M., Sembiring, E., & Handajani, M. (2022). The effect of filter media size and loading rate to filter performance of removing microplastics using rapid sand filter. *Journal of Engineering and Technological Sciences*, *54*(5). https://doi.org/10.5614/j.eng.technol.sci.2022.54.5.12

Girish, N., Parashar, N., & Hait, S. (2023). Coagulative removal of microplastics from aqueous matrices: Recent progresses and future perspectives. *Science of the Total Environment*, *899*, 165723. https://doi.org/10.1016/j.scitotenv.2023.165723

Hashemihedeshi, M., Haywood, E., Gatch, D. C., Jantunen, L., Helm, P. A., Diamond, M. L., Dorman, F. L., Cahill, L. S., & Jobst, K. J. (2024). Size-resolved identification and quantification of micro/nanoplastics in indoor air using pyrolysis gas chromatography–ion mobility mass spectrometry. *Journal of the American Society for Mass Spectrometry*, *35*(2), 275–284. https://doi.org/10.1021/jasms.3c00362

Hildebrandt, L., Voigt, N., Zimmermann, T., Reese, A., & Proefrock, D. (2019). Evaluation of continuous flow centrifugation as an alternative technique to sample microplastic from water bodies. *Marine Environmental Research*, *151*, 104768. https://doi.org/10.1016/j.marenvres.2019.104768

Jiang, B., Kauffman, A. E., Li, L., McFee, W., Cai, B., Weinstein, J., Lead, J. R., Chatterjee, S., Scott, G. I., & Xiao, S. (2020). Health impacts of environmental contamination of micro- and nanoplastics: A review. *Environmental Health and Preventive Medicine*, *25*(1), 29. https://doi.org/10.1186/s12199-020-00870-9

Jiang, R., Lu, G., Yan, Z., Liu, J., Wu, D., & Wang, Y. (2021). Microplastic degradation by hydroxy-rich bismuth oxychloride. *Journal of Hazardous Materials*, *405*, 124247. https://doi.org/10.1016/j.jhazmat.2020.124247

Jiang, S., Yin, M., Ren, H., Qin, Y., Wang, W., Wang, Q., & Li, X. (2023). Novel CuMgAlTi-LDH photocatalyst for efficient degradation of microplastics under visible light irradiation. *Polymers*, *15*(10). https://doi.org/10.3390/polym15102347

Juraij, K., Ammed, S. P., Chingakham, C., Ramasubramanian, B., Ramakrishna, S., Vasudevan, S., & Sujith, A. (2023). Electrospun polyurethane nanofiber membranes for microplastic and nanoplastic separation. *ACS Applied Nano Materials*, *6*(6), 4636–4650. https://doi.org/10.1021/acsanm.3c00112

Keawchouy, S., Na-Phatthalung, W., Keaonaborn, D., Jaichuedee, J., Musikavong, C., & Sinyoung, S. (2022). Enhanced coagulation process for removing dissolved organic matter, microplastics, and silver nanoparticles. *Journal of Environmental Science and Health, Part A*, *57*(13–14), 1084–1098. https://doi.org/10.1080/10934529.2022.2155419

Kundu, A., Shetti, N. P., Basu, S., Raghava Reddy, K., Nadagouda, M. N., & Aminabhavi, T. M. (2021). Identification and removal of micro- and nano-plastics: Efficient and cost-effective methods. *Chemical Engineering Journal*, *421*, 129816. https://doi.org/10.1016/j.cej.2021.129816

Lapointe, M., Farner, J. M., Hernandez, L. M., & Tufenkji, N. (2020). Understanding and improving microplastic removal during water treatment: Impact of coagulation and flocculation. *Environmental Science & Technology*, *54*(14), 8719–8727. https://doi.org/10.1021/acs.est.0c00712

Li, J., Wang, B., Chen, Z., Ma, B., & Chen, J. P. (2021). Ultrafiltration membrane fouling by microplastics with raw water: Behaviors and alleviation methods. *Chemical Engineering Journal*, *410*, 128174. https://doi.org/10.1016/j.cej.2020.128174

Li, L., Guo, W., Zhang, S., Guo, R., & Zhang, L. (2023). Electrospun nanofiber membrane: An efficient and environmentally friendly material for the removal of metals and dyes. *Molecules*, *28*(8). https://doi.org/10.3390/molecules28083288

Liao, Z., Nguyen, M. N., Wan, G., Xie, J., Ni, L., Qi, J., Li, J., & Schäfer, A. I. (2020). Low pressure operated ultrafiltration membrane with integration of hollow mesoporous carbon nanospheres for effective removal of micropollutants. *Journal of Hazardous Materials*, *397*, 122779. https://doi.org/10.1016/j.jhazmat.2020.122779

Lin, L., Yi, J., Wang, J., Qian, Q., Chen, Q., Cao, C., & Zhou, W. (2024). Enhancing microplastic degradation through synergistic photocatalytic and pretreatment approaches. *Langmuir*, *40*(43), 22582–22590. https://doi.org/10.1021/acs.langmuir.4c02124

Liu, S., He, Y., Yin, J., Zhu, Q., Liao, C., & Jiang, G. (2024). Neurotoxicities induced by micro/nanoplastics: A review focusing on the risks of neurological diseases. *Journal of Hazardous Materials*, *469*, 134054. https://doi.org/10.1016/j.jhazmat.2024.134054

Luo, D., Chu, X., Wu, Y., Wang, Z., Liao, Z., Ji, X., Ju, J., Yang, B., Chen, Z., Dahlgren, R., Zhang, M., & Shang, X. (2024). Micro- and nano-plastics in the atmosphere: A review of occurrence, properties and human health risks. *Journal of Hazardous Materials*, *465*, 133412. https://doi.org/10.1016/j.jhazmat.2023.133412

Martin, L. M. A., Sheng, J., Zimba, P. V., Zhu, L., Fadare, O. O., Haley, C., Wang, M., Phillips, T. D., Conkle, J., & Xu, W. (2022). Testing an iron oxide nanoparticle-based method for magnetic separation of nanoplastics and microplastics from water. *Nanomaterials*, *12*(14), 1–17. https://doi.org/10.3390/nano12142348

Masiá, P., Sol, D., Ardura, A., Laca, A., Borrell, Y. J., Dopico, E., Laca, A., Machado-Schiaffino, G., Díaz, M., & Garcia-Vazquez, E. (2020). Bioremediation as a promising strategy for microplastics removal in wastewater treatment plants. *Marine Pollution Bulletin*, *156*, 111252. https://doi.org/10.1016/j.marpolbul.2020.111252

Mogha, N. K., & Shin, D. (2023). Nanoplastic detection with surface enhanced Raman spectroscopy: Present and future. *TrAC Trends in Analytical Chemistry*, *158*, 116885. https://doi.org/10.1016/j.trac.2022.116885

Mohana, A. A., Farhad, S. M., Haque, N., & Pramanik, B. K. (2021). Understanding the fate of nano-plastics in wastewater treatment plants and their removal using membrane processes. *Chemosphere*, *284*, 131430. https://doi.org/10.1016/j.chemosphere.2021.131430

Murray, A., & Örmeci, B. (2020). water with separation processes used for water and wastewater treatment. *Water*, *12*, 635–649.

Nakazawa, Y., Abe, T., Matsui, Y., Shinno, K., Kobayashi, S., Shirasaki, N., & Matsushita, T. (2021). Differences in removal rates of virgin/decayed microplastics, viruses, activated carbon, and kaolin/montmorillonite clay particles by coagulation, flocculation, sedimentation, and rapid sand filtration during water treatment. *Water Research*, *203*, 117550. https://doi.org/10.1016/j.watres.2021.117550

Napi, N. Nor Amirah Mohd, Ibrahim, N., Adli Hanif, M., Hasan, M., Dahalan, F. A., Syafiuddin, A., & Boopathy, R. (2023). Column-based removal of high concentration microplastics in synthetic wastewater using granular activated carbon. *Bioengineered*, *14*(1). https://doi.org/10.1080/21655979.2023.2276391

Oza, H., & Pathak, T. (2024). A review on micro and nano plastics removal from aqueous environment through bionanomaterials. *Interactions*, *245*(1). https://doi.org/10.1007/s10751-024-02194-4

Pandey, R. P., Rasheed, P. A., Gomez, T., Azam, R. S., & Mahmoud, K. A. (2020). A fouling-resistant mixed-matrix nanofiltration membrane based on covalently cross-linked $Ti_3C_2T_X$ (MXene)/cellulose acetate. *Journal of Membrane Science*, *607*, 118139. https://doi.org/10.1016/j.memsci.2020.118139

Pervez, M. N., Mishu, M. R., Talukder, M. E., Stylios, G. K., Buonerba, A., Hasan, S. W., Cai, Y., Zhao, Y., Figoli, A., Zarra, T., Belgiorno, V., & Naddeo, V. (2022). Electrospun nanofiber membranes for the control of micro/nanoplastics in the environment. *Water Emerging Contaminants and Nanoplastics*, *1*(2), 1–6. https://doi.org/10.20517/wecn.2022.05

Peydayesh, M., Suta, T., Usuelli, M., Handschin, S., Canelli, G., Bagnani, M., & Mezzenga, R. (2021). Sustainable removal of microplastics and natural organic matter from water by coagulation–flocculation with protein amyloid fibrils. *Environmental Science & Technology*, *55*(13), 8848–8858. https://doi.org/10.1021/acs.est.1c01918

Pramanik, B. K., Pramanik, S. K., & Monira, S. (2021). Understanding the fragmentation of microplastics into nano-plastics and removal of nano/microplastics from wastewater using membrane, air flotation and nano-ferrofluid processes. *Chemosphere*, *282*, 131053. https://doi.org/10.1016/j.chemosphere.2021.131053

Prattichizzo, F., Ceriello, A., Pellegrini, V., La Grotta, R., Graciotti, L., Olivieri, F., Paolisso, P., D'Agostino, B., Iovino, P., Balestrieri, M. L., Rajagopalan, S., Landrigan, P. J., Marfella, R., & Paolisso, G. (2024). Micro-nanoplastics and cardiovascular diseases: Evidence and perspectives. *European Heart Journal*, *45*(38), 4099–4110. https://doi.org/10.1093/eurheartj/ehae552

Rahim, N. A. S. M., Islahudin, F., Tahrim, N. A. B. U., & Jasamai, M. (2022). Microplastics in cosmetics and personal care products: Impacts on aquatic life and rodents with potential alternatives. *Sains Malaysiana*, *51*(8), 2495–2506. https://doi.org/10.17576/jsm-2022-5108-12

Rai, P. K., Kumar, V., Sonne, C., Lee, S. S., Brown, R. J. C., & Kim, K.-H. (2021). Progress, prospects, and challenges in standardization of sampling and analysis of micro- and nano-plastics in the environment. *Journal of Cleaner Production*, *325*, 129321. https://doi.org/10.1016/j.jclepro.2021.129321

Rajala, K., Grönfors, O., Hesampour, M., & Mikola, A. (2020). Removal of microplastics from secondary wastewater treatment plant effluent by coagulation/flocculation with iron, aluminum and polyamine-based chemicals. *Water Research*, *183*, 116045. https://doi.org/10.1016/j.watres.2020.116045

Ramage, S. J. F. F., Pagaling, E., Haghi, R. K., Dawson, L. A., Yates, K., Prabhu, R., Hillier, S., & Devalla, S. (2022). Rapid extraction of high- and low-density microplastics from soil using high-gradient magnetic separation. *SSRN Electronic Journal*. https://doi.org/10.2139/ssrn.4003275

Ramirez Arenas, L., Ramseier Gentile, S., Zimmermann, S., & Stoll, S. (2021). Nanoplastics adsorption and removal efficiency by granular activated carbon used in drinking water treatment process. *Science of the Total Environment*, *791*, 148175. https://doi.org/10.1016/j.scitotenv.2021.148175

Rezaei Kahkha, M. R., Kahkha Zhaleh, M. R., Rezaei Kahkha, B., Khodadadi, M., & Faghihi-Zarandi, M. (2024). Removal of polypropylene nanoplastics from aqueous solution by biochar derived from Date palm fibers: Kinetics and isotherms studies. *Analytical Methods in Environmental Chemistry Journal*, *6*(4), 65–75. https://doi.org/10.24200/amecj.v6.i04.314

Rhein, F., Scholl, F., & Nirschl, H. (2019). Magnetic seeded filtration for the separation of fine polymer particles from dilute suspensions: Microplastics. *Chemical Engineering Science*, *207*, 1278–1287. https://doi.org/10.1016/j.ces.2019.07.052

Saravanan, A., Kumar, P. S., Vo, D.-V. N., Jeevanantham, S., Karishma, S., & Yaashikaa, P. R. (2021). A review on catalytic-enzyme degradation of toxic environmental pollutants: Microbial enzymes. *Journal of Hazardous Materials*, *419*, 126451. https://doi.org/10.1016/j.jhazmat.2021.126451

Sembiring, E., Fajar, M., & Handajani, M. (2021). Performance of rapid sand filter—single media to remove microplastics. *Water Supply*, *21*(5), 2273–2284. https://doi.org/10.2166/ws.2021.060

Shahi, N. K., Maeng, M., Kim, D., & Dockko, S. (2020). Removal behavior of microplastics using alum coagulant and its enhancement using polyamine-coated sand. *Process Safety and Environmental Protection, 141*, 9–17. https://doi.org/10.1016/j.psep.2020.05.020

Sharma, B., Dangi, A. K., & Shukla, P. (2018). Contemporary enzyme based technologies for bioremediation: A review. *Journal of Environmental Management, 210*, 10–22. https://doi.org/10.1016/j.jenvman.2017.12.075

Shen, M., Xiong, W., Song, B., Zhou, C., Almatrafi, E., Zeng, G., & Zhang, Y. (2022). Microplastics in landfill and leachate: Occurrence, environmental behavior and removal strategies. *Chemosphere, 305*, 135325. https://doi.org/10.1016/j.chemosphere.2022.135325

Shende, T., Andaluri, G., & Suri, R. (2021). Power density modulated ultrasonic degradation of perfluoroalkyl substances with and without sparging Argon. *Ultrasonics Sonochemistry, 76*, 105639. https://doi.org/10.1016/j.ultsonch.2021.105639

Singh, S., Kapoor, S., & Singh, J. P. (2024). Synergistic photocatalysis by α-MoO_3 nanostructures and SWCNT nanocomposites for efficient cross-linking and oxidative degradation of polystyrene nanoplastics. *ACS Applied Materials & Interfaces, 16*(31), 40914–40926. https://doi.org/10.1021/acsami.4c06616

Skaf, D. W., Punzi, V. L., Rolle, J. T., & Kleinberg, K. A. (2020). Removal of micron-sized microplastic particles from simulated drinking water via alum coagulation. *Chemical Engineering Journal, 386*, 123807. https://doi.org/10.1016/j.cej.2019.123807

Soylak, M., Ozalp, O., & Uzcan, F. (2021). Magnetic nanomaterials for the removal, separation and preconcentration of organic and inorganic pollutants at trace levels and their practical applications: A review. *Trends in Environmental Analytical Chemistry, 29*, e00109. https://doi.org/10.1016/j.teac.2020.e00109

Suhalim, N. S., Kasim, N., Mahmoudi, E., Shamsudin, I. J., Mohammad, A. W., Zuki, F. M., & Jamari, N. L. A. (2022). Rejection mechanism of ionic solute removal by nanofiltration membranes: An overview. *Nanomaterials, 12*(3). https://doi.org/10.3390/nano12030437

Sun, Y., Wu, Q., Li, X., Sun, W., Zhou, J., & Shah, K. J. (2023). Preparation of composite coagulant for the removal of microplastics in water. *Water Environment Research, 95*(12), e10969. https://doi.org/10.1002/wer.10969

Talvitie, J., Mikola, A., Koistinen, A., & Setälä, O. (2017). Solutions to microplastic pollution—Removal of microplastics from wastewater effluent with advanced wastewater treatment technologies. *Water Research, 123*, 401–407. https://doi.org/10.1016/j.watres.2017.07.005

Tang, Y., Zhang, S., Su, Y., Wu, D., Zhao, Y., & Xie, B. (2021). Removal of microplastics from aqueous solutions by magnetic carbon nanotubes. *Chemical Engineering Journal, 406*, 126804. https://doi.org/10.1016/j.cej.2020.126804

Thomas, C., Spatayeva, T., Yu, D., Loh, A., Yim, U. H., & Yoon, J.-Y. (2024). A comparison of current analytical methods for detecting particulate matter and micro/nanoplastics. *Applied Physics Reviews, 11*(1), 11313. https://doi.org/10.1063/5.0153106

Wang, R., Zhang, L., Chen, B., & Zhu, X. (2020a). Low-pressure driven electrospun membrane with tuned surface charge for efficient removal of polystyrene nanoplastics from water. *Journal of Membrane Science, 614*, 118470. https://doi.org/10.1016/j.memsci.2020.118470

Wang, Zhao, Crandall, C., Sahadevan, R., Menkhaus, T. J., & Fong, H. (2017). Microfiltration performance of electrospun nanofiber membranes with varied fiber diameters and different membrane porosities and thicknesses. *Polymer, 114*, 64–72. https://doi.org/10.1016/j.polymer.2017.02.084

Wang, Zhifeng, Lin, T., & Chen, W. (2020b). Occurrence and removal of microplastics in an advanced drinking water treatment plant (ADWTP). *Science of the Total Environment, 700*, 134520. https://doi.org/10.1016/j.scitotenv.2019.134520

Wolff, S., Weber, F., Kerpen, J., Winklhofer, M., Engelhart, M., & Barkmann, L. (2021). Elimination of microplastics by downstream sand filters in wastewater treatment. *Water, 13*(1), 33. https://doi.org/10.3390/w13010033

Wu, P., Tang, Y., Cao, G., Li, J., Wang, S., Chang, X., Dang, M., Jin, H., Zheng, C., & Cai, Z. (2020). Determination of environmental micro(nano)plastics by matrix-assisted laser desorption/ionization–time-of-flight mass spectrometry. *Analytical Chemistry, 92*(21), 14346–14356. https://doi.org/10.1021/acs.analchem.0c01928

Wulandari, M., Marpaung, K., Prasaningtyas, A., Yorika, R., Harfadli, M. M., & Zulfikar, A. (2024). Performance of rapid sand filter dual media for microplastic removal in the water: The effect of microplastic size and effective size of filter media. *Journal of Community Based Environmental Engineering and Management, 8*(8), 103–110. https://doi.org/10.23969/jcbeem.v8i1.12502

Xu, D., Zheng, J., Zhang, X., Lin, D., Gao, Q., Luo, X., Zhu, X., Li, G., Liang, H., & Van der Bruggen, B. (2022). Mechanistic insights of a thermoresponsive interface for fouling control of thin-film composite nanofiltration membranes. *Environmental Science & Technology*, *56*(3), 1927–1937. https://doi.org/10.1021/acs.est.1c06156

Xu, Q., Huang, Q.-S., Luo, T.-Y., Wu, R.-L., Wei, W., & Ni, B.-J. (2021). Coagulation removal and photocatalytic degradation of microplastics in urban waters. *Chemical Engineering Journal*, *416*, 129123. https://doi.org/10.1016/j.cej.2021.129123

Yahyanezhad, N., Bardi, M. J., & Aminirad, H. (2021). An evaluation of microplastics fate in the wastewater treatment plants: Frequency and removal of microplastics by microfiltration membrane. *Water Practice and Technology*, *16*(3), 782–792. https://doi.org/10.2166/wpt.2021.036

Yu, K.-F., Li, P., Li, H., Zhang, B., Yang, J., Huang, F.-Y., Li, R., & He, Y. (2021). Potential of coagulation to remove particle-associated and free-living antibiotic resistome from wastewater. *Journal of Hazardous Materials*, *406*, 124295. https://doi.org/10.1016/j.jhazmat.2020.124295

Zhang, Yongli, Diehl, A., Lewandowski, A., Gopalakrishnan, K., & Baker, T. (2020). Removal efficiency of micro- and nanoplastics (180 nm–125 μm) during drinking water treatment. *Science of the Total Environment*, *720*, 137383. https://doi.org/10.1016/j.scitotenv.2020.137383

Zhang, Yujian, Zhou, G., Yue, J., Xing, X., Yang, Z., Wang, X., Wang, Q., & Zhang, J. (2021). Enhanced removal of polyethylene terephthalate microplastics through polyaluminum chloride coagulation with three typical coagulant aids. *Science of the Total Environment*, *800*, 149589. https://doi.org/10.1016/j.scitotenv.2021.149589

Zhao, H., Wu, J., Su, F., & He, X. (2023). Removal of polystyrene nanoplastics from aqueous solutions by a novel magnetic zeolite adsorbent. *Human and Ecological Risk Assessment*, *29*(2), 327–346. https://doi.org/10.1080/10807039.2022.2071209

Zhou, Y., Kumar, M., Sarsaiya, S., Sirohi, R., Awasthi, S. K., Sindhu, R., Binod, P., Pandey, A., Bolan, N. S., Zhang, Z., Singh, L., Kumar, S., & Awasthi, M. K. (2022). Challenges and opportunities in bioremediation of micro-nano plastics: A review. *Science of the Total Environment*, *802*, 149823. https://doi.org/10.1016/j.scitotenv.2021.149823

Zhu, N., Yan, Q., He, Y., Wang, X., Wei, Z., Liang, D., Yue, H., Yun, Y., Li, G., & Sang, N. (2022). Insights into the removal of polystyrene nanoplastics using the contaminated corncob-derived mesoporous biochar from mining area. *Journal of Hazardous Materials*, *433*, 128756. https://doi.org/10.1016/j.jhazmat.2022.128756

Zodrow, K., Brunet, L., Mahendra, S., Li, D., Zhang, A., Li, Q., & Alvarez, P. J. J. (2009). Polysulfone ultrafiltration membranes impregnated with silver nanoparticles show improved biofouling resistance and virus removal. *Water Research*, *43*(3), 715–723. https://doi.org/10.1016/j.watres.2008.11.014

10 Application of Artificial Intelligence Tools in Plastic Waste Management

Jitendra Choudhary[1,2], Rahul Lanjewar[1], Karthikeyan C.[1], and Sankar Chakma[1]*

[1]Department of Chemical Engineering, Indian Institute of Science Education and Research, Bhopal, India

[2]Artie McFerrin Department of Chemical Engineering, Texas A&M University, College Station, Texas, USA

10.1 INTRODUCTION

Managing plastic waste has emerged as a significant global issue because of its severe environmental consequences. Since the mid-20th century, plastic production has surged dramatically, with more than 8.3 billion metric tons produced so far (Geyer et al., 2017). Unfortunately, only a tiny portion of this plastic has been recycled, resulting in billions of tons polluting our ecosystems and clogging landfills. The accumulation of plastic waste poses serious threats to living organisms, marine habitats, and human health (Dehghani et al., 2017; Eriksen et al., 2014; Kershaw et al., 2011). Moreover, other environmental effects, such as soil infertility and plastic entanglement, result in the formation of microplastics from waste plastics (Brennecke et al., 2016; Macfadyen et al., 2009).

In recent years, artificial intelligence (AI) has become a game-changing tool in revolutionizing waste management, providing innovative solutions to address the plastic waste crisis. Through the use of advanced algorithms, machine learning, and data analytics, AI is reshaping the processes of collecting, sorting, recycling, and repurposing plastic waste (*AI Revolution Shaping the Waste Industry in 2024 and Beyond*, n.d.-a). A significant and promising application of AI for plastic waste monitoring is the deployment of autonomous waste collection systems. These technologies not only eliminate harmful debris but also collect crucial data on waste patterns, which can guide future management strategies. Furthermore, AI is transforming the sorting processes at recycling facilities. Robots equipped with machine vision technology are capable of analyzing and categorizing waste with exceptional speed and accuracy, greatly exceeding the capabilities of human workers (*AI Revolution Shaping the Waste Industry in 2024 and Beyond*, n.d.-b)

Another innovative advancement is the use of smart recycling bins powered by AI. These bins can automatically sort and process plastic waste, employing sensors and machine learning to identify and separate different plastic types (Khan et al., 2021). By automating waste collection and sorting at the source, these systems greatly enhance recycling rates and minimize contamination. Beyond collection and sorting, AI is proving essential in uncovering new approaches for plastic disposal and recycling. The combination of AI and Internet of Things (IoT) technologies has led to the development of smart waste management systems by waste monitoring and tracking. These AI-driven solutions, using sensors and data analytics, provide real-time insights into patterns for generation of waste, enabling proactive strategies that augment collection routes and lower operational costs (Fang et al., 2023; Praveen et al., 2020).

* Corresponding author's email: schakma@iiserb.ac.in

DOI: 10.1201/9781003581000-13

Machine learning algorithms can quickly analyze vast amounts of data, pinpointing optimal solutions that may be missed by human researchers. Additionally, AI is being utilized to improve plastic pyrolysis processes, which transform plastic waste into valuable products like fuel (Wang et al., 2024a). By optimizing these processes, AI can boost resource recovery and support the transition to a circular economy. Various machine learning–based algorithms have been applied to model pyrolysis and gasification processes. These methods include support vector machines (SVM), tree-based algorithms like decision trees (DT), random forests (RF), and artificial neural networks (ANN) (Cheng et al., 2020; Mutlu & Yucel, 2018; Yaka et al., 2022). Furthermore, AI technology is revolutionizing the waste-to-energy (WTE) industry by offering innovative solutions for renewable energy generation and sustainable resource recovery. In light of increasing concerns about environmental sustainability and waste management, AI has become a powerful tool to optimize various aspects of the WTE process (Muntasir Shovon et al., 2024).

Integrating AI into plastic waste management holds immense promise for tackling one of today's most urgent environmental issues. AI technologies are revolutionizing waste collection, increasing sorting accuracy, optimizing recycling processes, and fostering innovation in disposal methods, paving the way for a more sustainable future. As advancements in this field progress, collaboration between technologists, environmental scientists, policymakers, and industry stakeholders will be essential in harnessing AI's full potential to transform plastic waste management on a global scale.

10.2 CHALLENGES IN PLASTIC WASTE MANAGEMENT

The management of plastic waste has emerged as a critical environmental challenge in recent years. With the continuous increase in global plastic production, the complexities surrounding waste disposal have grown significantly. This section delves into the primary obstacles in plastic waste management, as depicted in Figure 10.1.

Here are the current challenges in plastic waste management:

1. *Rising plastic production and consumption*: Global plastic production has surged dramatically in recent decades, reaching 8300 million metric tons (MMT) of waste plastic production to date (Geyer et al., 2017). By 2015, around 6300 MMT of plastic waste had been produced, with only 9% being recycled, 12% incinerated, and the remaining 79% ending up in landfills or the natural environment (Geyer et al., 2017). By 2020, the total production of primary plastics had reached approximately 9200 MMT, and around 12% of total global solid waste is composed of plastic (Alawa & Chakma, 2025). This rapid expansion in production and consumption has resulted in a significant rise in plastic waste generation, putting immense pressure on current waste management systems.
2. *Waste disposal*: A key factor contributing to the plastic waste issue is the lack of public awareness regarding proper waste disposal and recycling practices. Encouraging responsible consumption and shifting consumer behavior remains a persistent challenge in waste management initiatives (Xanthos & Walker, 2017).
3. *Environmental impact and microplastics*: The environmental consequences of waste disposal are significant, leading to air and water pollution, land degradation, and the release of greenhouse gases from landfills and incineration (Abubakar et al., 2022). These impacts contribute to climate change and present threats to both human health and ecosystems. When plastics enter the soil, they can harm soil structure, fertility, and biodiversity. Microplastics, in particular, may alter soil properties, introduce harmful pollutants, and impact soil organisms, potentially stimulating soil enzymes and influencing the behavior of other contaminants in the soil (Alimi et al., 2018; Gaylor et al., 2013; Hodson et al., 2017; Wang et al., 2015). Microplastics (MPs) from plastic waste accumulate in the environment and living organisms, posing serious risks to both human health and ecosystems (Dehghani et al., 2017). Marine animals ingest MPs, which can then enter the human

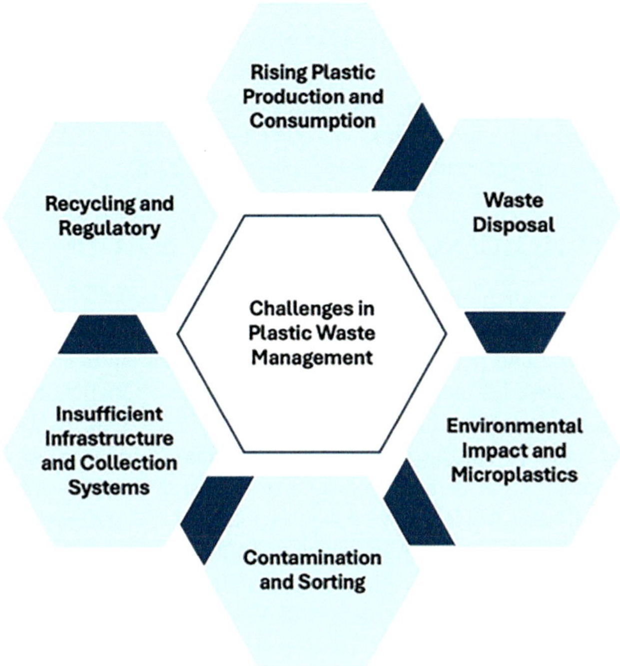

FIGURE 10.1 Flowchart depicting the challenges associated with plastic waste management.

food chain through seafood, potentially leading to health problems such as cancer and endocrine disruption. Plastic pollution severely affects marine environments, with 8–20 million tons of plastic waste entering the oceans each year, disrupting marine life and ecological processes (Eriksen et al., 2014; Jambeck et al., 2015; Mazhandu et al., 2020). Additionally, plastic entanglement poses a major threat to marine wildlife, with a 40% increase in affected species in the environment (Derraik, 2002).

4. *Contamination and sorting challenges*: The wide range of plastic types and compositions used in consumer products complicates the recycling process. Each type of plastic requires its own specific recycling method, and the absence of standardization makes sorting and processing more difficult (Hopewell et al., 2009). A significant obstacle in plastic recycling is the contamination of recyclable materials with non-recyclable items or mixed types of plastics. Such contamination can make entire batches of recyclables unusable, resulting in higher costs and decreased efficiency in recycling operations (Ragaert et al., 2017).
5. *Insufficient infrastructure and collection systems*: In many areas, especially in developing nations, there is a lack of adequate infrastructure for effective waste collection and management. This shortfall often results in improper disposal methods, such as open dumping or burning, which in turn contributes to environmental pollution and poses health risks (Jambeck et al., 2015).
6. *Challenges of recycling*: Despite the growing focus on recycling efforts, global plastic recycling rates remain extremely low, with only about 9% of all plastic waste ever being recycled (Geyer et al., 2017). Factors such as material contamination, inefficient sorting technology, and limited market demand for recycled plastics contribute to these poor recycling outcomes. Moreover, the financial viability of plastic recycling is often questioned because the costs involved in collecting, sorting, and processing plastic can exceed the value of the recycled materials. The market for recycled plastics faces significant

limitations due to concerns about quality and consistency, leading many industries to favor virgin plastics over recycled materials. This economic barrier discourages investment in recycling technologies and infrastructure (Hahladakis & Iacovidou, 2018).

7. *Regulatory challenges*: Additionally, regulatory challenges compound the issue, as inconsistent or insufficient regulations across various regions hinder the development of unified and effective waste management strategies (Dauvergne, 2018).

10.3 ARTIFICIAL INTELLIGENCE AS A POTENTIAL SOLUTION

AI developments in recent years have demonstrated encouraging potential to improve several facets of plastic waste management. Table 10.1 shows the comparison of traditional and AI-optimized plastic recycling processes. The percentage of efficiency increase indicates the potential benefits that AI technologies may have for managing plastic waste. These benefits include increased waste-to-energy conversion rates, better landfill management techniques, improved recycling and sorting procedures, optimized waste collection routes, and better monitoring accuracy. There is a fantastic chance to transform plastic waste management and transition to a more sustainable and effective strategy by utilizing AI. These developments have the potential to greatly reduce the negative environmental effects of plastic waste and improve the circular economy for plastics (Andeobu et al., 2022). Figure 10.2 shows the summary of plastic waste management strategies using AI tools.

Several conclusions can be made when comparing AI-enabled plastic waste management to conventional plastic waste management:

1. *Efficiency enhancement*: When compared to traditional techniques, AI-enabled plastic waste management shows significant efficiency gains in several areas. AI-enabled systems consistently beat conventional methods in all the listed categories, including trash monitoring, collection routes, generation forecast, recycling and sorting, waste-to-energy conversion, and landfill management.

TABLE 10.1
Comparison of Traditional and AI-Optimized Plastic Recycling Processes

Aspect	Traditional Recycling	AI-Optimized Recycling
Sorting efficiency	Depends on manual labor or basic mechanical sorting; prone to mistakes and contamination.	AI employs image recognition and machine learning for extremely accurate sorting while minimizing contamination.
Cost	Increased operating expenses owing to labor and inefficiency.	Higher initial expenditures, but lower long-term costs owing to labor savings and higher efficiency.
Scalability	Scaling-up necessitates additional workers, space, and resources.	Automation allows for easy scalability with minimum additional manpower or space.
Adaptability to new materials	Struggles in adapting to new plastic varieties or changes in recycling standards.	AI can quickly respond to new plastic kinds and changing recycling rules.
Environmental impact	Less efficient, resulting in more plastic waste in landfills or incineration.	A more effective recycling process decreases waste while increasing material recovery, which benefits the environment.
Labor requirements	High-labor intensity with the possibility of human mistake and injury.	Reduces the need for manual labor, increasing safety.

Sources: Ashwin et al., 2021; Ruj et al., 2015; Z. Wang, Ye, et al., 2024b.

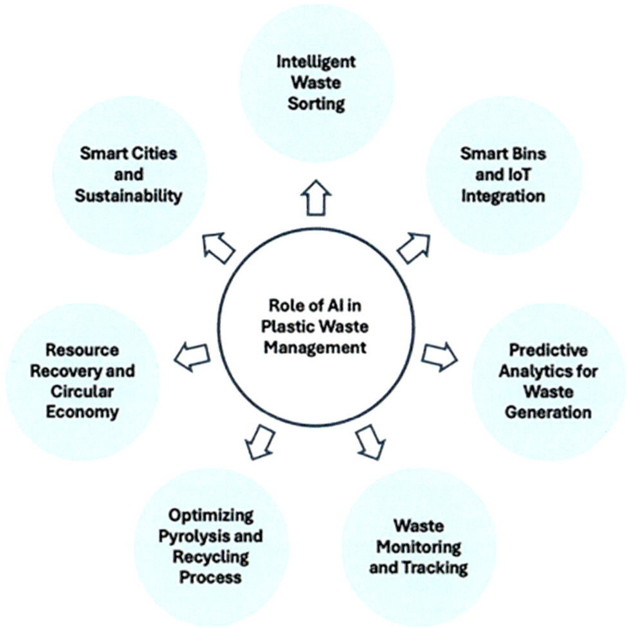

FIGURE 10.2 Summary of the areas in which AI can revolutionize plastic waste management strategies.

2. *Significant efficiency gain*: The percentages showing the efficiency gain indicate how much AI technologies may improve the handling of plastic trash. When compared to traditional methods, AI-enabled systems perform noticeably better, with efficiency gains ranging from 30% to 40%.
3. *Possibility of improved collection and monitoring*: Higher efficiency in trash monitoring and collection route optimization is demonstrated by AI-enabled plastic waste management. By using AI algorithms and technology, plastic garbage can be monitored more precisely, which improves data-driven decision-making in waste-collecting procedures.
4. *Improved garbage generation prediction*: AI-powered systems are very good at spotting trends in the production of plastic garbage. These systems can produce more accurate forecasts by utilizing machine learning algorithms and previous data analysis. This allows for resource allocation and proactive waste control techniques.
5. *Better recycling and sorting*: AI-powered plastic waste management provides notable improvements in recycling and sorting procedures. Cutting-edge AI technologies can automate and optimize sorting processes, increasing recycling rates and improving the quality of plastic-type segregation (Wilts et al., 2021).
6. *Enhanced waste-to-energy conversion*: AI-enabled methods show a discernible increase in the effectiveness of waste-to-energy conversion. More energy may be extracted from plastic garbage by using AI algorithms to optimize waste-to-energy operations like gasification or incineration, which helps to promote a more sustainable approach to waste management.
7. *Enhanced landfill management*: When compared to traditional approaches, AI-enabled landfill management systems provide significant efficiency advantages. AI approaches have high potential to optimize landfill operations, improve trash compaction, and enhance environmental monitoring, leading to substantial sustainability development and efficient organization of plastic waste in landfills.

Utilizing advanced technologies, AI has ability for transforming multiple aspects of waste management, ranging from collection to recycling, as follows:

1. *Intelligent waste sorting*: AI-powered systems can dramatically improve the precision and effectiveness of plastic waste arrangement. By integrating machine learning (ML) algorithms and computer vision, these techniques can help accurately in identifying and classifying various plastic types. For instance, AI-driven robotic sorting technologies utilize near-infrared spectroscopy and ML to sort plastics at high speeds, achieving sorting accuracies up to 99%, far surpassing manual methods. This technology addresses contamination issues that hinder current recycling efforts.
2. *Predictive analytics for waste generation*: AI can analyze historical data to forecast waste generation trends, enabling waste management authorities to optimize collection routes and allocate resources more efficiently. By factoring in variables such as population density, consumer behavior, and seasonal changes, machine learning models provide accurate waste generation predictions. This predictive approach results in more efficient waste collection strategies and reduced operational costs. AI-driven chatbots and personalized apps can educate individuals about proper waste disposal and recycling practices. These tools provide real-time guidance on how to dispose of specific items, encouraging better waste management habits at the consumer level.
3. *Smart bins and IoT integration*: AI-enabled smart bins, equipped with sensors, can monitor fill levels, detect contamination, and optimize waste collection schedules. These systems communicate in real time with waste collection vehicles, ensuring routes are adjusted based on actual needs rather than fixed schedules, significantly improving collection efficiency.
4. *Enhanced recycling processes*: AI can optimize recycling operations by analyzing and adjusting various parameters like temperature, pressure, and chemical composition in real time. Machine learning algorithms help recycling plants operate more efficiently, while advanced AI systems more effectively identify and remove contaminants, improving the quality and market value of recycled materials. AI algorithms analyze market trends to predict demand for recycled plastics, assisting recycling companies in aligning production with market needs and boosting the economic viability of recycling.
5. *Blockchain for traceability*: Combining AI with blockchain technology enhances the traceability of plastic waste along the supply chain. This integration aids in tracking contamination sources, ensuring regulatory compliance, and fostering accountability in waste management practices. AI-powered image recognition systems, deployed via drones or surveillance cameras, can detect and map litter in public areas. This technology helps identify locations that require immediate cleanup, preventing plastic waste from entering ecosystems and improving environmental cleanliness.
6. *Microplastics detection and monitoring*: AI-powered systems can detect and track microplastics in water bodies. Using machine learning to analyze water samples and satellite images helps researchers monitor the spread of microplastics and contribute to mitigation efforts.
7. *Policy and regulation support*: AI helps policymakers by processing large datasets to identify effective waste management strategies and forecast the impacts of new regulations. This data-driven approach allows for more informed and effective decision-making in policy creation.

10.4 MACHINE LEARNING ALGORITHMS

Machine learning is revolutionizing plastic waste management through the capacity to classify waste automatically, sort with efficiency, and predict waste generation. Most conventional waste management systems rely on manual sorting and conventional rule-based methods, which are error-prone

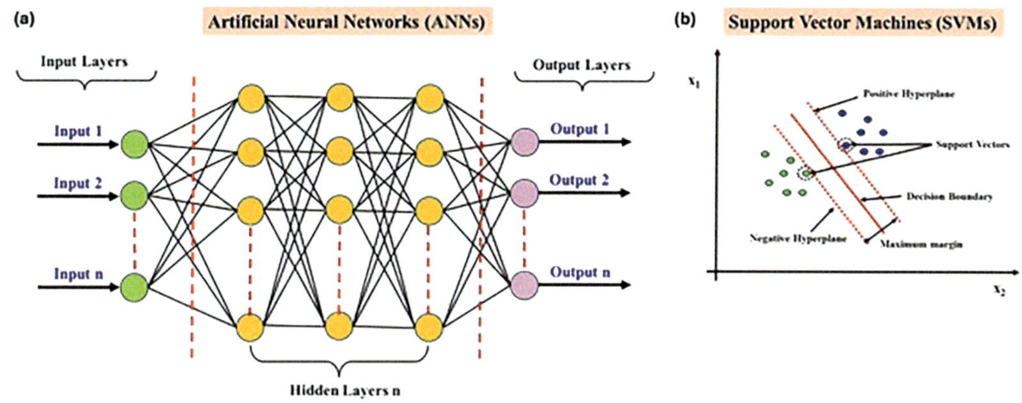

FIGURE 10.3 Schematic of architectures for supervised learning algorithms: (a) ANN, (b) SVMs.

and inefficient. Machine learning algorithms, on the other hand, can handle large datasets, identify patterns, and make accurate predictions, thereby improving recycling efficiency and sustainability. Using ML techniques, plastic waste can be segregated based on visual, chemical, and structural properties. Such methods facilitate effective separation of recyclable and non-recyclable materials, thus enhancing the recovery rate of waste. Further, predictive models through ML streamline the collection and processing of waste by forecasting patterns of waste generation, reducing the environmental impact, and increasing the circular economy. These machine learning architectures have advanced AI capabilities across many sectors, including plastic waste management, by enhancing data analysis, pattern recognition, and decision-making processes.

10.4.1 Supervised Learning Algorithms

Supervised learning is a type of ML where models are trained using labeled data; that is, input features have known output labels. These algorithms have numerous applications in plastic waste classification, sorting, and prediction. Supervised learning models primarily use architecture such as ANNs and SVMs as shown in Figure 10.3. Each of these models has distinct structures and functions as follows (Chakraborty et al., 2022):

1. *Artificial neural networks*: ANNs consist of multiple layers of interconnected neurons, including an input layer, one or more hidden layers, and an output layer. Through a training algorithm, typically backpropagation, ANN adjusts the weights of connections based on the error between predicted and actual outcomes. This architecture allows ANNs to detect complex patterns and nonlinear relationships, making them highly adaptable and capable of approximating a wide range of functions. ANNs are utilized to predict the composition of municipal solid waste, such as plastic fractions, from environmental data. For instance, a work utilized ANNs to forecast organic, paper, plastics, and textile waste stream percentages as input variables like temperature, wind speed, and humidity. The process aids in the understanding of the patterns of wastes and enhancing waste management strategies (Adeleke et al., 2021).
2. *Support vector machines*: These are supervised machine learning algorithms used for classification and regression. They find the optimal hyperplane that separates data points of different classes, maximizing the margin between the closest points (support vectors). SVMs handle both linear and non-linear data, using kernel functions like linear, polynomial, and radial basis functions (RBF) to transform non-linear data into a higher-dimensional space

for linear separation. Widely used in text categorization, image recognition, and bioinformatics, SVMs are valued for their accuracy and ability to manage complex relationships, though they can be computationally demanding for large datasets. SVMs are utilized for the classification of different plastic materials. By using features extracted from images, SVMs can classify different plastic types, for example, polyethylene terephthalate (PET) and non-PET bottles. This automatic classification enhances sorting in recycling facilities, leading to more efficient plastic waste management ("Support Vector Machines for Automated Classification of Plastic Bottles," 2010).

10.4.2 Deep Learning Algorithms

Deep learning is a sub-area of machine learning composed of neural networks with very deep layers that learn by themselves and extract high-level patterns from data. Deep models are particularly powerful in handling huge datasets for plastic waste classification and sorting. Deep learning models can be classified as convolutional neural networks (CNNs), recurrent neural networks (RNNs), graph neural networks (GNNs), and vision transformers (ViTs) as shown in Figure 10.4.

1. *Convolutional neural networks*: CNNs, frequently used in image recognition, are particularly effective for tasks such as image classification and segmentation. Their structure involves convolutional, pooling, and fully connected layers that automatically extract features from images, making them valuable for detecting visual elements like pests, plant diseases, or plastic waste in environmental applications. CNN architectures like LeNet, AlexNet, VGG16, InceptionV3, and ResNet have demonstrated success in various image-based tasks (Chakraborty et al., 2022). CNNs are mostly used for image-based classification of waste management. They allow plastic waste to be detected and sorted correctly according to visual characteristics. An example is the development of AI systems for waste classification that utilize CNNs to sort waste products and increase recycling efficiency (Yi & Kim, 2024).
2. *Recurrent neural networks*: RNNs are designed to handle sequential data and time-series problems by retaining memory of previous inputs, allowing them to process data in a temporal sequence. This ability makes them ideal for tasks involving historical data, such as predicting waste patterns based on past trends, enabling more informed decision-making. RNNs, particularly long short-term memory (LSTM) networks, find application in forecasting waste generation patterns over time (Chakraborty et al., 2022). With the analysis of temporal data, RNNs forecast future waste generation, enabling better planning and resource allocation in waste management systems. For example, LSTM models have been utilized in forecasting municipal solid waste generation to support the design of efficient collection and disposal systems (Solano Meza et al., 2023).
3. *Graph neural networks*: GNNs are a specialized deep model designed to work with graph-structured data. While CNNs are for working with grid-based data such as images and RNNs are for sequential data such as time series, GNNs are used where data can be represented as a network of relations, for example, among wastage items, recycling centers, or sensor networks (Hembert et al., 2024).
4. *Vision transformers*: A ViT is a deep learning model predominantly used for processing sequential data and has revolutionized natural language processing (NLP) and computer vision applications. In contrast to recurrent models, transformers use a self-attention mechanism to process input data in parallel rather than sequentially, thus making them more efficient and scalable for big datasets. It has been employed in image recognition applications for waste classification (Vaswani et al., 2017). They are better than traditional CNNs and RNNs in complex classification tasks as they capture long-range dependencies in images and take shorter time for training as transformers don't have any recurrence units.

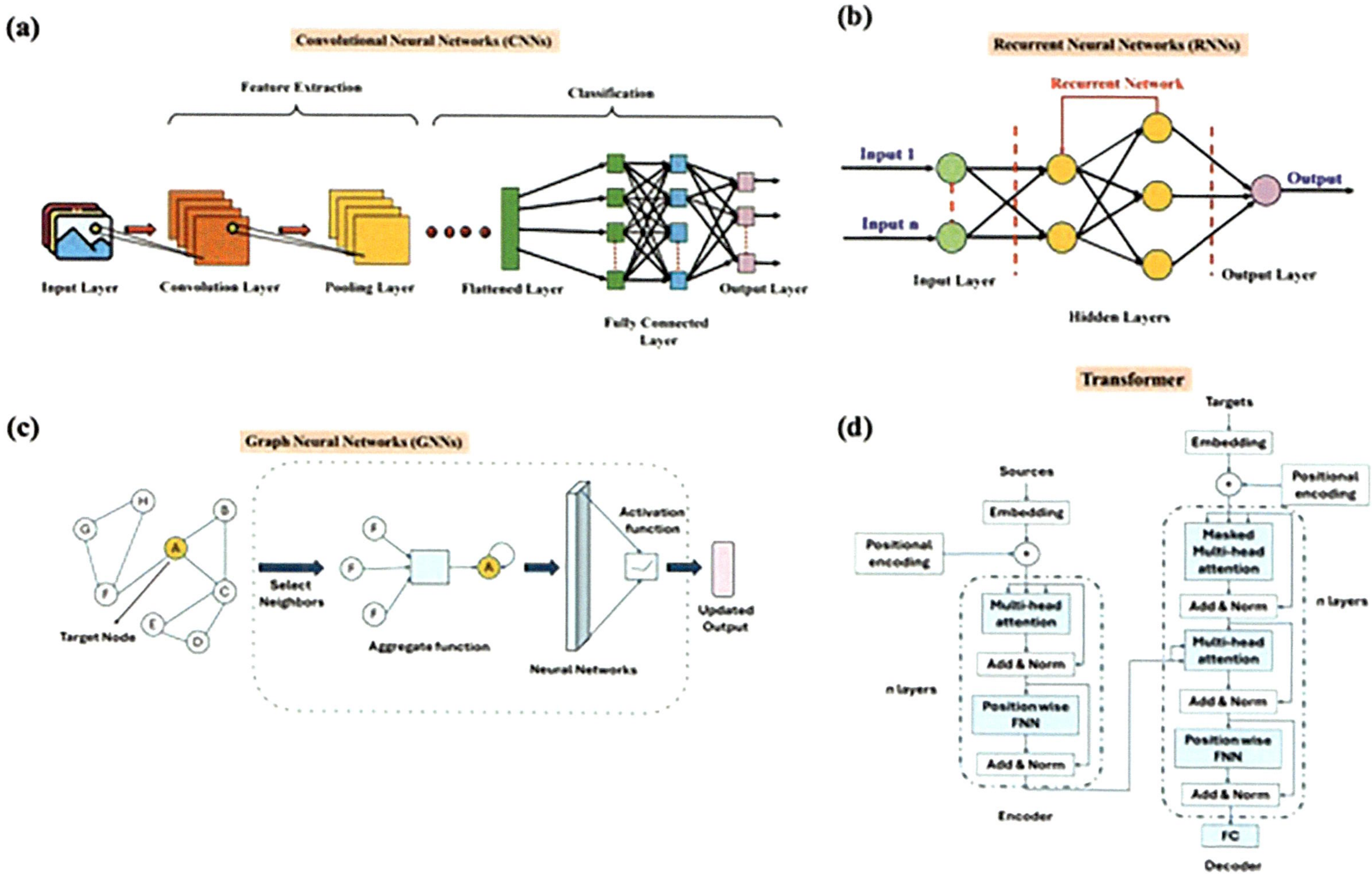

FIGURE 10.4 Architectures for deep learning algorithms: (a) CNNs, (b) RNNs, (c) GNNs, (d) ViTs.

10.4.3 Hybrid Algorithms

Hybrid models, or hybrid algorithms, combine various ML architectures such as the combination of two supervised architectures, two deep learning architectures, or a supervised architecture and a deep learning architecture to harness their respective strengths and compensate for their weaknesses. Nowadays, hybrid models are getting more popular because of their numerous advantages over individual ML algorithms, such as improved feature selection, better handling of complex relationships, and improved generalization and robustness in classification tasks. Hybrid models tend to blend CNNs with alternative architectures like transformers or GNNs to improve their performance and robustness. CNNs are well suited to identifying local spatial patterns and extracting specific features from an image. Transformer models, by contrast, employ self-attention to identify wider contextual relationships across an image. Similarly, a hybrid model of CNN and LSTM was developed for waste classification, which showed enhanced performance and efficiency than just CNN (Lilhore et al., 2024). Moreover, a CNN and a ViT-based hybrid model used for waste classification also showed enhanced accuracy than individual CNN and ViT models (Alrayes et al., 2023; Wang et al., 2024c) and a LSTM-GNN advanced hybrid model improved the predictive capabilities and reduced the computational complexity (Kumar et al., 2024). By integrating these complementary methods, researchers can build advanced hybrid models that leverage both precise local feature extraction and global image understanding, resulting in more robust and effective image recognition systems for waste sorting and classification (Alrayes et al., 2023; Dosovitskiy et al., 2021).

10.5 APPLICATION OF AI IN WASTE COLLECTION AND SORTING

The growing amount of plastic waste presents a substantial environmental issue. Moreover, the collection of waste plays a vital role in maintaining urban sanitation and public health. It ensures the timely and efficient removal of waste from residential, commercial, and industrial zones. Conventional waste management techniques are insufficient to handle the complexity and volume of today's waste streams. Traditional waste collection processes involve humans identifying the waste type and then disposing of it in the respective bins using the hand-picking method. However, AI-supervised trash and bins can be implemented and can be welcomed in the market, which can have the ability to complete the classification automatically and reduce the time and human involvement (Qu, 2022). AI-based waste-sorting robots utilize advanced technologies, including computer vision, machine learning, and robotics, to effectively identify, sort, and classify plastic waste. These sophisticated systems are specifically designed for use in recycling facilities, where they handle large volumes of waste with remarkable precision and speed. By automating the sorting process, these robots significantly reduce the need for manual labor, minimize human error, and enhance the overall efficiency of recycling operations. Several researchers have explored and reported AI-based waste collection systems using different AI models such as ANN, RNNs, SVMs, and CNNs. However, CNN models such as ResNet, visual geometric group (VGG), Inception, and Inception-ResNet are the most accurate due to their high accuracy in image recognition and classification (Adedeji & Wang, 2019; Nañez Alonso et al., 2021). Adedeji and Wang (2019) reported an intelligent waste classification system based on deep learning CNN, where the CNN model serves as an extractor and SVM, which is used to classify the waste into different types such as organic, plastic, and metal. These classifications were done based on the image dataset (around 2000 images) to train the CNN models and then SVM was implemented to classify the materials and an accuracy of 87% was reported. Similarly, Alonso et al. proposed a CNN using VGG and ResNet based on image classification and achieved a 90% reliability rate in classifying different types of wastes (Nañez Alonso et al., 2021).

Traditional waste-sorting methods are often labor intensive, time consuming, and error prone. Recent progress in AI has revolutionized waste management practices. In this section, we explore how AI-powered technologies are reshaping waste sorting and recycling processes. AI-powered waste-sorting robots leverage computer vision and ML algorithms to identify and categorize different types of waste (*AI Revolution Shaping the Waste Industry in 2024 and Beyond*, n.d.-a). These robots operate in recycling facilities and landfills and even at household level. Their accuracy in distinguishing between plastic, glass, paper, and other materials significantly improves recycling rates. Waste-sorting robots can process vast amounts of waste quickly. They analyze visual data from sensors and cameras, making real-time decisions about which conveyor belt to divert specific items to. This automation reduces human error and ensures consistent sorting, as shown in Figure 10.5. Studies are actively exploring disassembly techniques, often involving collaboration between robots and humans. Overall, the integration of robots in recycling and disassembly processes with a rate of 60–70 items/minute has the potential to transform what has traditionally been a labor-intensive industry (Bogue, 2019). Mohammed et al. proposed a system based on ANN and the fusion of different features to improve the accuracy of waste classification. The system aims to correctly separate waste into categories such as paper, glass, plastic, and metal. Key contributions include the use of ANN for classification, extraction of features from waste images, and the fusion of different feature vectors. The proposed approach achieves an impressive accuracy of 91.7%. However, challenges remain, such as computational complexity and the need for more diverse waste classes. Future work could explore genetic algorithms and multi-label image classification to enhance waste-sorting efficiency (Mohammed et al., 2023). Satav et al. (2023) discussed various algorithms used in robotic waste sorting, including deep learning (DL) models like CNNs and region-based CNNs. These algorithms, when combined with computer vision, enable robots to detect and classify waste materials accurately with an accuracy of 98.53% for ResNet models and >95% for SVM models.

Hybrid algorithms have become more popular due to their improved performance, robustness, low computational power, and reduced timing. Some researchers explored transformers and developed advanced hybrid ML algorithms such as CNN-ViT, CNN-LSTM, LSTM-GNN, and CNN-ViT, which used even larger datasets for waste classification and achieved classification accuracy of 95.8%, 98.76%, 96.54%, 95.45, respectively (Alrayes et al., 2023; Kumar et al., 2024; Lilhore et al., 2024; Wang et al., 2024c). These advanced hybrid algorithms performed better than their respective individual algorithms. Table 10.2 shows the summary of the performances of different ML and hybrid algorithms used in various studies.

Several companies and research institutions are at the forefront of developing and implementing AI-based waste-sorting robots. ZenRobotics, a leader in smart robotic recycling, has introduced

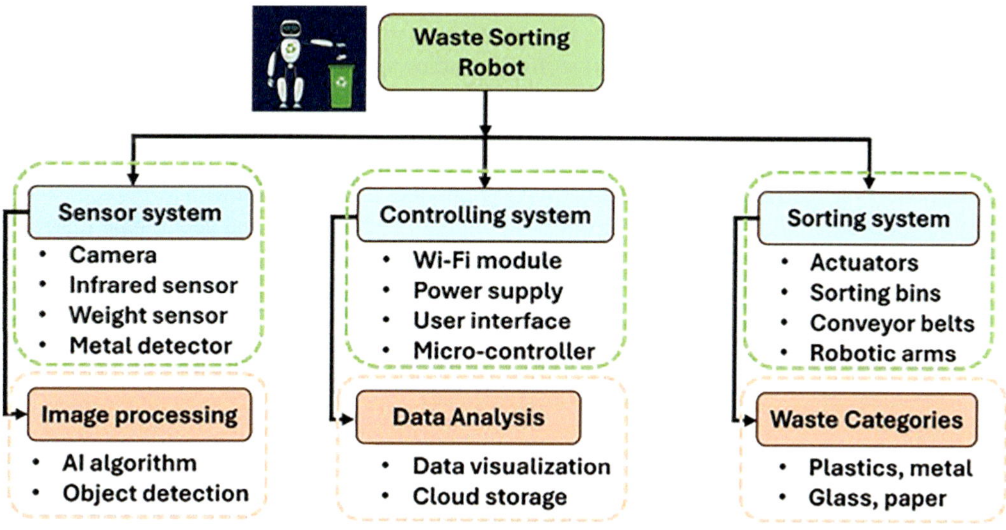

FIGURE 10.5 Flowchart showing the different components of waste-sorting robots.

TABLE 10.2
Summary of Algorithms and Their Performance Used for Waste Sorting Using Robotic Systems

Materials	Algorithm	Classification Accuracy (%)	Sorting Robot Accuracy (%)	Reference
Plastics, metal, and glass	ResNet-50 (CNN)	98.53	89.77	Sirawattananon et al., 2021
Seven different types of plastics	CNN	87	87	Bobulski & Kubanek, 2021
Plastics, metal, bio-waste, paper, and glass	EfficientNet-B2 (CNN)	75	70	Majchrowska et al., 2022
Plastic, metal, and glass	CNN	70	–	Abeygunawardhana et al., 2020
Plastic, paper, and metal	CNN, SVM	83	94.8	Sakr et al., 2016
Plastics, metal, bio-waste, paper, and glass	ResNet-50 (CNN)	87	–	Adedeji & Wang, 2019
Paper, glass, plastic, and metal	ANN	91.7	91.7	Mohammed et al., 2023
Glass, metal, paper, plastic, cloth, and organic foods	ResNet-50 (CNN)	99.5	–	Toğaçar et al., 2020
Metal, biological, glass, plastic, paper, clothes, and battery	ViT	95	–	Islam Prova, 2024
Glass, metal, paper, plastic, and cardboard	CNN-ViT	95.8	–	Alrayes et al., 2023
Textile, garden, metal, food, glass, plastic, paper, and others	LSTM-GNN	98.76	–	Kumar et al., 2024
Organic and recyclable waste	CNN-LSTM	95.45	–	Lilhore et al., 2024
Glass, metal, paper, plastic, cloth, cardboard, battery, and organic foods	CNN-ViT	96.54	–	Wang et al., 2024c

multiple generations of waste-sorting robots that have revolutionized recycling operations worldwide. Similarly, AMP Robotics has deployed its AI-guided sorting systems in hundreds of facilities, demonstrating significant improvements in sorting accuracy and recovery rates. Predictive models based on machine learning help municipalities and waste management companies anticipate future waste generation (*AI Revolution Shaping the Waste Industry in 2024 and Beyond*, n.d.-b). This information enables proactive planning, efficient resource allocation, and better waste management strategies. Machine learning and image-processing techniques are combined with technologies like radio frequency identification (RFID), global positioning systems (GPS), and geographic information systems (GIS) to automate container-level detection and improve waste management efficiency (Fang et al., 2023). This technological advancement not only streamlines the recycling process but also contributes to more sustainable waste management practices.

10.6 SMART BINS AND WASTE MONITORING

Smart bins represent a cutting-edge solution in waste management, utilizing advanced technologies such as AI and IoT to boost efficiency and sustainability. Traditional garbage bins only store waste, necessitating manual checks by sanitation workers to check the level of bins. This process is inefficient for regular waste disposal monitoring. Noiki et al. (2021) reported the inefficiency of traditional waste disposal methods such as open dumping, burial, and open-air dumping, which

FIGURE 10.6 Flowchart denoting the estimation and prediction of waste generation using smart bins.

are detrimental to the environment as well as public health. The other disadvantages of traditional methods are the responsibility of cleaners to empty waste bins periodically and the spilling of waste due to the significant rise of waste, which spreads diseases, odor, and unhealthy environments. Smart bins are outfitted with sensors and AI algorithms, enabling them to perform tasks like monitoring fill levels, sorting different types of waste, and alerting waste management teams when they need to be emptied as shown in Figure 10.6. This innovative approach not only optimizes waste collection schedules but also enhances overall resource management by minimizing human intervention and reducing operational costs (Fang et al., 2023).

A large amount of research has reported on intelligent and smart garbage bins that work on automatic waste classification and monitoring. A smart garbage bin can be created by using AI-based technologies like intelligent garbage bins, wireless sensors, classification robots, and AI algorithms, which can be used to monitor waste, predict waste, and optimize waste processing. Praveen et al. created an intelligent garbage bin by integrating the Espressif System chip, which can consist of gas, humidity, temperature, odor, sound, and infrared sensors. These smart garbage bins can detect objects and enable threshold limits and are connected to a computer system that can transmit the real-time collection and classification data, and further, transfer it for processing (Praveen et al., 2020). Khan et al. (2021) proposed a smart garbage bin monitoring system based on ultrasonic sensor networks to track municipal solid waste. Moreover, this study also proposed an architecture system to construct zone control units, that is, various smart bins which were able to track waste levels using a Global System for Mobile Communications (GSM) network controlled by computer and system software. Numerous cities and organizations have effectively adopted smart bins and waste-monitoring systems, as summarized in Table 10.3. In summary, smart bins and waste-monitoring systems signify a major leap forward in waste management technology. Utilizing real-time data and connectivity, these systems enhance efficiency, lower environmental impact, and promote sustainable waste management practices. However, to ensure successful implementation, it is crucial to address the associated challenges and costs effectively.

10.7 PLASTIC PYROLYSIS AND ENERGY CONVERSION

Optimization of thermo-chemical processes involves careful assessment of interrelated operating parameters, including catalyst types, dose, residence time, and operating temperature. Process optimization in chemical engineering has traditionally been approached through a series of laboratory investigations, one variable at a time. The optimal values for different operational parameters that impact a process's effectiveness or the ideal operational parameters for a process are found using conventional models, which are mathematical formulations (Chandel et al., 2024; Choudhary et al., 2023). Models are also helpful for efficiency, process management, and product yield forecasts (Xayachak et al., 2022). Plastic pyrolysis is currently modeled using global kinetic models and

TABLE 10.3
Highlights and Case Studies of Smart Bins Placed at Different Locations

Case Study/ Location	Type of Waste	Smart Bin Type	Algorithm/Method	Accuracy	Road Planning	Economic Feasibility	Reference
Singapore	Mixed waste (plastic, metal, paper, glass, and organic)	Camera-equipped bins	Deep learning (CNN)	93%	Not specified	High initial cost, efficiency gains	Chu et al., 2018
Santander, Spain	General waste (paper, cardboard, non-hazardous household waste)	IoT-enabled bins	Genetic algorithm	Not specified	Optimized collection schedules	High initial cost, long-term savings	Gutierrez et al., 2015
Seoul, South Korea	General waste (paper, cardboard, non-hazardous household waste)	GPS-enabled bins	K-means clustering	Not specified	Efficient bin placement	Moderate initial cost, improved efficiency	Hong et al., 2014
Barcelona, Spain	General waste (paper, cardboard, non-hazardous household waste)	Smart bins with fill-level sensors	Neural networks	91%	Predictive collection planning	High initial cost, operational savings	Cerchecci et al., 2018
Copenhagen, Denmark	General waste (paper, cardboard, non-hazardous household waste)	IoT-based smart bins	Decision trees	89%	Optimized collection frequency	Moderate initial cost, efficiency gains	Soltani et al., 2015

semi-detailed kinetic models, including kinetic Monte Carlo, method-of-moment, and deterministic models. Regretfully, the use of these models is still limited due to their lack of mass and heat transfer, catalyst, and feedstock type integration (Ureel et al., 2022). By applying computational tools based on AI techniques, the efficiency of these processes can be increased and the time and money needed to synthesize and test them can be decreased (Armenise et al., 2021; Chen et al., 2020; David et al., 2020).

Machine learning presents several benefits over conventional methods in various areas. In recent years, ML has garnered considerable research attention for its applications in yield prediction, parameter optimization, and the monitoring of pyrolytic processes (Wang et al., 2024a). ML is a potent tool for modeling complicated processes by mapping inputs with matching outputs without involving underlying mechanisms to derive mathematical correlations among them (Ascher et al., 2022). ML could make predictions without doing experiments, saving time and money on experiment load. It also demonstrates the capacity to handle complicated, large, and high-dimensional data (Li et al., 2023). The steps involved in the ML processes are shown in Figure 10.7 and described later: (1) data acquisition: data collection from literature or experiment; (2) data preprocessing: normalization and dimensionality reduction for data; (3) model screening: model selection and hyper parameters; (4) model training: hyper-parameters optimization by cross-validation based on training dataset; (5) model testing: evaluate the model performance based on a testing dataset; (6) model interpretation: analyze the relationship between inputs and outputs to enhance model explainability and interpretability.

Many algorithms have been used to represent pyrolysis or gasification processes. These methods include SVM, tree-based algorithms such as DTs, RF, and ANN (Cheng et al., 2023; Mutlu & Yucel, 2018; Yaka et al., 2022). The summary of the application of machine learning to the pyrolysis of plastic waste is displayed in Table 10.4. The most popular machine learning methodology is the ANN method. Choudhary et al. (2024) used TGA data to optimize and predict thermodynamic parameters in the co-pyrolysis of banana peel and waste plastics using the Advanced Isoconversional Model (AIC) model and ANN modeling. Using TGA data, some researchers have developed ANN models to forecast the pyrolytic behavior of particular materials' thermal breakdowns (Al-Yaari & Dubdub, 2020; Naqvi et al., 2018). An ANN outperformed a real-gas equilibrium model in the biomass gasification process by incorporating tar, char, and permanent gas interactions (Baruah & Baruah, 2014). Cheng et al. (2020) suggested combining life cycle evaluation and economic analysis with an RF-based predictive model. Cao et al. (2016) aimed to achieve a comprehensive evaluation holistically with different pyrolysis feedstocks. SVM is frequently employed in applications involving the prediction of pyrolysis. For instance, SVM outperformed ANN in estimating the yield of pyrolysis biochar at R^2 and RMSE. Li et al. (2023) used a hybrid model of the artificial neural network-genetic algorithm (ANN-GA) to predict and optimize the co-pyrolysis oil yield due to the

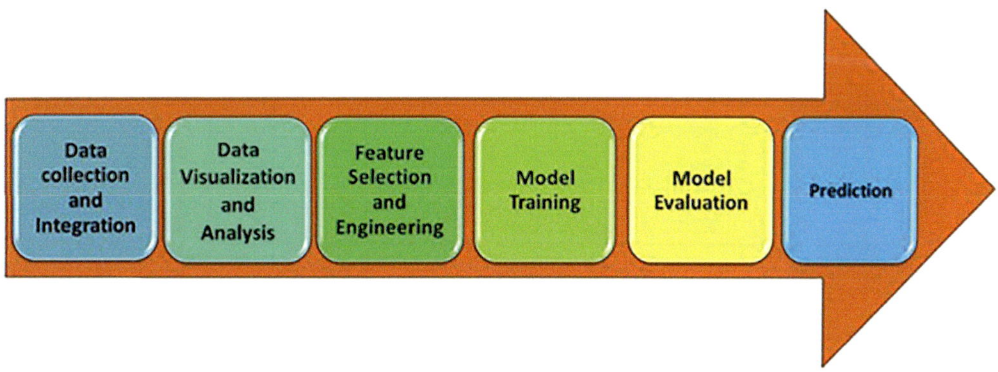

FIGURE 10.7 Steps involved in the ML process and implementation.

TABLE 10.4
Summary of Machine Learning Applications in Plastic Waste Pyrolysis and Co-Pyrolysis

Feed	Conversions	Input Variables	Output Targets	ML Model	Data Size	Best R^2	Reference
PS	Microwave-assisted pyrolysis	Mass of KOH and PS	Yield of char, gas, oil, pyrolysis time, heating rate, power, energy, conversion efficiency of specific microwave, and conductive heat loss	SVM	13	0.98–1.00	Terapalli et al., 2022
Mixed plastic	Pyrolysis	Reaction temperature, catalyst ratio, and reaction time	Liquid and gaseous yield	ANN, SVM	1048–1130	0.99	Mighani et al., 2017
PP, PE, PVC, PS, PET	Pyrolysis and co-pyrolysis	Plastic type, ultimate analysis, particle size, feed intake, pyrolysis temperature, residence time	Yield of oil gas, char, liquid, and gas compositions	SVM, DT, ANN	274–301	0.61–0.94	Y. Cheng et al., 2023
A mixture of HDPE, LDPE, PP, and PS	Co-pyrolysis	Ratios of different plastics	Yield of char, oil, and gas	ANN	24	0.95–0.99	Abnisa et al., 2019
Mixed plastics	Co-pyrolysis	Temperature, heating rate	Weight loss	ANN	358–752	0.99	Al-Yaari & Dubdub, 2020

Abbreviations: DT, Decision Tree; LDPE, Low-Density Polyethylene; NN, Neural Networks; PE, Polyethylene; PET, polyethylene terephthalate; PLA, polylactic acid; PP, polypropylene; PS, polystyrene; PVC, polyvinyl chloride; SVM, Support Vector Machine.

complex interactions of the waste polyethylene and polystyrene (WPE/WPS) mixture composition and operating conditions. Figure 10.8 shows the schematics of different ML algorithm architectures used for plastic waste pyrolysis processes.

However, pyrolysis is a complex, nonlinear, and multi-scale process affected by various operating factors, including catalyst composition, temperature gradients, heating rate, and residence time. Moreover, insufficient experimental data often results in a mismatch between the ML model and the physical process, especially for feedstocks that are not present in the training data set, making it difficult to develop credible ML models for biomass pyrolysis and gasification. In recent publications, alternative approaches such as physics-informed neural networks (PINNs) or Bayesian optimization are explored to enhance parameter tuning, uncertainty quantification, and more reliable and efficient process modeling.

Figure 10.9 shows the basic flow diagram of PINNs and Bayesian optimization algorithms. Bayesian optimization is a ML-based method that globally optimizes black-box objective functions (Alruqi & Sharma, 2023; Dogu et al., 2025). Bayesian optimization techniques have two essential components: a probabilistic surrogate model, which approximates the objective function, and an acquisition function, which selects the next evaluation point. Dogu et al. (2023) used a tree-based kinetic Monte Carlo with parameter adjusting based on Bayesian optimization to estimate the pyrolysis pathways of PS (Polystyrene). It was discovered that the production and selectivity of styrene monomers could be efficiently maximized in PS pyrolysis and reused as a feedstock for styrene polymerization. Alruqi and Sharma (2023) used the Bayesian Approach for Hyperparameter Optimization for Gaussian Process Regression to manufacture biomethane from a mixture of sugarcane vinasse, solid waste, and spent tea waste. This study showed that the contemporary Bayesian methodology for hyperparameter tuning for Gaussian process regression is an effective way of

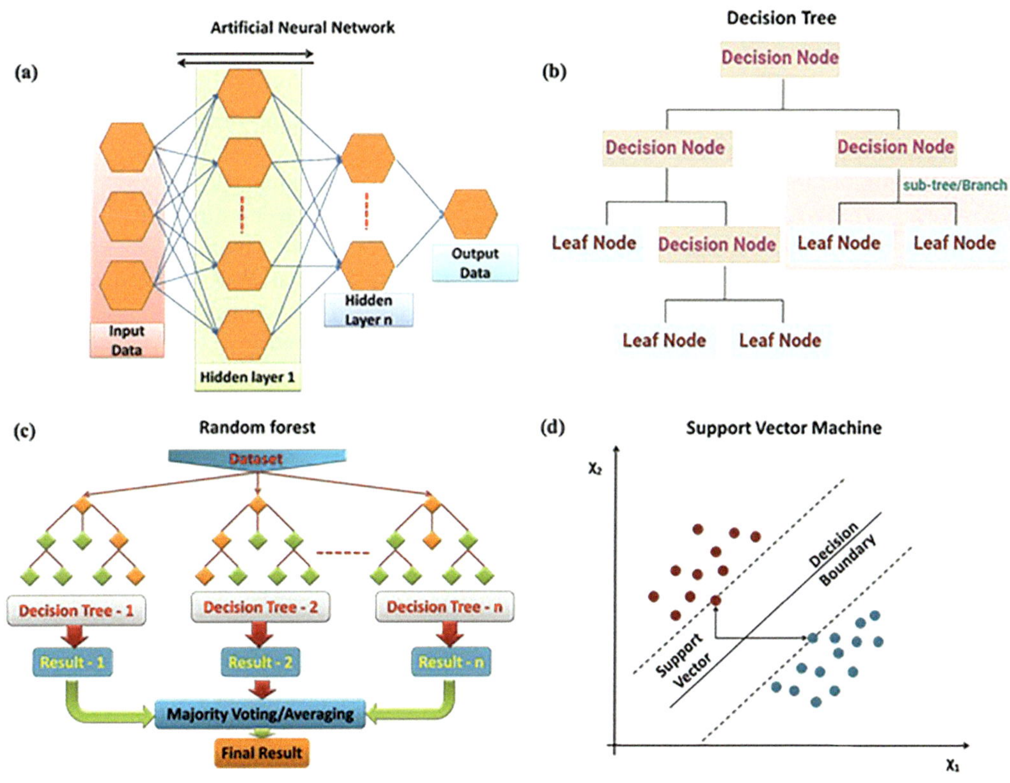

FIGURE 10.8 Schematics of ML algorithms: (a) ANN, (b) RF, (c) DT, and (d) SVM.

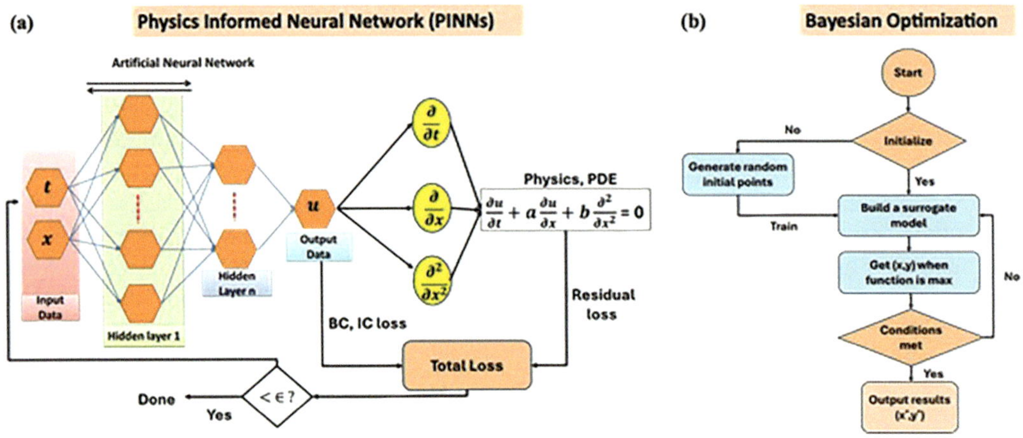

FIGURE 10.9 Schematics of (a) PINNs and (b) Bayesian optimization algorithms.

model prediction. Similarly, Kaya et al. (2024) developed new predictive models for hydro char's high heating value utilizing Gaussian Process Regression (GPR), Ensemble, DT algorithms, and Bayesian Optimization (BO). The Bayesian Optimization technique minimizes prediction errors by combining GPR, Ensemble, and DT with BO. These results revealed that BO-GPR converged during training with the lowest Mean Absolute Error (0.1783) compared to BO-Ensemble (0.5128) and BO-DT (0.7430). Kanthasamy et al. (2023) employed Bayesian optimal multilayer perceptron neural network modeling to generate biochar and syngas from biomass-derived wastes and the results reveal that Bayesian-optimized models allow the correct prediction of the biochar and syngas yields.

PINN techniques embed reliable physical knowledge into the model training process to reduce the demand for training samples while improving model generalization performance and interpretability (Fernández de la Mata et al., 2023; Wang et al., 2022). Wu et al. (2023) reported that PINN models with limited training data outperformed fully data-driven ML models in terms of prediction accuracy. Sharma et al. (2023) proposed a physics-informed deep learning model that uses mild penalty constraints to enforce physics while successfully predicting temperature and velocity fields in the melting field with only a modest amount of training data. van Wyk (2025) developed a revolutionary physics-based machine learning model for enhanced thermochemical waste conversion. The physics-informed model was created to improve data-driven machine learning models for waste gasification/thermal cracking applications. This model covered a wide range of feedstocks, making it applicable to a variety of processes, including biomass conversion for the manufacture of renewable fuels and chemicals, as well as chemical recycling of plastic waste. It reliably predicts the whole carbon product slate (desirable and undesired products) while maintaining scientific consistency and provided economic and environmental benefits by optimizing process settings to optimize product yields while minimizing waste streams. Iannello et al. (2025) explored the dynamic behavior of single polypropylene particles in a lab-scale bubbling fluidized bed reactor during pyrolysis at temperatures ranging from 500°C to 650°C. A mechanistic one-dimensional axial segregation model and two alternative physics-informed machine learning (PIML) models based on the PINN framework were presented and confirmed using experimental data. The augmented version of the PINN (augPINN) outperformed the mechanistic model and the normal PINN in terms of predictive ability. When it comes to describing the axial mixing/segregation of polypropylene particles, the augPINN outperforms both the mechanistic and the ordinary PINN models. Furthermore, the resulting value P (Interphase Distribution Parameter) was determined to be physically interpretable. Ghaderi et al. (2022) utilized collocation training to create PINN architectures that predicted temperature distributions and degree of burning during pyrolysis in a variety of one- and two-dimensional cases. The applied model predicted the system's performance loss by estimating the char formation pattern and the localized degree of burning at each continuity. Therefore, PINNs or Bayesian optimization can be used for more reliable and efficient process modeling of plastic pyrolysis processes.

10.8 RESOURCE RECOVERY AND CIRCULAR ECONOMY

According to data from the Organization for Economic Co-operation and Development, the lifespan of plastics currently contributes 1.8 giga tons (Gt) of greenhouse gas emissions (GHGs); but due to rising plastic usage and waste, this will more than double by 2060, reaching >4.3 Gt CO_2e (OECD, 2022). Zheng & Suh (2019) anticipated that worldwide emissions from plastics at all stages of their life cycle would rise from 1.7 Gt CO_2e in 2015 to 6.5 Gt CO_2e by 2050 if the plastics industry keeps up its current trend of demand without implementing any changes to the way plastics are disposed. Several systems have been developed and discussed for the high-value, safe disposal of plastic trash. The main method for disposing of plastic waste in large quantities is incineration with energy recovery, although this method has drawn criticism due to its significant carbon dioxide and air pollution emissions (Zhou et al., 2023). It has been demonstrated that pyrolysis of mixed plastic wastes to polyolefins produces 185% less CO_2 emissions than conversion for the production of heat and power (Creadore & Castaldi, 2022). When compared to the conventional incineration of plastic waste, Park et al. (2024) proposed the waste plastic pyrolysis oil method for Ewha Woman's University in South Korea to decrease 0.8 Mt CO_2 e/year. The environmentally optimal option(s) for managing plastic garbage can be carefully compared and identified using the life cycle assessment (LCA) approach. Analyzing the potential environmental effects of a substance, process, activity, or product is achievable with the LCA approach. Currently, LCA stands as the most scientific and practical method for assessing a process's or product's environmental impact (Aryan et al., 2018).

Global interest in recovering energy from solid waste as liquid fuel is growing as a means of addressing the issues of energy security and environmental protection. Compared to other materials, plastics are almost always in contact with humans. Concerning the growing amount of plastic garbage in the environment, land is developing on a global scale. Plastic solid waste mismanagement may worsen in the absence of an innovative shift in management approach, eliminating opportunities to recover these priceless waste resources. Pyrolysis is an advanced and substitute method for managing plastic waste (Alawa et al., 2023). Syngas, bioliquid, and char/residues are the main by-products of the pyrolysis process and are utilized as alternative fuels to produce energy (Praveenkumar et al., 2024). Figure 10.10 shows plastic waste pyrolysis, and their products and applications are shown here:

1. *Gas*: The content of the feedstock material affects the composition of the gas generated during the pyrolysis of waste plastic. Gases derived from the pyrolysis of various plastic feedstock types include mostly methane (34%), hydrogen (20%), nitrogen (17.5%), carbon monoxide (13%), carbon dioxide (11%), ethylene (3%), and oxygen (1.5%). The volume of gas produced increases as the pyrolysis process's temperature rises. It has been shown that in the pyrolysis process, long residence time and high temperature maximize gas output. The gas generated during the pyrolysis process has a higher thermal value and can be utilized in gas turbines to generate electricity or in boilers for heating with no need for flue gas treatment.
2. *Char*: Char is a by-product of pyrolysis, which is the unburned plastic that remains in the reactor. At very low temperatures, char production maximizes with a moderate heating rate whereas it is minimal with rapid pyrolysis. Most char is made up of fixed carbon and volatile matter (97%) with trace amounts of ash and moisture. Because char has a low sulfur level, using it as fuel is convenient. Char can be applied to a variety of environmental problems, including the removal of heavy metals from hazardous gases and wastewater from industry and municipalities. Char can be utilized as a building material, as an energy

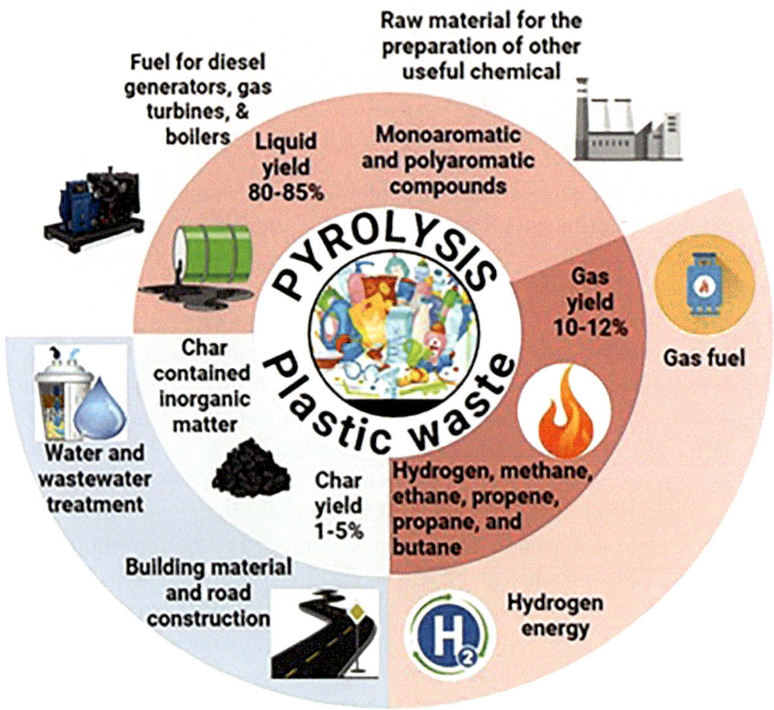

FIGURE 10.10 Plastic waste pyrolysis, their valuable products, and applications.

source in boilers, or as a feedstock for activated carbon, in road surfaces, as an absorbent in water treatment to remove heavy metals by upgrading treatment, and in other applications (Faisal et al., 2024; Maqsood et al., 2021).

3. *Pyro-oil*: The pyrolysis process produces either liquid or waxy plastic oil. The resulting wax has a high boiling point, is yellow, and is more viscous at room temperature. After plastic waste is depolymerized through pyrolysis, hundreds of chemical components combine to form plastic oil, a complex mixture of various hydrocarbons. Because pyrolysis oil shares some significant characteristics with diesel, such as density, viscosity, and calorific value, it has the potential to replace conventional liquid fuel. Benzene, toluene, xylene, limonene, and styrene are the constituents of pyrolysis oil and can be utilized as chemical raw materials in the chemical industry (Chang, 2023; Radhakrishnan et al., 2023; Zhang et al., 2024). Some examples of plastic oil, char, and gas derived from the pyrolysis of different single (HDPE, LDPE, PP, PS, and PET) plastic wastes are as given in Table 10.5.

The basic concepts of the circular economy date back hundreds of years, and they are rooted in traditional practices like recycling, reuse, and repair. The utilization of cutting-edge technologies and business models to apply these concepts in a contemporary setting and achieve a more efficient and sustainable use of resources is novel. The basis of the circular economy concept is based on the idea of closing the material loop or extending the useful life of resources while reducing waste and pollution. This is done by employing a range of techniques, such as designing durable and easily repairable items, utilizing renewable and recycled materials, and advocating for circular business models like sharing platforms and product-as-a-service (Seyyedi et al., 2023; Yang et al., 2023). An intelligent economic system always seeks novel ways to reduce costs while improving

TABLE 10.5
Product Yield from the Pyrolysis of Plastic Waste

Plastic Type	Temperature (°C)	Heating Rate (°C/min)	Yield (wt%) Liquid	Char	Gas	Reference
LDPE	500	20	97.3	0.4	2.3	Jaafar et al., 2022
HDPE	500	20	99.4	–	1.9	
PP	500	20	95.9	–	4.2	
PS	500	20	100	–	0.0	
MP	500	20	95.3	–	4.7	
LDPE	500	17	94	0.0	4	Genuino et al., 2023
HDPE	500	17	96	0.0	4	
PP	500	17	96	0.0	4	
PS	500	17	94	4	2	
PET	500	17	54	32	9	

the effectiveness and sustainability of actions that affect the entire system. Under these conditions, AI-driven technologies are prepared to usher in new industrial paradigms (Noman et al., 2022). AI technology is transforming the WTE industry by providing game-changing solutions for renewable energy production and sustainable resource recovery. AI has emerged as a controlling tool to improve many components of the WTE process in response to growing concerns about environmental sustainability and waste management. Additionally, there has been a persistent push in recent years to integrate artificial intelligence with every facet of the circular economy (Muntasir Shovon et al., 2024). Figure 10.11 shows the general schemes of AI-integrated plastic waste management toward a circular economy.

Thanks to AI, powered garbage sorting devices, robotic sensors, and machine learning algorithms, waste segregation is now more accurate and efficient, optimizing resource recovery from a variety of waste streams (Andeobu et al., 2022). Moreover, AI enables predictive maintenance by monitoring data from sensors, reducing downtime, and enhancing overall plant performance (Liu et al., 2022). AI finds valuable materials by examining the composition of garbage, which encourages recycling and the circular economy. With the potential to increase productivity, sustainability, and profitability, artificial intelligence is quickly changing the WTE industry. In the WTE sector, AI-powered solutions are being created to address a variety of issues (Muntasir Shovon et al., 2024; Pathan et al., 2023).

1. *Waste sorting*: AI can be used to create sophisticated waste-sorting systems that accurately identify and separate a variety of waste materials, including plastics, metals, and organic materials. This can lessen the quantity of waste that ends up in landfills and increase the efficiency of WTE units.
2. *Energy generation*: To ensure that waste is burned at the ideal temperature and oxygen content, AI can be utilized to optimize the combustion process in WTE plants. Emissions may be decreased, and energy output raised as a result.
3. *Process monitoring and control*: Currently, AI can be utilized to control and monitor the WTE process. This can aid in the early detection and resolution of such issues, averting expensive downtime and interruptions.
4. *Predictive maintenance*: AI may be used to forecast when WTE plant equipment is most likely to break. By doing so, maintenance may be scheduled more efficiently, and unscheduled downtime can be avoided.

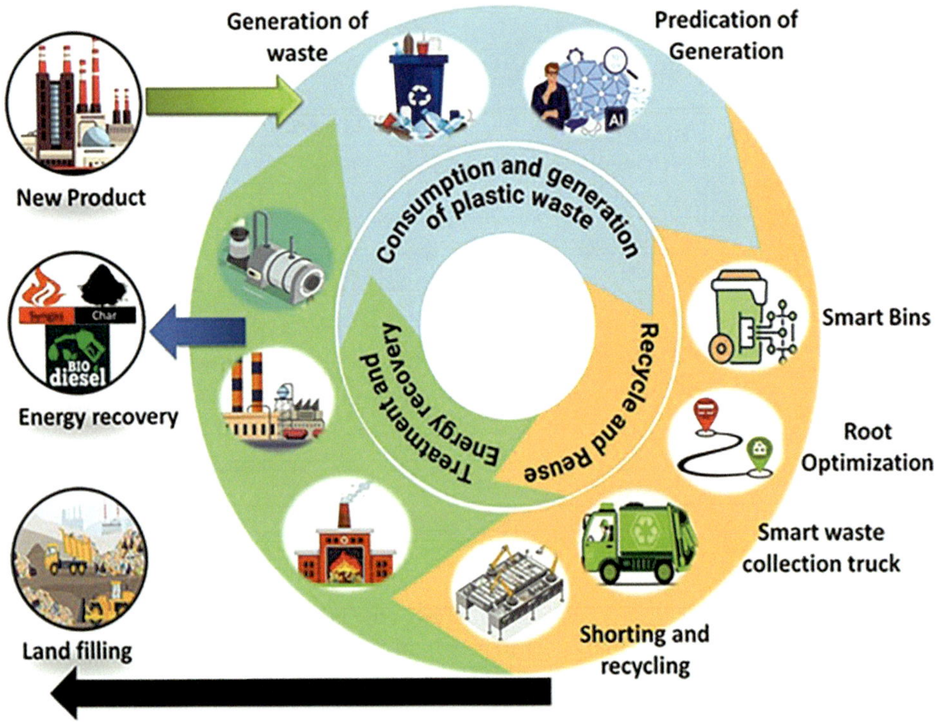

FIGURE 10.11 AI-integrated plastic waste management toward a circular economy.

10.8.1 Highlight of Success Stories and Case Studies

ML has been used to predict municipal solid waste generation and diversion in Canada (Kannangara et al., 2018). The work predicted the production of municipal solid waste (MSW) and diversion in Canada using ML based on socioeconomic and demographic factors. Models were created using decision trees and neural networks on data from 220 municipal areas in Ontario. The models used data from the Canadian Census to map residential MSW volumes along with socioeconomic and demographic factors. Creating MSW inventory for the entire country of Canada was the aim. According to the study, decision trees fared worse than neural networks when it came to developing models with 72% accuracy for MSW production prediction. However, due to the absence of variables that explain recycling behavior in the dataset, the work's diversion models exhibited low accuracy, ranging from 32% to 36%. The study concluded that if enough socioeconomic data are employed, machine learning may provide trash prediction models that are accurate.

Mohammed et al. (2023) created an automated system based on the fusion of several features and an ANN to lessen the impact of inappropriate garbage disposal, particularly household waste such as glass, paper, plastic, and organic waste. Waste can be appropriately divided into many categories using the automated approach that is being suggested. With a 91.7% accuracy rate, the ANN and fusion of different features is one of the approaches that can effectively tackle this problem, according to the data. However, compared to other conventional techniques, ANN and fusion of different features seem to be more computationally expensive. Furthermore, in the future, strategies like augmentation and fine-tuning may be used to increase the accuracy of ANN systems. Furthermore, ANNs perform better when there is greater data available. All things considered, this analysis points to the possibility that a digitally empowered circular economy vision could enhance garbage sorting services and recycling choices made along the smart city value chain.

10.9 SMART CITIES AND SUSTAINABLE PRACTICES

Smart cities offer a revolutionary approach to urban development by harnessing technology and data to enhance quality of life, streamline city operations, and foster sustainability. Unlike conventional cities, smart cities integrate cutting-edge technologies such as IoT, AI, and big data analytics to create interconnected urban networks. These networks enable the collection and analysis of real-time data, optimizing various aspects of city life, including traffic management, energy use, public services, and waste management. Table 10.6 shows the differences between smart cities and traditional urban cities.

Prospective sustainable routes and suggestions for the management of plastic waste are as follows:

1. *Progressing research and innovation*: To effectively handle the intricate problems associated with managing plastic trash, ongoing research and innovation are crucial. Research and development expenditures can result in the creation of novel materials, technology for recycling, and long-term substitutes for single-use plastics. To promote research projects and foster innovation in the industry, governments, corporations, and academic institutions should work together.
2. *Increasing international cooperation*: The global problem of plastic waste calls for increased international cooperation. To provide uniform guidelines, rules, and best practices for managing plastic trash, governments, groups, and interested parties should work together. Cross-border exchange of information, insights, and resources can hasten the transition to a plastic circular economy.
3. *Infrastructure development and investment*: The construction and modernization of waste management infrastructure are essential to the efficient management of plastic trash. To increase the capability and effectiveness of managing plastic trash, the public and private sectors should make investments in recycling centers, waste-to-energy facilities, and collection systems. Sufficient infrastructure will make it possible to properly separate, recycle, and dispose of plastic trash, minimizing its negative effects on the environment.
4. *Encouragement of circular business models*: Companies should adopt circular business models, which emphasize recycling, waste minimization, and the usage of recycled resources. This entails implementing take-back initiatives, designing items for durability and recyclability, and implementing sustainable packaging options. Encouraging companies to incorporate the ideas of the circular economy into their operations will accelerate the shift to a system that is more resource-efficient and sustainable.

TABLE 10.6
Comparison of the Characteristics of Smart Cities and Traditional Urban Cities

Characteristic	Smart Cities	Traditional Urban Cities
Infrastructure and connectivity	Employ IoT, sensors, and high-speed Internet for integrated urban systems and real-time data analysis	Use traditional methods and systems lacking real-time connectivity and integration
Sustainability and resource management	Focus on reducing carbon footprints through smart grids, renewable energy, and efficient waste management	Outdated infrastructure leading to inefficiencies and higher environmental impacts
Public services and citizen engagement	Enhanced access to services through digital platforms and encourages participatory governance	Citizen engagement occurs through traditional means like town halls with less digital emphasis
Economic opportunities and innovation	Attract businesses and startups by fostering an innovative environment	

5. *Customer education and modification of behavior*: Changing consumer behavior requires consumer-focused education and awareness efforts. Governments, nongovernmental organizations, and corporations ought to fund educational initiatives that encourage conscientious consumption, waste minimization, and appropriate recycling procedures. More sustainable plastic waste management will result from giving people the information they need to make educated decisions and from offering easy access to sustainable substitutes.
6. *Rules and regulations*: To encourage environmentally friendly methods of managing plastic garbage, governments should pass and implement thorough rules and regulations. This entails putting in place plastic reduction objectives, extended producer responsibility programs, plastic bans or limitations, and financial incentives for the use of recycled materials. Businesses and individuals can implement sustainable practices with the help of a framework that is supportive of clear and consistent policies.
7. *Working together with the private sector*: Public–private alliances are essential for promoting investment, innovation, and the execution of plastic waste management programs. To take advantage of the private sector's knowledge, assets, and market power, governments ought to interact with it. Innovation in problem-solving, technical breakthroughs, and the expansion of sustainable practices can all result from collaboration.
8. *Involving local communities*: Local communities are essential to the management of plastic trash. Communities should be involved by governments and organizations using participatory methodologies, and they should be given the resources, infrastructure, and information needed to actively participate in waste management procedures. Long-term habit change and sustainability can be achieved by supporting waste collection programs, encouraging community-led projects, and cultivating a sense of ownership.

By concentrating on these suggestions and future paths, we can strive toward a more efficient and sustainable method for managing plastic garbage. Using a blend of policy interventions, technological innovations, stakeholder cooperation, and behavioral adjustments, we may establish a future in which plastic trash is reduced, recycled, and appropriately handled, promoting a more sustainable environment and circular economy (Dris et al., 2015).

10.10 TRENDS AND CHALLENGES

The future of AI in plastic waste management is highly promising. As technology continues to evolve, we can anticipate even more advanced AI applications that will further enhance waste collection, boost recycling efficiency, and drive innovation in plastic alternatives and circular economy practices. In conclusion, AI is emerging as a powerful tool in the battle against plastic pollution. By improving waste collection, increasing sorting accuracy, optimizing recycling processes, and fostering innovative disposal methods, AI is helping pave the way toward a more sustainable future. As the global plastic waste crisis persists, integrating AI into waste management strategies offers a hopeful path toward more effective solutions and a cleaner environment. As the field evolves, new directions and possibilities emerge. Figure 10.12 depicts the current trends and the important areas of focus.

1. *Integration of AI and Internet of Things*: The combination of AI and IoT allows for the development of intelligent waste management systems that can monitor, analyze, and optimize garbage collection, categorization, reprocessing, and disposal operations automatically (Mousavi et al., 2023). This connection will enable efficient waste management operations, predictive maintenance of waste management systems, and enhanced real-time monitoring and control.
2. *Developments in machine learning and deep learning*: Waste categorization, sorting, and predictive modelling skills are being improved using machine learning techniques. Data-driven algorithms are revolutionizing trash image recognition, material identification, and

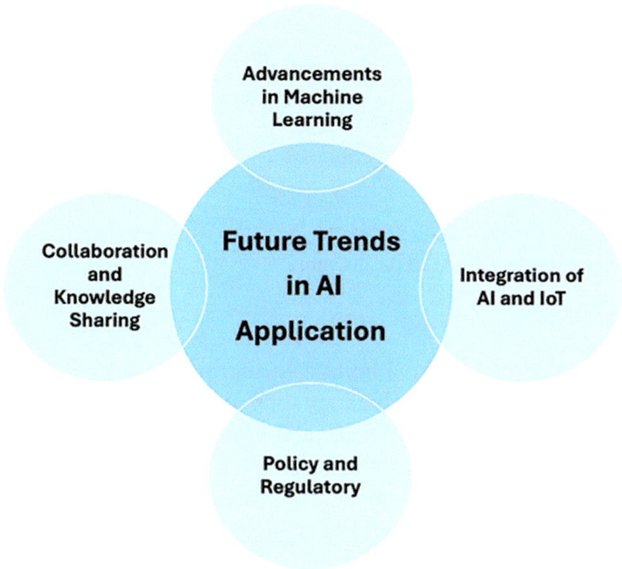

FIGURE 10.12 Trends of AI applications in plastic waste management.

quality control processes. ML algorithms can handle complicated waste streams, adapt to dynamic waste management applications, and utilize large-scale data for enhanced decision-making and optimization (Oruganti et al., 2023; Adeleke et al., 2021; Li et al., 2023).

3. *Collaboration and knowledge sharing*: Collaboration among stakeholders, including waste management agencies, technology suppliers, and legislators, is essential for the appropriate deployment of AI in trash management (Kurniawan et al., 2024). Working together, stakeholders may share their expertise, resources, and best practices, stimulating innovation and creating beneficial outcomes for both the environment and society. This partnership might result in the creation of bespoke AI solutions that suit the special needs of trash management while adhering to legislation and ethical norms. It can also help to make AI systems more open and responsible, fostering public confidence and adoption. Finally, collaboration among stakeholders can speed up the implementation of AI in trash management, leading to more efficient, sustainable, and environmentally friendly practices (Olawade et al., 2024).
4. *Policy and monitoring frameworks*: Legislative frameworks and industry guidelines are critical for ensuring the accountable and ethical utilization of AI in waste controlling. Policy measures can help promote data sharing, incentivize AI usage, and facilitate interoperability. Regulatory frameworks may help drive responsible waste-monitoring practices, confirming that AI algorithms emphasize environmental sustainability and social equality. Governments and regulatory organizations play an important role in defining these frameworks, since they may create an enabling climate that encourages innovation while protecting public interests (Olawade et al., 2024).

While AI has enormous promise to alter waste management processes, various obstacles and constraints must be overcome for successful deployment. Current trends point to the following significant areas of concern, as shown in Figure 10.13.

1. *Data availability*: One of the most significant issues in AI-driven waste administration is data accessibility and quality control. AI systems rely largely on huge, diversified datasets to learn and forecast accurately. To address these problems, efforts are underway to

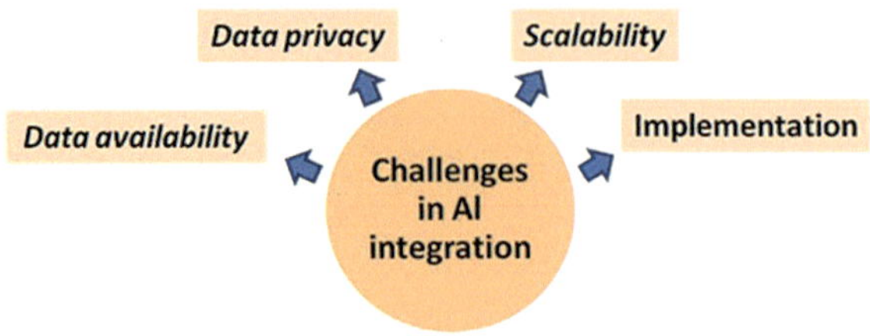

FIGURE 10.13 Future challenges related to AI integration.

enhance data-gathering methods and build data-sharing protocols. Collaborations among waste management organizations, technology suppliers, and researchers must strive to standardize data formats, build data platforms, and promote data-sharing practices (Olawade et al., 2024).

2. *Data privacy*: The use of AI in trash managing presents privacy and security concerns. AI systems frequently need access to sensitive data, such as trash creation trends, collection routes, and user behavior (Edinov & Fauzi, 2023; Pournaghi et al., 2020). To tackle these issues, privacy-improving methods like data anonymization and encryption are being used. Furthermore, legislative frameworks are growing to ensure appropriate data-processing procedures and preserve individuals' privacy rights. Striking the correct balance between data accessibility for AI algorithms and data privacy and security is a crucial issue that requires continual attention.
3. *Scalability*: Scalability is one of the obstacles of using machine learning to municipal trash management. Machine learning algorithms may need to be scaled up or down based on the quantity of garbage created in a given location. This might be difficult since it may necessitate major investments as well as adjustments to existing systems and procedures. Scaling machine learning models is crucial because it allows systems to manage vast volumes of data and adapt to changing conditions (Lin et al., 2022; Munir et al., 2023). For example, in locations with expanding populations, the amount of garbage created may rise. Machine learning models must be able to manage the increased data volume and react to variations in waste composition.

10.11 CONCLUSIONS AND PERSPECTIVES

AI tools hold immense potential in revolutionizing plastic waste management, addressing critical challenges related to public health, environmental protection, economic benefits, energy efficiency, and regulatory compliance. However, the existing waste management infrastructure faces significant hurdles, such as a lack of real-time monitoring, inefficient sorting systems, suboptimal collection routes, insufficient predictive maintenance, and limited public engagement. AI and IoT offer promising solutions to these challenges, providing technologies for real-time tracking, automated sorting, route optimization, smart bins, and targeted consumer education. Despite these advancements, maximizing AI's potential in waste management requires further research and development. AI systems rely heavily on large datasets and a deep understanding of both the data and the problem being addressed. Additionally, AI alone cannot solve the complex issue of plastic waste management—it must be implemented alongside other strategies, including effective policies, regulations, and community initiatives, to create a sustainable and integrated waste management ecosystem.

Cost–benefit analysis, scalability, and the feasibility of deploying AI solutions are critical factors for success. The biggest challenges in leveraging AI for plastic waste administration are the availability and quality of data. To address this, improving data-collection infrastructure, standardizing formats, ensuring data integrity, collaborating across sectors, and monitoring data quality are essential steps. Synthetic data generation and robust data governance strategies can further enhance AI's effectiveness. AI offers clear benefits to various stakeholders in waste management. For municipalities, it can streamline garbage collection and disposal, monitor waste containers, and predict future waste needs. Waste-managing companies can optimize gathering routes, track waste levels, and identify recycling and waste reduction opportunities. For residents and businesses, AI reduces the frequency of waste collection, improves recycling accuracy, and ensures cleaner waste containers. Environmental groups also benefit from AI's ability to identify waste reduction strategies, enhance recycling efforts, and reduce greenhouse gas emissions. Ultimately, AI has the potential to contribute to a cleaner, healthier, and more sustainable world.

To fully realize this potential, ongoing research and collaboration are vital. Comparative studies of AI-driven waste management systems have demonstrated impressive results, from optimizing collection routes to improving recycling efficiency through robotic sorting systems. However, addressing global waste management challenges requires collective action. The international community must prioritize sharing knowledge, technology transfer, and cooperation between governments, businesses, and local communities.

A unified global effort involving research institutions, industries, and policymakers can create standardized, adaptable frameworks that ensure AI-driven waste management solutions are sustainable, inclusive, and globally applicable. With concerted efforts, AI can play a key role in closing the gap between developed and developing countries and contributing to a healthier, more sustainable planet.

REFERENCES

Abeygunawardhana, A. G. D. T., Shalinda, R. M. M. M., Bandara, W. H. M. D., Anesta, W. D. S., Kasthurirathna, D., & Abeysiri, L. (2020). AI—driven smart bin for waste management. *2020 2nd International Conference on Advancements in Computing (ICAC)*, *1*, 482–487. https://doi.org/10.1109/ICAC51239.2020.9357151

Abnisa, F., Anuar Sharuddin, S. D., bin Zanil, M. F., Wan Daud, W. M. A., & Indra Mahlia, T. M. (2019). The yield prediction of synthetic fuel production from pyrolysis of plastic waste by Levenberg–Marquardt approach in feedforward neural networks model. *Polymers*, *11*(11), Article 11. https://doi.org/10.3390/polym11111853

Abubakar, I. R., Maniruzzaman, K. M., Dano, U. L., AlShihri, F. S., AlShammari, M. S., Ahmed, S. M. S., Al-Gehlani, W. A. G., & Alrawaf, T. I. (2022). Environmental sustainability impacts of solid waste management practices in the global south. *International Journal of Environmental Research and Public Health*, *19*(19), 12717. https://doi.org/10.3390/ijerph191912717

Adedeji, O., & Wang, Z. (2019). Intelligent waste classification system using deep learning convolutional neural network. *Procedia Manufacturing*, *35*, 607–612. https://doi.org/10.1016/j.promfg.2019.05.086

Adeleke, O., Akinlabi, S. A., Jen, T.-C., & Dunmade, I. (2021). Application of artificial neural networks for predicting the physical composition of municipal solid waste: An assessment of the impact of seasonal variation. *Waste Management & Research*, *39*(8), 1058–1068. https://doi.org/10.1177/0734242X21991642

AI Revolution Shaping the Waste Industry in 2024 and Beyond: Part One. (n.d.-a). Retrieved June 1, 2024, from https://www.waste360.com/industry-insights/ai-revolution-shaping-the-waste-industry-in-2024-and-beyond-part-one

AI Revolution Shaping the Waste Industry in 2024 and Beyond: Part Two. (n.d.-b). Retrieved June 1, 2024, from https://www.waste360.com/industry-insights/ai-revolution-shaping-the-waste-industry-in-2024-and-beyond-part-two

Alawa, B., & Chakma, S. (2025). A review on feasibility and techno-economic analysis of hydrocarbon liquid fuels production via catalytic pyrolysis of waste plastic materials. *Carbon Capture Science & Technology*, *14*, 100337. https://doi.org/10.1016/j.ccst.2024.100337

Alawa, B., Choudhary, J., & Chakma, S. (2023). Discernment of synergism in co-pyrolysis of HDPE and PP waste plastics for production of pyro-oil: Mechanistic investigation with economic analysis and health risk assessment. *Process Safety and Environmental Protection*, *169*, 107–131. https://doi.org/10.1016/j.psep.2022.10.077

Alimi, O. S., Farner Budarz, J., Hernandez, L. M., & Tufenkji, N. (2018). Microplastics and nanoplastics in aquatic environments: Aggregation, deposition, and enhanced contaminant transport. *Environmental Science & Technology*, *52*(4), 1704–1724. https://doi.org/10.1021/acs.est.7b05559

Alrayes, F. S., Asiri, M. M., Maashi, M. S., Nour, M. K., Rizwanullah, M., Osman, A. E., Drar, S., & Zamani, A. S. (2023). Waste classification using vision transformer based on multilayer hybrid convolution neural network. *Urban Climate*, *49*, 101483. https://doi.org/10.1016/j.uclim.2023.101483

Alruqi, M., & Sharma, P. (2023). Biomethane production from the mixture of sugarcane vinasse, solid waste and spent tea waste: A Bayesian approach for hyperparameter optimization for gaussian process regression. *Fermentation*, *9*(2), Article 2. https://doi.org/10.3390/fermentation9020120

Al-Yaari, M., & Dubdub, I. (2020). Application of artificial neural networks to predict the catalytic pyrolysis of HDPE using non-isothermal TGA data. *Polymers*, *12*(8), Article 8. https://doi.org/10.3390/polym12081813

Andeobu, L., Wibowo, S., & Grandhi, S. (2022). Artificial intelligence applications for sustainable solid waste management practices in Australia: A systematic review. *Science of the Total Environment*, *834*, 155389. https://doi.org/10.1016/j.scitotenv.2022.155389

Armenise, S., Wong, S., Ramírez-Velásquez, J. M., Launay, F., Wuebben, D., Nyakuma, B. B., Rams, J., & Muñoz, M. (2021). Application of computational approach in plastic pyrolysis kinetic modelling: A review. *Reaction Kinetics, Mechanisms and Catalysis*, *134*(2), 591–614. https://doi.org/10.1007/s11144-021-02093-7

Aryan, V., Font-Brucart, M., & Maga, D. (2018). A comparative life cycle assessment of end-of-life treatment pathways for photovoltaic backsheets. *Progress in Photovoltaics: Research and Applications*, *26*(7), 443–459. https://doi.org/10.1002/pip.3003

Ascher, S., Watson, I., & You, S. (2022). Machine learning methods for modelling the gasification and pyrolysis of biomass and waste. *Renewable and Sustainable Energy Reviews*, *155*, 111902. https://doi.org/10.1016/j.rser.2021.111902

Ashwin, M., Alqahtani, A. S., & Mubarakali, A. (2021). IoT based intelligent route selection of wastage segregation for smart cities using solar energy. *Sustainable Energy Technologies and Assessments*, *46*, 101281. https://doi.org/10.1016/j.seta.2021.101281

Baruah, D., & Baruah, D. C. (2014). Modeling of biomass gasification: A review. *Renewable and Sustainable Energy Reviews*, *39*, 806–815. https://doi.org/10.1016/j.rser.2014.07.129

Bobulski, J., & Kubanek, M. (2021). Vehicle for plastic garbage gathering. *2021 International Conference on Electrical, Computer, Communications and Mechatronics Engineering (ICECCME)*, 1–5. https://doi.org/10.1109/ICECCME52200.2021.9591096

Bogue, R. (2019). Robots in recycling and disassembly. *Industrial Robot: The International Journal of Robotics Research and Application*, *46*(4), 461–466. https://doi.org/10.1108/IR-03-2019-0053

Brennecke, D., Duarte, B., Paiva, F., Caçador, I., & Canning-Clode, J. (2016). Microplastics as vector for heavy metal contamination from the marine environment. *Estuarine Coastal and Shelf Science*, *178*, 189–195. https://doi.org/10.1016/j.ecss.2015.12.003

Cao, H., Xin, Y., & Yuan, Q. (2016). Prediction of biochar yield from cattle manure pyrolysis via least squares support vector machine intelligent approach. *Bioresource Technology*, *202*, 158–164. https://doi.org/10.1016/j.biortech.2015.12.024

Cerchecci, M., Luti, F., Mecocci, A., Parrino, S., Peruzzi, G., & Pozzebon, A. (2018). A low power IoT sensor node architecture for waste management within smart cities context. *Sensors*, *18*(4), Article 4. https://doi.org/10.3390/s18041282

Chakraborty, S. K., Chandel, N. S., Jat, D., Tiwari, M. K., Rajwade, Y. A., & Subeesh, A. (2022). Deep learning approaches and interventions for futuristic engineering in agriculture. *Neural Computing and Applications*, *34*(23), 20539–20573. https://doi.org/10.1007/s00521-022-07744-x

Chandel, A., Kumar, V., Choudhary, J., Upadhyay, P., Baidya, D., Bhattacharjee, M., & Chakma, S. (2024). Sono-extracted crystalline nanocellulose thin film as a supporting material for fabrication of highly sensitive and stable flexible sensors for thermal applications. *ACS Applied Engineering Materials*, *2*(1), 76–86. https://doi.org/10.1021/acsaenm.3c00574

Chang, S. H. (2023). Plastic waste as pyrolysis feedstock for plastic oil production: A review. *Science of the Total Environment*, *877*, 162719. https://doi.org/10.1016/j.scitotenv.2023.162719

Chen, A., Zhang, X., Chen, L., Yao, S., & Zhou, Z. (2020). A machine learning model on simple features for CO_2 reduction electrocatalysts. *The Journal of Physical Chemistry C*, *124*(41), 22471–22478. https://doi.org/10.1021/acs.jpcc.0c05964

Cheng, F., Luo, H., & Colosi, L. M. (2020). Slow pyrolysis as a platform for negative emissions technology: An integration of machine learning models, life cycle assessment, and economic analysis. *Energy Conversion and Management*, *223*, 113258. https://doi.org/10.1016/j.enconman.2020.113258

Cheng, Y., Ekici, E., Yildiz, G., Yang, Y., Coward, B., & Wang, J. (2023). Applied machine learning for prediction of waste plastic pyrolysis towards valuable fuel and chemicals production. *Journal of Analytical and Applied Pyrolysis*, *169*, 105857. https://doi.org/10.1016/j.jaap.2023.105857

Choudhary, J., Alawa, B., & Chakma, S. (2023). Insight into the kinetics and thermodynamic analyses of co-pyrolysis using advanced isoconversional method and thermogravimetric analysis: A multi-model study of optimization for enhanced fuel properties. *Process Safety and Environmental Protection*, *173*, 507–528. https://doi.org/10.1016/j.psep.2023.03.033

Choudhary, J., Kumar, A., Alawa, B., & Chakma, S. (2024). Optimization and prediction of thermodynamic parameters in co-pyrolysis of banana peel and waste plastics using AIC model and ANN modeling. *Energy Nexus*, *14*, 100302. https://doi.org/10.1016/j.nexus.2024.100302

Chu, Y., Huang, C., Xie, X., Tan, B., Kamal, S., & Xiong, X. (2018). Multilayer hybrid deep-learning method for waste classification and recycling. *Computational Intelligence and Neuroscience*, *2018*(1), 5060857. https://doi.org/10.1155/2018/5060857

Creadore, L. T., & Castaldi, M. J. (2022). Quantitative comparison of life cycle assessments of advanced recycling technologies for end-of-life plastics. *Journal of Energy Resources Technology*, *145*(042201). https://doi.org/10.1115/1.4055843

Dauvergne, P. (2018). Why is the global governance of plastic failing the oceans? *Global Environmental Change*, *51*, 22–31. https://doi.org/10.1016/j.gloenvcha.2018.05.002

David, L., Thakkar, A., Mercado, R., & Engkvist, O. (2020). Molecular representations in AI-driven drug discovery: A review and practical guide. *Journal of Cheminformatics*, *12*(1), 56. https://doi.org/10.1186/s13321-020-00460-5

Dehghani, S., Moore, F., & Akhbarizadeh, R. (2017). Microplastic pollution in deposited urban dust, Tehran metropolis, Iran. *Environmental Science and Pollution Research International*, *24*(25), 20360–20371. https://doi.org/10.1007/s11356-017-9674-1

Derraik, J. G. B. (2002). The pollution of the marine environment by plastic debris: A review. *Marine Pollution Bulletin*, *44*(9), 842–852. https://doi.org/10.1016/s0025-326x(02)00220–5

Dogu, O., Conka, R., Akin, O., Withoeck, D., D'hooge, D. R., Van Steenberge, P. H. M., & Van Geem, K. M. (2025). Advancing polyethylene pyrolysis kinetics: A tree-based kinetic Monte Carlo approach with Bayesian optimization. *Chemical Engineering Journal*, *507*, 160686. https://doi.org/10.1016/j.cej.2025.160686

Dogu, O., Eschenbacher, A., John Varghese, R., Dobbelaere, M., D'hooge, D. R., Van Steenberge, P. H. M., & Van Geem, K. M. (2023). Bayesian tuned kinetic Monte Carlo modeling of polystyrene pyrolysis: Unraveling the pathways to its monomer, dimers, and trimers formation. *Chemical Engineering Journal*, *455*, 140708. https://doi.org/10.1016/j.cej.2022.140708

Dosovitskiy, A., Beyer, L., Kolesnikov, A., Weissenborn, D., Zhai, X., Unterthiner, T., Dehghani, M., Minderer, M., Heigold, G., Gelly, S., Uszkoreit, J., & Houlsby, N. (2021). An image is worth 16x16 words: Transformers for image recognition at scale. *arXiv* arXiv:2010.11929. https://doi.org/10.48550/arXiv.2010.11929

Dris, R., Gasperi, J., Rocher, V., Saad, M., Renault, N., & Tassin, B. (2015). Microplastic contamination in an urban area: A case study in Greater Paris. *Environmental Chemistry*, *12*(5), 592–599. https://doi.org/10.1071/EN14167

Edinov, S., & Fauzi, R. (2023). Community behavior in artificial intelligence-based waste management. *Formosa Journal of Sustainable Research*, *2*(2), 341–350. https://doi.org/10.55927/fjsr.v2i2.2993

Eriksen, M., Lebreton, L. C. M., Carson, H. S., Thiel, M., Moore, C. J., Borerro, J. C., Galgani, F., Ryan, P. G., & Reisser, J. (2014). Plastic pollution in the world's oceans: More than 5 trillion plastic pieces weighing over 250,000 tons afloat at sea. *PLOS ONE*, *9*(12), e111913. https://doi.org/10.1371/journal.pone.0111913

Faisal, F., Rasul, M. G., Chowdhury, A. A., & Jahirul, M. I. (2024). Optimisation of process parameters to maximise the oil yield from pyrolysis of mixed waste plastics. *Sustainability*, *16*(7), Article 7. https://doi.org/10.3390/su16072619

Fang, B., Yu, J., Chen, Z., Osman, A. I., Farghali, M., Ihara, I., Hamza, E. H., Rooney, D. W., & Yap, P.-S. (2023). Artificial intelligence for waste management in smart cities: A review. *Environmental Chemistry Letters*, *21*(4), 1959–1989. https://doi.org/10.1007/s10311-023-01604-3

Fernández de la Mata, F., Gijón, A., Molina-Solana, M., & Gómez-Romero, J. (2023). Physics-informed neural networks for data-driven simulation: Advantages, limitations, and opportunities. *Physica A: Statistical Mechanics and Its Applications*, *610*(C). https://ideas.repec.org//a/eee/phsmap/v610y2023ics0378437122009736.html

Gaylor, M. O., Harvey, E., & Hale, R. C. (2013). Polybrominated diphenyl ether (PBDE) accumulation by earthworms (Eisenia fetida) exposed to biosolids-, polyurethane foam microparticle-, and Penta-BDE-amended soils. *Environmental Science & Technology*, *47*(23), 13831–13839. https://doi.org/10.1021/es403750a

Genuino, H. C., Pilar Ruiz, M., Heeres, H. J., & Kersten, S. R. A. (2023). Pyrolysis of mixed plastic waste: Predicting the product yields. *Waste Management*, *156*, 208–215. https://doi.org/10.1016/j.wasman.2022.11.040

Geyer, R., Jambeck, J. R., & Law, K. L. (2017). Production, use, and fate of all plastics ever made. *Science Advances*, *3*(7), e1700782. https://doi.org/10.1126/sciadv.1700782

Ghaderi, A., Akbari, R., Chen, Y., & Dargazany, R. (2022). A knowledge-driven physics-informed neural network model; pyrolysis and ablation of polymers. *arXiv* arXiv:2209.11749. https://doi.org/10.48550/arXiv.2209.11749

Gutierrez, J. M., Jensen, M., Henius, M., & Riaz, T. (2015). Smart waste collection system based on location intelligence. *Procedia Computer Science*, *61*, 120–127. https://doi.org/10.1016/j.procs.2015.09.170

Hahladakis, J. N., & Iacovidou, E. (2018). Closing the loop on plastic packaging materials: What is quality and how does it affect their circularity? *Science of the Total Environment*, *630*, 1394–1400. https://doi.org/10.1016/j.scitotenv.2018.02.330

Hembert, P., Ghnatios, C., Cotton, J., & Chinesta, F. (2024). Assessing sensor integrity for nuclear waste monitoring using graph neural networks. *Sensors*, *24*(5), Article 5. https://doi.org/10.3390/s24051580

Hodson, M. E., Duffus-Hodson, C. A., Clark, A., Prendergast-Miller, M. T., & Thorpe, K. L. (2017). Plastic bag derived-microplastics as a vector for metal exposure in terrestrial invertebrates. *Environmental Science & Technology*, *51*(8), 4714–4721. https://doi.org/10.1021/acs.est.7b00635

Hong, I., Park, S., Lee, B., Lee, J., Jeong, D., & Park, S. (2014). IoT-based smart garbage system for efficient food waste management. *Scientific World Journal*, *2014*(1), 646953. https://doi.org/10.1155/2014/646953

Hopewell, J., Dvorak, R., & Kosior, E. (2009). Plastics recycling: Challenges and opportunities. *Philosophical Transactions of the Royal Society B: Biological Sciences*, *364*(1526), 2115–2126. https://doi.org/10.1098/rstb.2008.0311

Iannello, S., Friso, A., Galvanin, F., & Materazzi, M. (2025). A hybrid physics–machine learning approach for modeling plastic–bed interactions during fluidized bed pyrolysis. *Energy & Fuels*. https://doi.org/10.1021/acs.energyfuels.4c05870

Islam Prova, N. N. (2024). Garbage intelligence: Utilizing vision transformer for smart waste sorting. *2024 Second International Conference on Intelligent Cyber Physical Systems and Internet of Things (ICoICI)*, 1213–1219. https://doi.org/10.1109/ICoICI62503.2024.10696177

Jaafar, Y., Abdelouahed, L., Hage, R. E., Samrani, A. E., & Taouk, B. (2022). Pyrolysis of common plastics and their mixtures to produce valuable petroleum-like products. *Polymer Degradation and Stability*, *195*, 109770. https://doi.org/10.1016/j.polymdegradstab.2021.109770

Jambeck, J. R., Geyer, R., Wilcox, C., Siegler, T. R., Perryman, M., Andrady, A., Narayan, R., & Law, K. L. (2015). Plastic waste inputs from land into the ocean. *Science*, *347*(6223), 768–771. https://doi.org/10.1126/science.1260352

Kannangara, M., Dua, R., Ahmadi, L., & Bensebaa, F. (2018). Modeling and prediction of regional municipal solid waste generation and diversion in Canada using machine learning approaches. *Waste Management*, *74*, 3–15. https://doi.org/10.1016/j.wasman.2017.11.057

Kanthasamy, R., Almatrafi, E., Ali, I., Hussain Sait, H., Zwawi, M., Abnisa, F., Choe Peng, L., & Victor Ayodele, B. (2023). Bayesian optimized multilayer perceptron neural network modelling of biochar and syngas production from pyrolysis of biomass-derived wastes. *Fuel*, *350*, 128832. https://doi.org/10.1016/j.fuel.2023.128832

Kaya, E. Y., Ali, I., Ceylan, Z., & Ceylan, S. (2024). Prediction of higher heating value of hydrochars using Bayesian optimization tuned Gaussian process regression based on biomass characteristics and process conditions. *Biomass and Bioenergy*, *180*, 106993. https://doi.org/10.1016/j.biombioe.2023.106993

Kershaw, P., Katsuhiko, S., Lee, S., & Woodring, D. (2011). *Plastic Debris in the Ocean*. United Nations Environment Programme.

Khan, A. A., Sajib, A. A., Shetu, F., Bari, S., Zishan, Md. S. R., & Shikder, K. (2021). Smart waste management system for Bangladesh. *2021 2nd International Conference on Robotics, Electrical and Signal Processing Techniques (ICREST)*, 659–663. https://doi.org/10.1109/ICREST51555.2021.9331159

Kumar, N., Rajeswari, P., Jeya Priya, D., & Uma Maguesvari, M. (2024). Forecasting municipal solid waste generation using advanced transformer and multi-layer perceptron techniques. *Clean Technologies and Environmental Policy*. https://doi.org/10.1007/s10098-024-03091-8

Kurniawan, A., Saragih, V. C., Wiranti, A., Rahmadi, D., & Poerbo, H. W. (2024, September). Revealing the future of urban design: Exploring human and artificial intelligence collaboration in planning and design for the sustainable built environment (Case Study: IKN Nusantara). In *IOP Conference Series: Earth and Environmental Science* (Vol. 1394, No. 1, p. 012032). IOP Publishing. https://doi.org/10.1088/1755-1315/1394/1/012032

Li, J., Yu, D., Pan, L., Xu, X., Wang, X., & Wang, Y. (2023). Recent advances in plastic waste pyrolysis for liquid fuel production: Critical factors and machine learning applications. *Applied Energy, 346*, 121350. https://doi.org/10.1016/j.apenergy.2023.121350

Lilhore, U. K., Simaiya, S., Dalal, S., & Damaševičius, R. (2024). A smart waste classification model using hybrid CNN-LSTM with transfer learning for sustainable environment. *Multimedia Tools and Applications, 83*(10), 29505–29529. https://doi.org/10.1007/s11042-023-16677-z

Lin, J. Y., Liu, H. T., & Zhang, J. (2022). Recent advances in the application of machine learning methods to improve identification of the microplastics in environment. *Chemosphere, 307*, 136092. https://doi.org/10.1016/j.chemosphere.2022.136092

Liu, Z., Sun, Y., Xing, C., Liu, J., He, Y., Zhou, Y., & Zhang, G. (2022). Artificial intelligence powered large-scale renewable integrations in multi-energy systems for carbon neutrality transition: Challenges and future perspectives. *Energy and AI, 10*, 100195. https://doi.org/10.1016/j.egyai.2022.100195

Macfadyen, G., Huntington, T., & Cappell, R. (2009). *Abandoned, Lost or Otherwise Discarded Fishing Gear (No. 523, 115pp).* Food and Agriculture Organization of the United Nations (FAO).

Majchrowska, S., Mikołajczyk, A., Ferlin, M., Klawikowska, Z., Plantykow, M. A., Kwasigroch, A., & Majek, K. (2022). Deep learning-based waste detection in natural and urban environments. *Waste Management, 138*, 274–284. https://doi.org/10.1016/j.wasman.2021.12.001

Maqsood, T., Dai, J., Zhang, Y., Guang, M., & Li, B. (2021). Pyrolysis of plastic species: A review of resources and products. *Journal of Analytical and Applied Pyrolysis, 159*, 105295. https://doi.org/10.1016/j.jaap.2021.105295

Mazhandu, Z. S., Muzenda, E., Mamvura, T. A., Belaid, M., & Nhubu, T. (2020). Integrated and consolidated review of plastic waste management and bio-based biodegradable plastics: Challenges and opportunities. *Sustainability, 12*(20), Article 20. https://doi.org/10.3390/su12208360

Mighani, M., Shahi, A., & Antonioni, G. (2017). *Catalytic Pyrolysis of Plastic Waste Products: Time Series Modeling Using Least Square Support Vector Machine and Artificial Neural Network.* In 16th International Conference on Sustainable Energy Technologies, Bologna.

Mohammed, M. A., Abdulhasan, M. J., Kumar, N. M., Abdulkareem, K. H., Mostafa, S. A., Maashi, M. S., Khalid, L. S., Abdulaali, H. S., & Chopra, S. S. (2023). Automated waste-sorting and recycling classification using artificial neural network and features fusion: A digital-enabled circular economy vision for smart cities. *Multimedia Tools and Applications, 82*(25), 39617–39632. https://doi.org/10.1007/s11042-021-11537-0

Mousavi, S., Hosseinzadeh, A., & Golzary, A. (2023). Challenges, recent development, and opportunities of smart waste collection: A review. *Science of the Total Environment, 886*, 163925. https://doi.org/10.1016/j.scitotenv.2023.163925

Munir, M. T., Li, B., & Naqvi, M. (2023). Revolutionizing municipal solid waste management (MSWM) with machine learning as a clean resource: Opportunities, challenges and solutions. *Fuel, 348*, 128548. https://doi.org/10.1016/j.fuel.2023.128548

Muntasir Shovon, S., Ahamed Akash, F., Abdur Rahman, Md., Rahman, W., Chakraborty, P., Uddin Monir, M., Md Rezaul Karim, K., Ahsan Habib, Md., & Forrukh Hossain Khan, M. (2024). The potential for sustainable waste management and energy recovery in Bangladesh: A review. *Sustainable Energy Technologies and Assessments, 64*, 103705. https://doi.org/10.1016/j.seta.2024.103705

Mutlu, A. Y., & Yucel, O. (2018). An artificial intelligence based approach to predicting syngas composition for downdraft biomass gasification. *Energy, 165*, 895–901. https://doi.org/10.1016/j.energy.2018.09.131

Nañez Alonso, S. L., Reier Forradellas, R. F., Pi Morell, O., & Jorge-Vazquez, J. (2021). Digitalization, circular economy and environmental sustainability: The application of artificial intelligence in the efficient self-management of waste. *Sustainability, 13*(4), Article 4. https://doi.org/10.3390/su13042092

Naqvi, S. R., Tariq, R., Hameed, Z., Ali, I., Taqvi, S. A., Naqvi, M., Niazi, M. B. K., Noor, T., & Farooq, W. (2018). Pyrolysis of high-ash sewage sludge: Thermo-kinetic study using TGA and artificial neural networks. *Fuel, 233*, 529–538. https://doi.org/10.1016/j.fuel.2018.06.089

Noiki, A., Afolalu, S. A., Abioye, A. A., Bolu, C. A., & Emetere, M. E. (2021). Smart waste bin system: A review. *IOP Conference Series: Earth and Environmental Science, 655*(1), 012036. https://doi.org/10.1088/1755-1315/655/1/012036

Noman, A. A., Akter, U. H., Pranto, T. H., & Haque, A. B. (2022). Machine learning and artificial intelligence in circular economy: A bibliometric analysis and systematic literature review. *Annals of Emerging Technologies in Computing, 6*(2), 13–40. https://doi.org/10.33166/AETiC.2022.02.002

OECD. (2022). *Global Plastics Outlook: Policy Scenarios to 2060.* Organisation for Economic Co-operation and Development. https://www.oecd-ilibrary.org/environment/global-plastics-outlook_aa1edf33-en

Olawade, D. B., Fapohunda, O., Wada, O. Z., Usman, S. O., Ige, A. O., Ajisafe, O., & Oladapo, B. I. (2024). Smart waste management: A paradigm shift enabled by artificial intelligence. *Waste Management Bulletin, 2*(2), 244–263. https://doi.org/10.1016/j.wmb.2024.05.001

Oruganti, R. K., Biji, A. P., Lanuyanger, T., Show, P. L., Sriariyanun, M., Upadhyayula, V. K., Gadhamshetty, V., & Bhattacharyya, D. (2023). Artificial intelligence and machine learning tools for high-performance microalgal wastewater treatment and algal biorefinery: A critical review. *Science of the Total Environment, 876*, 162797. https://doi.org/10.1016/j.scitotenv.2023.162797

Park, H., Kim, K., Yu, M., Yun, Z., & Lee, S. (2024). Economic analysis of the circular economy based on waste plastic pyrolysis oil: A case of the university campus. *Environment, Development and Sustainability, 26*(3), 6293–6313. https://doi.org/10.1007/s10668-023-02963-1

Pathan, M. S., Richardson, E., Galvan, E., & Mooney, P. (2023). The role of artificial intelligence within circular economy activities—A view from Ireland. *Sustainability, 15*(12), Article 12. https://doi.org/10.3390/su15129451

Pournaghi, S. M., Bayat, M., & Farjami, Y. (2020). MedSBA: A novel and secure scheme to share medical data based on blockchain technology and attribute-based encryption. *Journal of Ambient Intelligence and Humanized Computing, 11*(11), 4613–4641. https://doi.org/10.1007/s12652-020-01710-y

Praveen, A., Radhika, R., Rammohan, M. U., Sidharth, D., Ambat, S., & Anjali, T. (2020). IoT based smart bin: A Swachh-Bharat Initiative. *2020 International Conference on Electronics and Sustainable Communication Systems (ICESC)*, 783–786. https://doi.org/10.1109/ICESC48915.2020.9155626

Praveenkumar, T. R., Sekar, M., Pasupuleti, R. R., Gavurová, B., Arun Kumar, G., & Vignesh Kumar, M. (2024). Current technologies for plastic waste treatment for energy recovery, it's effects on poly aromatic hydrocarbons emission and recycling strategies. *Fuel, 357*, 129379. https://doi.org/10.1016/j.fuel.2023.129379

Qu, D. (2022). Application of artificial intelligence in waste classification management at university. In H. Sharma, V. K. Vyas, R. K. Pandey, & M. Prasad (Eds.), *Proceedings of the International Conference on Intelligent Vision and Computing (ICIVC 2021)* (pp. 330–343). Springer International Publishing, https://doi.org/10.1007/978-3-030-97196-0_27

Radhakrishnan, K., Senthil Kumar, P., Rangasamy, G., Praveen Perumal, L., Sanaulla, S., Nilavendhan, S., Manivasagan, V., & Saranya, K. (2023). A critical review on pyrolysis method as sustainable conversion of waste plastics into fuels. *Fuel, 337*, 126890. https://doi.org/10.1016/j.fuel.2022.126890

Ragaert, K., Delva, L., & Van Geem, K. (2017). Mechanical and chemical recycling of solid plastic waste. *Waste Management, 69*, 24–58. https://doi.org/10.1016/j.wasman.2017.07.044

Ruj, B., Pandey, V., Jash, P., & Srivastava, V. K. (2015). Sorting of plastic waste for effective recycling. *International Journal of Applied Sciences and Engineering Research, 4*(4). https://doi.org/10.6088.ijaser.04058

Sakr, G. E., Mokbel, M., Darwich, A., Khneisser, M. N., & Hadi, A. (2016). Comparing deep learning and support vector machines for autonomous waste sorting. *2016 IEEE International Multidisciplinary Conference on Engineering Technology (IMCET)*, 207–212. https://doi.org/10.1109/IMCET.2016.7777453

Satav, A. G., Kubade, S., Amrutkar, C., Arya, G., & Pawar, A. (2023). A state-of-the-art review on robotics in waste sorting: Scope and challenges. *International Journal on Interactive Design and Manufacturing (IJIDeM), 17*(6), 2789–2806. https://doi.org/10.1007/s12008-023-01320-w

Seyyedi, S. Reza, Kowsari, E., Ramakrishna, S., Gheibi, M., & Chinnappan, A. (2023). Marine plastics, circular economy, and artificial intelligence: A comprehensive review of challenges, solutions, and policies. *Journal of Environmental Management, 345*, 118591. https://doi.org/10.1016/j.jenvman.2023.118591

Sharma, R., Raissi, M., & Guo, Y. (2023). Physics-informed deep learning of gas flow-melt pool multi-physical dynamics during powder bed fusion. *CIRP Annals, 72*(1), 161–164. https://doi.org/10.1016/j.cirp.2023.04.005

Sirawattananon, C., Muangnak, N., & Pukdee, W. (2021). Designing of IoT-based smart waste sorting system with image-based deep learning applications. *2021 18th International Conference on Electrical Engineering/Electronics, Computer, Telecommunications and Information Technology (ECTI-CON)*, 383–387. https://doi.org/10.1109/ECTI-CON51831.2021.9454826

Solano Meza, J. K., Orjuela Yepes, D., Rodrigo-Ilarri, J., & Rodrigo-Clavero, M.-E. (2023). Comparative analysis of the implementation of support vector machines and long short-term memory artificial neural networks in municipal solid waste management models in megacities. *International Journal of Environmental Research and Public Health, 20*(5), 4256. https://doi.org/10.3390/ijerph20054256

Soltani, A., Hewage, K., Reza, B., & Sadiq, R. (2015). Multiple stakeholders in multi-criteria decision-making in the context of Municipal Solid Waste Management: A review. *Waste Management (New York, N.Y.), 35*, 318–328. https://doi.org/10.1016/j.wasman.2014.09.010

Support Vector Machines for automated classification of plastic bottles. (2010). *2010 6th International Colloquium on Signal Processing & Its Applications*, 1–5. https://doi.org/10.1109/CSPA.2010.5545265

Terapalli, A., Kamireddi, D., Sridevi, V., Tukarambai, M., Suriapparao, D. V., Rao, C. S., Gautam, R., & Modi, P. R. (2022). Microwave-assisted *in-situ* catalytic pyrolysis of polystyrene: Analysis of product formation and energy consumption using machine learning approach. *Process Safety and Environmental Protection*, *166*, 57–67. https://doi.org/10.1016/j.psep.2022.08.016

Toğaçar, M., Ergen, B., & Cömert, Z. (2020). Waste classification using AutoEncoder network with integrated feature selection method in convolutional neural network models. *Measurement*, *153*, 107459. https://doi.org/10.1016/j.measurement.2019.107459

Ureel, Y., Dobbelaere, M. R., Akin, O., Varghese, R. J., Pernalete, C. G., Thybaut, J. W., & Van Geem, K. M. (2022). Active learning-based exploration of the catalytic pyrolysis of plastic waste. *Fuel*, *328*, 125340. https://doi.org/10.1016/j.fuel.2022.125340

van Wyk, S. (2025). Development of a novel physics-informed machine learning model for advanced thermochemical waste conversion. *Chemical Engineering Journal Advances*, *21*, 100699. https://doi.org/10.1016/j.ceja.2024.100699

Vaswani, A., Shazeer, N., Parmar, N., Uszkoreit, J., Jones, L., Gomez, A. N., Kaiser, Ł. ukasz, & Polosukhin, I. (2017). Attention is all you need. *Advances in Neural Information Processing Systems*, *30*. https://proceedings.neurips.cc/paper/2017/hash/3f5ee243547dee91fbd053c1c4a845aa-Abstract.html

Wang, F., Shih, K. M., & Li, X. Y. (2015). The partition behavior of perfluorooctanesulfonate (PFOS) and perfluorooctanesulfonamide (FOSA) on microplastics. *Chemosphere*, *119*, 841–847. https://doi.org/10.1016/j.chemosphere.2014.08.047

Wang, S., Yu, X., & Perdikaris, P. (2022). When and why PINNs fail to train: A neural tangent kernel perspective. *Journal of Computational Physics*, *449*, 110768. https://doi.org/10.1016/j.jcp.2021.110768

Wang, W., Chang, J.-S., & Lee, D.-J. (2024a). Machine learning applications for biochar studies: A mini-review. *Bioresource Technology*, *394*, 130291. https://doi.org/10.1016/j.biortech.2023.130291

Wang, Z., Ye, L., Chen, F., Zhou, T., & Zhao, Y. (2024b). Multi-category sorting of plastic waste using Swin Transformer: A vision-based approach. *Journal of Environmental Management*, *370*, 122742. https://doi.org/10.1016/j.jenvman.2024.122742

Wang, Z., Zhou, W., & Li, Y. (2024c). Garbage FusionNet: A deep learning framework combining ResNet and Vision Transformer for waste classification. *Research Square*. https://doi.org/10.21203/rs.3.rs-4708918/v1

Wilts, H., Garcia, B. R., Garlito, R. G., Gómez, L. S., & Prieto, E. G. (2021). Artificial intelligence in the sorting of municipal waste as an enabler of the circular economy. *Resources*, *10*(4), Article 4. https://doi.org/10.3390/resources10040028

Wu, G., Yion, W. T. G., Dang, K. L. N. Q., & Wu, Z. (2023). Physics-informed machine learning for MPC: Application to a batch crystallization process. *Chemical Engineering Research and Design*, *192*, 556–569. https://doi.org/10.1016/j.cherd.2023.02.048

Xanthos, D., & Walker, T. R. (2017). International policies to reduce plastic marine pollution from single-use plastics (plastic bags and microbeads): A review. *Marine Pollution Bulletin*, *118*(1–2), 17–26. https://doi.org/10.1016/j.marpolbul.2017.02.048

Xayachak, T., Haque, N., Parthasarathy, R., King, S., Emami, N., Lau, D., & Pramanik, B. K. (2022). Pyrolysis for plastic waste management: An engineering perspective. *Journal of Environmental Chemical Engineering*, *10*(6), 108865. https://doi.org/10.1016/j.jece.2022.108865

Yaka, H., Insel, M. A., Yucel, O., & Sadikoglu, H. (2022). A comparison of machine learning algorithms for estimation of higher heating values of biomass and fossil fuels from ultimate analysis. *Fuel*, *320*, 123971. https://doi.org/10.1016/j.fuel.2022.123971

Yang, K., Wu, K., Li, F., Jia, L., Wang, S., & Zhang, H. (2023). Investigation on the co-pyrolysis of bamboo sawdust and low-density polyethylene via online photoionization mass spectrometry and machine learning methods. *Fuel Processing Technology*, *240*, 107579. https://doi.org/10.1016/j.fuproc.2022.107579

Yi, C. J., & Kim, C. F. (2024). AI-powered waste classification using convolutional neural networks (CNNs). *International Journal of Advanced Computer Science and Applications (IJACSA)*, *15*(10), Article 10. https://doi.org/10.14569/IJACSA.2024.0151009

Zhang, M., Qi, Y., Zhang, W., Wang, M., Li, J., Lu, Y., Zhang, S., He, J., Cao, H., Tao, X., Xu, H., & Zhang, S. (2024). A review on waste tires pyrolysis for energy and material recovery from the optimization perspective. *Renewable and Sustainable Energy Reviews*, *199*, 114531. https://doi.org/10.1016/j.rser.2024.114531

Zheng, J., & Suh, S. (2019). Strategies to reduce the global carbon footprint of plastics. *Nature Climate Change*, *9*(5), 374–378. https://doi.org/10.1038/s41558-019-0459-z

Zhou, X., He, P., Peng, W., Zhou, J., Jiang, M., Zhang, H., & Dong, W. (2023). A value-added and carbon-reduction approach to upcycle mixed plastic waste into methane and carbon microspheres. *Resources, Conservation and Recycling*, *193*, 106988. https://doi.org/10.1016/j.resconrec.2023.106988

Index

N

O

P

V

W

Z